Who Rules the Earth?

Who Rules the Earth?

How Social Rules Shape Our Planet and Our Lives

PAUL F. STEINBERG

OXFORD
UNIVERSITY PRESS

Oxford University Press is a department of the
University of Oxford. It furthers the University's objective
of excellence in research, scholarship, and education
by publishing worldwide.

Oxford   New York
Auckland   Cape Town   Dar es Salaam   Hong Kong   Karachi
Kuala Lumpur   Madrid   Melbourne   Mexico City   Nairobi
New Delhi   Shanghai   Taipei   Toronto

With offices in
Argentina   Austria   Brazil   Chile   Czech Republic   France   Greece
Guatemala   Hungary   Italy   Japan   Poland   Portugal   Singapore
South Korea   Switzerland   Thailand   Turkey   Ukraine   Vietnam

Oxford is a registered trade mark of Oxford University Press
in the UK and certain other countries.

Published in the United States of America by
Oxford University Press
198 Madison Avenue, New York, NY 10016

© Paul F. Steinberg 2015

All rights reserved. No part of this publication may be reproduced,
stored in a retrieval system, or transmitted, in any form or by any means,
without the prior permission in writing of Oxford University Press,
or as expressly permitted by law, by license, or under terms agreed with
the appropriate reproduction rights organization. Inquiries concerning
reproduction outside the scope of the above should be sent to the
Rights Department, Oxford University Press, at the address above.

You must not circulate this work in any other form
and you must impose this same condition on any acquirer.

Library of Congress Cataloging-in-Publication Data
Steinberg, Paul F.
Who rules the earth? : how social rules shape our planet and our lives / Paul F. Steinberg.
   pages cm
ISBN 978-0-19-989661-5 (hardback)
1. Environmentalism—Social aspects.   2. Environmental policy—Social aspects.   I. Title.
GE195.S776   2015
363.7—dc23   2014026033

3 5 7 9 8 6 4 2
Printed in the United States of America
on acid-free paper

*To my son
Benjamin Steinberg
Who rules my heart*

Contents

Acknowledgments   ix

I. SEARCHING FOR SOLUTIONS

1. Recycling Is Not Enough   5
2. Strings Attached   19
3. Feasible Worlds   35

II. WHO OWNS THE EARTH?

4. A Perilous Journey   63
5. The Big Trade   95
6. A Planet of Nations   127

III. TRANSFORMATIONS

7. Scaling Up   161

8	Scaling Down	183
9	Keep the Change	211

IV. LEVERAGE

10	Super Rules	245
11	Paper, Plastic, or Politics?	263
	Notes	281
	Index	317

Acknowledgments

Although I will play the role of author, and you of reader, in the pages to come, I suspect that anyone who opens a book with a title like this one is no stranger to grappling with ideas and expressing a point of view, whether in writing or some other form. So let me confide, as one merchant of words and ideas to another, that I find writing to be a peculiar (if deeply enriching) experience. It is an intensely solitary task, requiring the author to spend long hours sequestered from social distractions, silently scribbling on notepads, arranging and rearranging Post-it notes, and tapping away on the keyboard late at night with little accompaniment beyond a lightbulb and some music to keep the pace. At the same time, writing is the most social of activities because all writers draw from—and are therefore indebted beyond imagination to—the insights of countless people, past and present. That is certainly true of this book, and among the many people who have helped me along the way, a few deserve special mention.

My research assistants—Laurie Egan, Noah Proser, Ratik Asokan, and Thomas Carey—were undaunted in tackling the questions I sent their way, no matter how difficult or obscure. Whether tracking down the latest data on global water shortages, or figuring out what species of bird feather adorned the headdresses of Incan emperors, their good humor and great research instincts were indispensible throughout.

I began work on this book during a sabbatical at the Department of Environmental Science, Policy, and Management at the University of California at Berkeley. Kate O'Neill and Nancy Peluso were gracious hosts, welcoming me into their intellectual community. To Louise Fortmann, I owe thanks for helping me to navigate the voluminous literature on property rights.

Given the historical and geographic scope of this book, I often found it necessary to reach out to experts in diverse fields, soliciting their help in locating historical materials, interpreting the data, or simply making sure I got the story right. Especially generous in this regard were Evan Ringquist, Patrick Angel, Malcolm Lewis, David Gottfried, Mike Italiano, David Vogel, Ronald Mitchell, Richard Mehlinger, Kristin Dobbin, Paul Tukey, Sara Lowe, Claudia Olazábal, Jeffrey Sellers, and Matthew Lyons. Joseph Nye, George Lakoff, and Neil Shubin shared valuable insights into the craft of writing a book that is moored in the research, yet accessible to broader reading audiences. Thanks also go to those who took the time to read and comment on early drafts of chapters, including James Meadowcroft, Ken Conca, George Somogyi, Paul Stamler, Eugene Bardach, Robert Asher, Carol Williams, Sam Arenson, Barbara Steinberg, Frank Lossy, Kathryn Hochstetler, Javiera Barandiaran, and Lia Fox. David McBride of Oxford University Press struck a perfect balance of critique and support throughout the writing process. The visual content of the book was made possible thanks to the efforts of librarians, photographers, archivists, and scholars including Nazli Choucri, Sidney Gauthreaux, Margaret Taylor, Miriam Gago, Erik Nelson, Evan Johnson, Phil Sandlin, Simon Elliott, Michelle Levers, and Niccolo Tognarini.

Who Rules the Earth? is accompanied by a multimedia education and outreach effort called The Social Rules Project, which can be accessed at www.rulechangers.org. This was the result of work by more than 100 students at the Claremont Colleges and the California Institute of the Arts. Adrienne Luce and Steve Prince marshaled the resources that enabled the students to shine, and in the process became close colleagues and friends. To the student animators, videogame designers, website developers, environmental analysts, and others who took part: We imagined that the concept of institutions might be represented in a visually compelling way; it was your talent and dedication that turned this idea into a reality.

Finally, and most important, my wife Jennifer and son Benjamin were a constant source of support and inspiration. Ben, you were six years old when I started this project, and twelve when I finished. This book is dedicated to you.

Who Rules the Earth?

PART I

Searching for Solutions

1

Recycling Is Not Enough

Faced with an endless stream of alarming news about the environment—rising temperatures and declining water supplies, population growth and species extinction, oil spills and cancer clusters—people increasingly want to know what can actually be done to address these problems. Concerned parents comb through websites late at night in search of safer products for their children. Students pack lecture halls in hundreds of environmental studies programs that have popped up on college campuses across the globe. Our grocery aisles and magazine stands are filled with advertisements promising that sustainability is just one more purchase around the corner.

The major current of environmental thinking today emphasizes the small changes we can make as individuals, which (we are told) will add up to something big. Michael Maniates, a political scientist at Allegheny College, observes that the responsibility for confronting these issues too often "falls to individuals, acting alone, usually as consumers."[1] Yet solutions that promote green consumerism and changes in personal lifestyles strike many of us as strangely out of proportion with enormous problems like climate change, urban air pollution, and the disappearance of tropical forests. We learn that glaciers are melting and sea levels are expected to rise due to global warming—and in response we are advised to ride a bicycle to work. Scientists tell us that one out of every five mammal species in the world is threatened with extinction, and we react by switching coffee brands. Is it any wonder that people despair that real solutions are not within their grasp?

You may suspect that tackling these gargantuan problems will require something more—but what? The answer, it turns out, can be found in a mountain of books and research articles published by thousands of social scientists over the past quarter century. But their discoveries have remained largely hidden from public view. This book is an attempt to distill insights from that research and to share the findings with those who need them most: intelligent readers who are concerned about the environment and eager to learn what they can do beyond tossing a bottle in the recycling bin—an action that produces a strange, conflicted sensation of knowing you are doing the right thing, yet wondering if it really makes any difference at all. So what more might we do to move the world onto a more sustainable path? The answer can be found in the story of a country doctor in the small Canadian town of Hudson, who decided it was time to do something extraordinary: She changed the rules.

A NEW LANDSCAPE

Dr. June Irwin tended to one of her patients while, a few miles away, the votes were counted. It was May 6, 1991, and the town council of Hudson, Quebec, was considering a proposal to ban all nonessential pesticide use from homes and public spaces throughout their municipality. The move was unprecedented, but local officials had been swayed by Dr. Irwin's relentless research on the potential dangers to children. "If down the road, science shows that we are wrong," declared a town councilor, "then all that's happened because of our actions is a few more dandelions. But if in fact we're right, how many people did we save?"[2] Ultimately the council voted in favor of the ban—a decision that would have consequences far greater than local officials could possibly have imagined.[3]

Hudson is a picturesque community of some five thousand residents, nestled along a stretch of the Ottawa River thirty-five miles to the west of Montreal. In 1985, Dr. Irwin began showing up regularly at town council meetings, where she pleaded with her elected officials to put a stop to the practice of spraying pesticides on lawns and gardens. With a red-lipsticked smile that complements her 70-something years, June Irwin cuts quite a character. Most days she can be found tending a flock of sheep on her farm, wearing her signature sun hat and long skirt—conjuring an image of a biblical shepherd more than that of a dermatologist with a bustling private practice. Throughout the 1980s, Dr. Irwin grew increasingly concerned as her patients complained of ailments ranging from skin rashes to immune system disorders. She suspected that the culprit might be chemicals like 2,4-D that were routinely applied in home gardens and public parks to control weeds. She began her investigation by asking patients to provide tissue samples to test for pesticide residues. The results revealed that pesticides were present in the blood, hair, semen, and breast milk of the good citizens of Hudson.

Dr. Irwin's findings were consistent with data collected in large-scale "body burden" studies run by the US Centers for Disease Control, which show that our bodies contain a complex brew of pesticides and other industrial toxins. We are exposed to thousands of man-made chemicals on a daily basis, and few of these have undergone rigorous testing for health effects. Medical researchers do know, however, that many pesticides affect the brain, liver, and other organs. Children are most susceptible to these noxious effects because their growing bodies rely on internal chemical cues for the normal development of the nervous system and other vital functions.[4]

As she pored through the medical journals, June Irwin soon reached the conclusion that it is madness to routinely expose children to poisons just to maintain the cosmetic appearance of lawns. Town council members listened patiently as she offered lengthy discourses on pesticides and health, comparing her data on local body burdens with the latest findings from the medical journals. She wrote a steady stream of letters to the local *Hudson Gazette* in an effort to rally the community. But four years into her one-woman campaign, there was little to show for her efforts. This all changed in November 1989, when one of the town council members who had endured Dr. Irwin's lectures, a local carpenter by the name of Michael Elliot, was elected mayor. Six months after his election, Mayor Elliott pushed for approval of By-law 270, banning all nonessential pesticides from homes and public spaces in the quiet little town of Hudson.

What happened next would change the North American landscape, both physically and politically.

It began with an aggressive response from the pesticide companies, who moved quickly to quell Hudson's small act of defiance. In the fall of 1993, as the town's maples and aspens glowed bright in their full autumn color, ChemLawn and SprayTech, representing Canada's billion-dollar lawn care industry, sued Hudson in the Quebec Superior Court, arguing that the town had no legal right to regulate pesticides. They worried that if local communities could take it upon themselves to enact environmental rules stronger than those of the Canadian provinces, things could quickly spin out of control. At a deeper level, there were cultural norms at stake. The pesticide industry relied on the idea that a proper home lawn consists of a uniform stretch of green with no weeds whatsoever—a feat that requires applying poison to the grass. If picturesque Hudson could make do without pesticides, this would challenge the golf course aesthetic that has generated handsome profits for the industry since World War II.

"Nobody thought that we could win this—never," explained the town clerk to documentary filmmaker Paul Tukey, who followed the story in his film *A Chemical Reaction*. In the courtroom, a ChemLawn representative showed up with a bottle of pesticides that he intended to drink in front of the judge in a show of confidence. Before ChemLawn's man had a chance to ingest the poison, the judge demanded that he remove it from the courtroom and then ruled that Hudson was well within its rights in regulating pesticide use.

The publicity generated by the court case caught the attention of other communities. "If they can do it, why can't we?" asked Merryl Hammond, founder of Citizens for Alternatives to Pesticides. The movement soon spread across Quebec, as one town after the next banned nonessential pesticides (often making exceptions, as did Hudson, for agriculture and golf courses). On the defensive now, the pesticide industry took the case to the Canadian Supreme Court. This time they didn't attempt to drink pesticides in the courtroom, but repeated the argument that local governments have no right to decide whether chemicals are sprayed in their communities. On June 28, 2001, in an austere gray court building surrounded by an expansive green lawn, the Canadian Supreme Court justices convened in their traditional red and white gowns and ruled 9-0 in favor of the town of Hudson. The ruling electrified reformers throughout the country. In 2009, the province of Ontario passed even stricter rules than those adopted in Quebec. "When it comes to our homes, playgrounds, schoolyards, and the like," explained Dalton McGinty, the premier of Ontario, "we think that we have a special shared responsibility owed to the youngest generation." A year after the implementation of Ontario's new rules, concentrations of common pesticides in the province's waterways dropped by half.[5] By 2010, three-fourths of all Canadian citizens were covered by some form of protective legislation based on the Hudson model.[6]

MEANWHILE IN AMERICA

In the United States, the story unfolded very differently. Stunned by the unprecedented turn of events in Canada, the pesticide industry moved quickly to shore up their interests south of the border. "The activists plain outworked us up there," said Allen James, president of the lobbying group Responsible Industry for a Sound Environment (RISE). "We clearly have lost the battle in Canada for the most part.... We cannot allow this to happen in the U.S."[7] In June 1991, the US Supreme Court had affirmed that local communities have the right to implement pesticide regulations stronger than those of the federal government. The individual states, however, retained the right to preempt local authorities. Seizing the opportunity, a month after the decisions by the US and Canadian supreme courts, RISE joined a coalition of 180 industry organizations that reads like a who's who of pesticide lobbyists, including the National Agricultural Chemical Association and the US Chamber of Commerce. The Coalition for Sensible Pesticide Policy, as they called themselves, traveled from one state capital to the next, pressing lawmakers to pass new state preemption rules to prevent cities and counties from attempting to regulate pesticides.

The first time I encountered the story of Hudson, and the spread of preemption rules in the United States, was at the annual convention of the International Studies Association, where thousands of social scientists converge on a different city each year, sharing early research results with peers before submitting these for publication in the professional journals. I had agreed to serve in the role of discussant, offering feedback on draft papers written by a panel of graduate students and professors. One of these was Sarah Pralle of Syracuse University, who was investigating the astonishing divergence in outcomes between Canada and the United States. As I read her paper on the flight from Los Angeles, I was struck by Pralle's observation that in the United States, "the pesticide industry was far better organized than anti-pesticide activists." This hit home because, as it happens, I was one of those activists. In early 1992, when the state-by-state lobbying effort was in full swing, I was a young researcher at the San Francisco offices of Pesticide Action Network International. In the course of a given day, dozens of news items, campaign materials, and research reports from the pesticide reform community came across my desk; I don't recall seeing anything about an industry lobbying campaign to prevent local communities from regulating pesticides. American environmental groups were simply caught unaware.

For the pesticide industry, the stealth lobbying strategy worked like a charm. They scored their first successes in 1992 when preemption laws were adopted by state legislatures in Georgia, Kansas, Kentucky, Missouri, New Mexico, Tennessee, Virginia, Oklahoma, and Florida. The following year these states were joined by Alabama, Arkansas, New Hampshire, North Dakota, Montana, Nebraska, Texas, Illinois, and Wisconsin. Massachusetts, Iowa, Michigan, and Idaho were the next to go, and soon would be joined by many others (Figure 1.1).[8] Today all but a handful of American states have preemption rules in place. While Canadian children play at parks and homes that are largely pesticide free, American kids roll around on lawns that are drenched with 127 million pounds of pesticides every year.[9] The strategy was so successful, it was later copied by the tobacco industry, which lobbied for preemption rules in a number of states to prevent municipalities from banning cigarette smoking in public spaces.

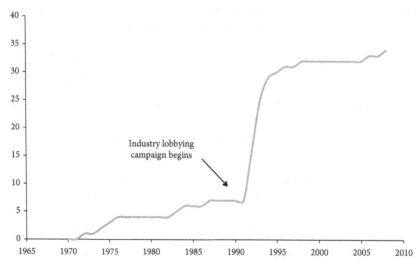

FIGURE 1.1 American states with preemption rules forbidding local control of pesticides

See note 8 for information sources.

CHANGING THE RULES

The fate of pesticide reform efforts in the United States and Canada carries a larger lesson for anyone who wants to get serious about promoting sustainability. Although they worked on opposite sides of the issue, June Irwin and pesticide industry leaders recognized something very profound: To bring about lasting change requires modifying the very rules that societies live by. For Dr. Irwin and her adversaries, the relevant rules were city by-laws as well as higher-order rules that decide how power is shared among local, regional, and national governments. But social rules are not limited to government laws and policies. In some cases they are encoded in private contracts, like when a business instructs its lawncare service to spray for weeds on the first of each month. Corporations rely on reams of written agreements to run their day-to-day operations, and these rules have a direct impact on the earth. Take Walmart, for example. In 2006, in response to pressure from marine conservation experts, the corporate giant put in place a new rule specifying that its fish products must come from sustainably harvested sources. This rule soon reverberated throughout the global economy, ricocheting along chains of buyers, distributors, and wholesalers, until it ultimately changed the harvesting practices of fishing fleets plowing through the high seas of the Bering Strait. In other cases, these rules take the form of widely understood social norms, like the golf course aesthetic that leads homeowners to spray their lawns to the nodding approval of their neighbors. Whether they take the form of national regulations or the most minute technical design standards, social rules are like thousands of invisible threads tugging at us as we go about our daily routines, shaping our decisions and determining how we relate to one another and to the planet we share.

Although they go by many names, these rules—which social scientists call *institutions*—are the machinery that makes coordinated social activity possible.[10] Douglass North, who won the Nobel Prize in Economics for his work on social rules, shows that the historical expansion of capitalism, from its humble origins as small-scale personal exchanges to today's complex global markets, was made possible by ever-expanding sets of rules governing everything from cargo insurance to banking practices. Social rules enable societies to function. They are also the source of our most recalcitrant *dys*functions, driving us blithely down paths that no rational society would choose to follow. In this book I will argue that these rules are the big levers that will ultimately decide whether we can reconcile the pursuit of prosperity with thoughtful environmental stewardship. Drawing on insights from the latest social science research, my goal is to explain what social rules are and why they matter for your personal well-being and for that of future generations. By exploring the enduring foundations of the environmental crisis, my hope is to provide you with a deeper and ultimately more satisfying understanding of what it will take to put society on a more sustainable path. To accomplish this, we will take a series of journeys, from a familiar stroll on the beach to a 3,000-mile odyssey following an endangered bird on its annual flight from the Peruvian rainforests to the mountains of West Virginia. Along the way we will see how communities, corporations, and countries are reinventing the rules they live by to achieve such seemingly simple yet elusive goals as clean air and water, vibrant urban spaces, and healthy food.

EXPLAINING THE PUZZLE

To change the rules, we must first become aware of them. Yet there is a reason why society's rules so often escape our attention—they're supposed to! When rules are routinely followed (as they must be, if they are to have an effect) we internalize them as habits, routines, and standard operating procedures. We take them for granted as part of the natural order of things.[11] (*Of course* I have the right to speak my mind without fear of imprisonment. *Obviously* your neighbor cannot pick your apples without permission. *Naturally* women are allowed to attend college.) Social rules also escape notice because, unlike the usual subjects of environmental science, you cannot place a rule in a child's hand for inspection, point it out to a group of tourists on safari, or mix it in a test tube. The rules we live by are invisible to our most powerful satellites and microscopes alike. Yet once we know what to look for, we are in a position to critically examine the powerful social structures shaping our planet and our lives.

Armed with this perspective, we can make sense of otherwise beguiling puzzles: If solar energy imposes fewer costs on society than fossil fuels, why then is solar expensive and oil cheap? The rules in place do not require the oil industry to pay for the environmental costs of its products, instead passing those costs to others in the form of global warming. Why are there dangerous amounts of toxic lead in older homes throughout the United States but not in Europe? A century ago, American officials chose to ignore a League of Nations resolution to ban lead from interior paints. Why is Costa Rica doing an outstanding job of protecting its forests while other tropical countries are collectively wiping out 22 million acres of forest habitat every year?[12] Costa Ricans have put in place innovative rules that pay farmers to leave trees standing on their property.

What makes this perspective so empowering is that rule changes are well within the reach of ordinary citizens. I will not promise you that change is easy. But when people join together, as did the citizens of Hudson, significant change is possible—often with less individual time and effort than is required to tackle a challenging college course. Frankly, we can't afford to wait. This book is born of the idea that the environmental problems we face today, from dwindling water supplies to disappearing coral reefs, from urban squalor to toxic waste dumps, are so large, and the social processes driving them are so powerful, that we need to think big—and soon. We need solutions whose power and scope match the severity and pace of the problems unfolding before us. We need new rules. To ride a bicycle to work is terrific; to lobby for a city ordinance putting in more bike lanes is even better. Constructing a "green" campus building is laudable. Creating new standards for campus construction, or catalyzing a change in the entire building industry (a feat discussed in chapter 3), is transformational. Placing a solar panel on your home is a positive step; placing a requirement for renewable energy in government legislation is an outright sprint.

WINDOWS INTO POLITICS

Political scientists are accustomed to thinking about the big forces that move society—the origins of revolutions, how new ideas spread, what determines who participates in politics (and who doesn't), and other engines of large-scale social change. As an active researcher in this field, I have spent the past twenty years trying to answer one question: What does it take to bring about social change to protect the environment?

When my colleagues in disciplines like conservation biology do their field research, they are likely to be found in lovely outdoor settings taking water samples, tracking grizzly bears, or observing the behavior of pollinators to open a window into the workings of the natural world. I am more likely to be found in a room with an influential policymaker somewhere in the developing world, straining to pick up the nuances of a foreign-language conversation above the sound of a tropical downpour on a tin roof, waiting for that golden moment when my interview subject opens up and reveals something stunning about how social change works in that part of the world. On one particular occasion in December 1997, in the town of Santa Cruz de la Sierra, Bolivia, I was interviewing Francisco Kempff, a forestry official who is the son of one of South America's most important conservationists, Noel Kempff Mercado. The elder Kempff pushed for the creation of national parks in the 1950s and '60s, at a time when few North Americans or Europeans had ever heard of rainforests, and when Bolivia was still in the grip of a dictatorship. Tragically, Francisco's father was murdered in a remote jungle in 1986 when his research team stumbled upon a landing strip used by Brazilian drug runners. Although I was eager to learn about his father's early efforts at reform, which had never been studied in detail, I decided that out of respect, I would not ask Francisco about his father's life unless he brought up the subject. The interview lasted an hour or so, and as we were finishing up I briefly mentioned his father's name when paying a compliment to the family. "A funny thing," he said, "Papa kept a copy of every letter he ever sent or received in his life. No one has ever asked to see them."

With only three days remaining in my six-month stay in Bolivia, I wiped my schedule clear and visited the Kempff family home, where his mother and her maid brought out box after dusty box of yellowed letters, portfolios, and notebooks in which Noel Kempff described his efforts to change the rules governing the Bolivian forests. The results were spectacular. We know very little about how rulemaking works under dictatorships; official records are scarce and researchers are often denied access to decision makers. Digging through the family archives, I discovered correspondence from the early 1970s between Noel Kempff and his brother Rolando, who had somehow managed to secure a position in the Agriculture Ministry and advised him on strategies for convincing the generals to create protected areas for bird conservation. Many years later, a similar sort of rulemaking savvy was deployed when news spread of Noel Kempff's death and Bolivians took to the streets demanding justice and accusing the government of complicity in the drug trade. Taking advantage of the opportunity, local activists convinced the city government of Santa Cruz de la Sierra to provide funding for what is today Noel Kempff National Park—an area so rich in natural variety that it provides habitat for more than five percent of the planet's bird species.

My field research has given me the opportunity to study how social change works in diverse political settings around the globe, and I will share findings from these investigations in the chapters to follow. But the essence of research—and the reason why political scientists and other research communities gather by the thousands in convention halls each year like salmon returning to the stream—is a recognition that none of us are smart enough to figure out the answers on our own. Fortunately, many other researchers have been asking the same questions I have about social change and sustainability from the perspective of fields including economics, sociology, anthropology, public policy, and law. Hundreds of researchers have fanned out across the globe to learn from communities that have crafted rules to sustainably manage their forests, water resources, and fisheries over long time horizons.[13] Other researchers roam the halls of government, exploring what it takes to bring about changes in public policies and why some countries are leaders, and others laggards, on issues like climate change and species conservation.[14] Economists have shown how rules governing property rights shape decisions about how much to pollute and whether to conserve agricultural topsoil.[15] Still others explore the conditions under which citizens' movements flourish and whether they have an impact on government policy.[16]

This sort of work requires a devotion to the craft that can assume comical proportions. In her work on the politics of nature protection in Siberia, Melinda Herrold-Menzies recounts how Russian tradition requires repeated rounds of drinks with interview subjects—the trick being to gain valuable research insights before losing consciousness.[17] Ronie Garcia-Johnson, whose book *Exporting Environmentalism* upended the conventional wisdom by showing that multinational corporations were actually raising environmental standards in Mexico, once told me that she used to stuff a pillow in her dress for the morning subway commute in Mexico City, feigning pregnancy to deter male hecklers on her way to interviews. My own research has had me navigating around logging trucks on windy mountain roads of Northern California in the middle of the night, haggling with bureaucrats in Costa Rica to access rare archival materials, and shamelessly sprinting across the floor of the UN General Assembly to share an elevator ride with a particularly knowledgeable diplomat.

The results of this research lead to the same overarching conclusion: The transition to sustainability requires transforming the rules we live by. Unfortunately, these findings have not been shared widely beyond insular groups of research specialists and our students. Those of us working in the trenches of environmental social science have simply not done as good a job as our counterparts in the natural sciences at communicating our discoveries to audiences outside of our fields—pulling insights out of the arcane language and inscrutable equations of research journals and sharing them in ways that discerning readers can appreciate. While the African savannahs and Amazonian rainforests blaze in full color on our television screens, the social forces that will determine whether these splendors survive into the next century remain invisible.

To address this problem, I have collaborated with over 100 students at six universities on an educational initiative called the Social Rules Project. At www.rulechangers.org, you will find free multimedia materials including a short animated film, an educational videogame, and Facebook group links exploring the themes raised in this book. The goal of the Social Rules Project is to foster public understanding and action on the institutional dimensions of sustainability. As part of that effort, this book is written for a diverse audience of concerned readers—students, scientists, parents, business entrepreneurs, community leaders, environmental professionals, and others from all walks of life—who wish to move beyond the headlines covering the latest ecological disaster and take a closer look at how the rules we make shape the course we take.

PAIRING RESEARCH AND ACTION

A few years ago I was discussing the concept for this book with a colleague at Harvey Mudd College when she asked me pointedly, "What kind of book are you writing—is it about research or activism?" My response, then and now, is *yes*. My colleague's question reflects a collective angst that runs throughout academia. Many scholars fear that their research reputations will be tarnished if they associate themselves too closely with efforts to change the world. This may strike many readers as odd, but it is a tension that university professors know well. I confess that the distinction between research and action has never made much sense to me. When I was an undergraduate studying biology at the University of California, Santa Barbara, in the 1980s, I joined with other students to launch a group called Scientists and Engineers for Responsible Technology. My time was divided between studying the mind-blowing elegance of the natural world, and speaking out in public forums on issues like nuclear weapons proliferation and offshore oil drilling. Moving between the science of nature and the politics that threaten to undermine it, I came to understand that rigorous research can serve as a powerful weapon against poorly construed public policies.

Some prominent researchers have bucked the anti-activism trend. Particularly within the natural sciences, the roster of scientists pairing knowledge and public advocacy is a distinguished one. It includes no less than Rachel Carson, the federal wildlife biologist whose book *Silent Spring* helped launch the modern environmental movement. It also includes Carl Sagan, who not only popularized physics and astronomy for a generation of stargazers, but also spoke out against the environmental consequences of nuclear weapons.[18] These intellectual giants and others like them—the ecologist Norman Myers (on extinction), atmospheric scientist Stephen Schneider (on climate change), zoologist Theo Colburn (on endocrine disruptors), and the physicist John Holdren (on alternative energy)—have never been content to stay put in their offices and laboratories.

Unfortunately, these outspoken figures are outliers. Their activities are at odds with the expectations guiding professional practice in most university departments. Within the social sciences in particular, there is widespread unease with research that is tagged as "normative," meaning it not only describes the world as it is, but ventures to say something about how it should be. To my mind, the normative epithet and its scarlet letter connotations rest on a sloppy conceptual distinction. Normative concerns lie just below the surface of most scientific research. Behind the most dispassionate effort by researchers to discover how flu viruses infect human cells, we find the normative position that human health should indeed be protected. Engineering is taught in universities because of an underlying normative belief that we should not live downstream from poorly designed dams.

A more useful distinction, I believe, is whether a researcher is committed to a cause (in the sense of caring about the world and acting on that concern) or so bound to a particular organizational or ideological agenda that the person relinquishes the practice of open inquiry. The litmus test is whether the researcher not only tolerates but actively solicits alternative points of view, and remains open to findings sharply at odds with expectations. Commitment to a social cause, and associated efforts at public advocacy, represent no threat to intellectual integrity. Botanists studying the fate of an endangered tree species may be motivated by a deep concern for its survival, but this does not prevent them from reaching objective conclusions about its future prospects. To surrender the right to critical inquiry, however, is something altogether different. A researcher working for a conservative think tank is not free to argue for more stringent environmental regulation, no matter what the evidence suggests. A journalist at an avowedly leftist magazine does not have the liberty to write a series of reports on the role of markets in alleviating poverty. The academic's greatest asset as a participant in public debate is not mere expertise, but freedom of inquiry. To squander that freedom by erecting barriers between the worlds of research and action strikes me as counterproductive and, frankly, a disservice to the cause of well-informed democratic dialogue.

In this book I try to combine the activist's sense of urgency with the scholar's professional skepticism, in the hope of providing the reader with something that is both relevant and reliable. I believe people are eager for this sort of synthesis. The social leaders whom I have had the privilege of knowing over the course of my career—businesspeople, nonprofit managers, community activists, policymakers, and philanthropists—are hungry for new knowledge to inform their efforts. They simply need it translated into a form that makes sense. For students, many enjoy the intellectual ride of their coursework, but wonder *How does this matter?* Is it really wrong for students to demand that their education relates to the world of action? Likewise, people with a predilection for action have the right to know where their information is coming from; to that end, throughout the book I provide notes where readers will find references and expansions on specific points raised in the body of the text.

After reading the following chapters, my hope is that you will not only come away with a fresh perspective on the world, but that it will be impossible to view it the same way again. By exploring the deeper foundations of the environmental crisis, a rule-based perspective marks a radical departure from the solutions we encounter in the popular media. But then, you always knew it would take something more than driving a hybrid, carrying around a reusable coffee cup, and hoping for the best. So let's begin with a political scientist's equivalent of an archaeological dig, revealing how social rules shape our everyday existence—and in the process helping us to answer the question *Who rules the earth?*

2

Strings Attached

Imagine for a moment that you are taking a leisurely walk on a favorite beach. As the calming sound of the waves and the wide horizon clear your mind and heighten your senses, you begin to notice the things around you. A group of birds floating on the wind. The play of light through scattered clouds. The remarkable process whereby stones too strong to break by hand have been transformed by time and ocean currents into countless sand particles crunching under your feet.

These and other aspects of the natural world capture our attention and inspire natural scientists to discover their secrets. But there are other realities here that go unseen by the untrained eye, and have yet to enter into the colorful documentaries provided by scientists, journalists, and other chroniclers of the natural world. These are the social rules that pattern this physical reality. Sometimes these rules take the form of laws. In other instances they appear as building codes or product design standards. Voting rules, property rights, and constitutional guarantees count among our most powerful social rules, which also include unwritten but widely recognized principles of right and wrong that guide our actions. Our task in this chapter is to make these social rules more visible—to help you "see" the rules shaping your everyday activities, to understand something of their political origins, and to appreciate why these rules matter for the future of our planet. To begin, let us return to our stroll on the beach and see what traces of politics and power we find amid the shells and stones.

INVISIBLE WORLDS

First consider what is missing from the beach. Why are there no fences? Why can we walk on this beach at all? If our social rules specified that the surf and sand were available to the highest bidder, or belonged to the first party to stake a claim, we would have no more right to swim in the ocean than we would to plunge uninvited into a neighbor's pool. In fact public access to the shoreline differs markedly from one country to the next, depending on the rules in place. In Scotland, the Land Reform Act of 2003 ensures that the beaches are widely available to all who wish to enjoy them.[1] You can watch the seals at Tentsmuir Sands on the east coast, or dip a toe in the chilly waters of Clashnessie Bay far in the north, unimpeded. In Ireland, in contrast, coastal access is a stingy affair. There public access is purely at the discretion of those who own property adjacent to the shore. As a result of this rule, much of the coastline is off-limits to Ireland's children, birdwatchers, joggers, and other devotees of what Rachel Carson called "that great mother of life, the sea."[2]

In the United States, the ability to enjoy a day at the beach is ensured by our embrace of an ancient legal principle known as the public trust doctrine. First spelled out by the Roman Emperor Justinian, and later passed down through Spanish, British, and American colonial law, the public trust doctrine holds that any waterways suitable for travel by boat, and the land underneath them, cannot be owned privately, but are instead held in trust by the government for public use.[3] The weight of this rule in the American legal system was confirmed in a famous Supreme Court case in 1892. The case concerned a jaw-dropping decision made by the state of Illinois, where legislators voted to sell Lake Michigan to a private corporation. The legislature granted the Illinois Central Railroad Company much of the state's portion of the great lake, including exclusive access to a huge swath of the Chicago harbor—over 1,000 acres in all. Fortunately, the US Supreme Court ruled that the state of Illinois was in violation of the public trust doctrine. The Court's decision quoted Andrew Kirkpatrick, a New Jersey Supreme Court judge who argued in 1821 that to grant a private party exclusive control over state waters would be "divesting all the citizens of their common right. It would be a grievance which never could be long borne by a free people."[4] A century and a half after Kirkpatrick penned these words, coastal access was enshrined in the US Coastal Zone Management Act of 1972, a landmark law that enabled comprehensive coastal protection and planning throughout the country.[5]

If the story of public access to beaches, and of social rules generally, were simply one of fair-minded rules fairly applied, there would be little reason for us to take interest in where these rules come from. But this story raises a larger point. The fact is that many of the simple pleasures we take for granted today, such as a walk on the beach, are possible only because others before us scrutinized the existing order of things, found it wanting, and changed the rules. The sight of a toddler wobbling toward the shore with a bucket of sand in tow seems far removed from the clash and clang of politics. Yet it is political engagement that made this innocent scene possible. The Coastal Zone Management Act did not come about as the inevitable result of a society coming to terms with the side effects of economic growth. It was the product of public protests stretching from Oregon to New Jersey, where citizens voiced concerns about the rapid development of coastal areas and declining public access.[6] Fences were bulldozed, city land use plans were modified, access paths were opened, and the beach was made available to all (with some notable holdouts, the focus of continuing public advocacy today).[7]

Public access to American beaches also required overturning rules inherited from a previous era that were deliberately designed to keep people out. Most notoriously, as recently as the 1960s, hundreds of state and local ordinances made it illegal for African Americans and other people of color to enjoy a day at the beach. During the Civil Rights era, organizers challenged the old rules using techniques such as "wade-ins" at whites-only beaches stretching from St. Augustine, Florida, to Biloxi, Mississippi (Figure 2.1). They often met with hostility and brutal violence. At a wade-in event in Chicago on August 28, 1960—along the same stretch of coastline that was almost sold off to a railroad company a century earlier—protesters were attacked by a mob of 1,000 stone-throwing segregationists. In the southern states, several wade-in activists were murdered. The remarkable sacrifices of these young Americans were ultimately successful in changing the rules and making a day at the beach a right available to all.[8]

22 WHO RULES THE EARTH?

FIGURE 2.1 Protesters are attacked during a "wade-in" at a segregated beach in St. Augustine, Florida, 1964

Russell Yoder/UPI.

TIES THAT BIND

As we continue our political excavation of the shoreline, we soon discover that the rules governing public access are only the beginning. The chemical composition of the ocean itself has been shaped by social rules—notably an international treaty that banned the practice of tankers intentionally dumping excess oil at sea, which by the 1970s had led to the discharge of a million tons of oil every year.[9] As you take a deep breath of the ocean breeze, the physical quality of the air filling your lungs is very much a function of social rules, specifically clean air regulations that have dramatically reduced pollution levels since the 1970s.[10] (Those readers wearing nail polish may be pleased to know that it contains fewer smog-producing chemicals now as a result of those rules.) The abundance and diversity of fish and other sea creatures are shaped by government policies and by private agreements among fishermen. In Maine, the lobster fishing community has designed its own sustainable harvesting agreements, enforced by local patrols of "harbor gangs" who ensure compliance with the rules.[11] Up the coast in Newfoundland, in contrast, ineffective rules governing offshore fishing practices led to the complete collapse of Canada's famous northern cod fishery by the early 1990s.[12]

Even the rays of the sun warming your face are affected by social rules, as implausible as this might seem. The level of ultraviolet radiation reaching your skin is more dangerous now than it used to be, due to the presence of man-made chemicals in the atmosphere that have thinned the earth's protective ozone layer. But it is safer now than it would have been (and over time will grow safer still) due to the Montreal Protocol, a set of international rules that successfully phased out the use of these substances.[13] The sunscreen you carry in your bag was available for purchase on the store shelf only because the company producing it can protect its invention through patents—rules of use that are enforceable in a court of law. As a consumer, in turn, your assurance that the product provides the level of protection advertised on the bottle is a function of rules issued by national regulatory authorities. (As I write, the US Food and Drug Administration is releasing a long-awaited revision to its sunscreen bottle regulations amid a flurry of public comments.)

As you walk past a "Do Not Litter" sign, you notice a large freight ship in the distance. The vessel spews a stream of toxic pollutants into the air, the result of low-grade bunker fuel that these ships use to power their massive diesel engines. The American Chemical Society estimates that air pollution from marine transport results in the premature deaths of 60,000 people every year in port cities around the world.[14] How can such a shocking situation endure with the full knowledge of our scientists and lawmakers? The answer is that the ship is exempt from domestic air quality rules because of its status as an international carrier. You might decide then and there to create a citizens' group to demand that your elected officials address the situation with an international treaty—but only if you happen to live in a country where constitutional rules protect the right of citizens to speak out and organize.

Of course, you don't need to visit a beach for examples of how social rules shape our planet and our lives. Chances are none of your clothes are from Cuba, due to international rules enforcing a trade embargo. But many of your accessories are from China, following a change of rules by the Communist regime that promoted market growth and facilitated China's entry into the rulemaking body known as the World Trade Organization. Your blood contains less of the pesticide DDT than was the case for someone reading a book about the environment forty years ago, before DDT was banned in industrialized countries. If you are in a public building with a restroom stall large enough to accommodate a wheelchair, this is not the result of goodwill on the part of the building owners. It is the end product of a protracted political struggle led by people with disabilities, whose efforts over two decades culminated in new rules such as the Americans with Disabilities Act of 1990. Whether we choose to notice them or not, social rules pervade every aspect of our lives (Figure 2.2).

RULES AND FREEDOM

All this talk of rules hiding in every crack and crevice of our existence may strike some readers as a little, well, creepy. Are these not a threat to individual liberty? Doesn't each rule, with its parade of dos and don'ts, erode a bit more of our freedom of action? Perhaps after reading this chapter you decide to liberate yourself from the shackles of social rules. If you happen to live in the United States, you might jump in your car (which you can drive because you meet age and competency requirements), start the engine (likely assembled in Mexico as a result of the North American Free Trade Agreement), and dart down the highway (thanks to the Federal-Aid Highway Act of 1956), driving, one would hope, on the required side of the road. Shaking these thoughts from your mind, you leave the city and head up a forested road, your car insurance contract rattling around in the glove compartment. Weary of these reminders of social rules, upon reaching the summit you dash out of your car and jettison all of your material belongings, until finally you come to rest on a precipice, poised in a state of pure nature, surveying the undisturbed wilderness around you. This meditative moment would be an opportune time to reflect on the US Wilderness Act of 1964, which made this peaceful respite possible by protecting the remnants of wild landscapes untrammeled by shopping malls. That is, the very absence of human intervention requires social rules. The same is true of freedom, including the freedom to travel up a mountain road, as the founders of the United States were acutely aware when they put in place social rules designed to constrain the exercise of government power.

Social rules are an indispensable and inescapable part of our existence. Just as individuals can't survive for long without societies, civilizations (and the school systems, airports, pizza joints, and soccer leagues that comprise them) cannot function without rules to guide the interactions

FIGURE 2.2 Social rules shape our world

among participants. The rules we live by shape our rivers, our skies, and the type and amount of energy we use. They determine whether our forests are clearcut until they resemble desolate moonscapes, or instead include intact ecosystems in which wildlife can flourish. At their best, social rules protect human rights and promote long-term prosperity. At their worst, social rules comprise elaborate systems for the subjugation of entire peoples and promote the pursuit of the quick buck regardless of the cost to our economy and our ecology.

Most important, the rules we live by can be changed. This may appear to be a daunting task. Every day we hear news stories about political gridlock, a polarized electorate, and the influence of money and power deployed in faraway places. Little wonder that so many of us assume we are powerless to change our world for the better. Yet it is often the case that the more we understand about a thing, the less fixed and immutable it appears. This is true of our rivers and grasslands, which despite their seemingly timeless character have changed dramatically over the centuries. And it is equally true of our political structures, which may appear permanent and unyielding on a daily basis, but are in fact prone to major shifts within a single human lifetime, punctuating long periods of stability with moments of sweeping social change. The election of an African American president, gay couples' right to marry, the collapse of the Soviet Union, and China's rocket-fueled economic growth are among the more visible examples of situations that in the span of a few years turned from unthinkable to unstoppable. But far away from the headlines that capture history's most dramatic changes in course, efforts are underway by reformers working behind the scenes, in rich and poor countries alike, to reconcile economic growth with environmental quality. Often these efforts are successful, as we will see in the next chapter. For now, we need to take a look under the hood to understand exactly how social rules work.

THE WORLD'S LARGEST MACHINE

What are the distinguishing characteristics of social rules? They are *social* in the sense that they shape interactions among people. Like DNA guiding the blizzard of chemical activity in a human cell, these social blueprints serve an essential coordinating function, preventing cars from crashing at intersections and promoting complex forms of joint activity like staging a rock concert, forming a corporation, or deploying an army. We are most interested here in rules that operate at scales larger than that of a family. If the Garcia family has a rule that TV is not allowed during dinner, this is not a "social" rule in the sense used in this book. If they attend church on Sunday mornings, their actions are guided by a social rule in the sense that it affects the activities of many families in predictable ways.

To have any meaning, a rule must be understood and followed, even if imperfectly, by those bound by it. A national law that no one knows about carries little relevance for this discussion, no matter how much fanfare may have surrounded its passage. Consider the rainforests. In tropical countries, if you flip through the musty old volumes weighing down the shelves of law libraries, you will find that most of these countries have had clear prohibitions against the destruction of forests since the colonial era. But it is only recently that reformers have begun to turn these ceremonial gestures into enforceable rules guiding the decisions of landowners and timber companies. It's not just what is written on paper that matters, but what we carry around in our heads. This is why social taboos, such as cutting in line at an airport, or speaking disrespectfully to a village elder, carry the force of social rules even if they are nowhere written down.

Every social rule assumes a common form. First, it clearly specifies a number of distinct roles. Next, it spells out the rights and responsibilities attached to anyone who occupies these roles. Thus a rental contract defines the roles of landlord and tenant and describes the obligations and expected benefits for each. British parliamentary rules specify that the person occupying the role of Speaker of the House has the right to choose which lawmakers speak in debates and in what order. The Speaker also carries the obligation to act with impartiality, resigning from his or her political party and no longer socializing with fellow legislators. Whether we are dealing with international treaties or department store return policies, all social rules can be understood in terms of these three R's: roles ("As a customer, ..."), rights ("you can return this item..."), and responsibilities ("in good condition within 30 days with proof of purchase"). The world's nations are now haggling over the three R's with respect to climate change, debating which countries have an obligation to control emissions of carbon dioxide and how this responsibility is weighed against rights to economic development and national sovereignty.

This same basic structure can be found within the unwritten rules guiding our actions. Consider the case of an elderly woman boarding a city bus.

In many cultures, those occupying a certain role—seated passengers who are young and/or male and are in close proximity to the elderly passenger—have associated with their role the duty to promptly offer their seat. Rules that lay quietly under the surface of things may become glaringly visible when broken. A teenager who neglects to offer his seat will learn this lesson quickly when confronted with a sea of scornful stares and a sharp comment from the bus driver. Often these unwritten rules eventually become codified in the law. The process resembles a well-trodden footpath cutting across a college lawn that campus authorities eventually acknowledge and turn into a paved walkway. The most effective rules combine formal written regulations with unwritten but widely shared understandings that give them legitimacy and force.

When rules catch our attention, they do so in a one-off manner. Sign here. Silence your cell phone during the performance. Hold the door for others. Entries will be judged based on originality and technique. Employees must greet customers as they enter the store. You have the right to an attorney. Place your recyclables in the blue bin.

To appreciate the true power of social rules, however, we need to take in the whole picture. As these rules build on one another, they interlock and intertwine, forming formidable structures—agglomerations of corporate contracts, traffic regulations, cultural norms, and dos and don'ts enforced by opinion leaders, judges, priests, neighbors, referees, bosses, voters, and friends. Huge networks of rules underlie and perpetuate these things we call Korean culture, the Interstate Highway System, Chevron, or Boston, Massachusetts. When social scientists use the language of "institutions," we are trying to draw attention to these large interconnected systems as well as the individual rules that comprise them. Sometimes in harmony, at other times in conflict, rapidly changing or stubbornly steadfast, our social rules privilege certain agendas over others, direct resources this way instead of that, and set the ground rules for economic growth and political change. They are what enable civilizations to flourish or implode. We reproduce our kind through biology, but our ways through institutions. In their totality, social rules make up the world's largest machine.

THE GHOST OF POLITICS

If there was ever a man who aspired to rule the earth, it was surely Napoleon Bonaparte. Napoleon knew a thing or two about power. At the height of his reign, the diminutive emperor ruled over 44 million subjects in 130 departments encompassing most of Western Europe. When the British and Prussians finally defeated Napoleon's Grande Armée at the Battle of Waterloo in 1815, they were so wary of his influence that they forced him into exile on the remote island of St. Helena, a speck of land in the South Atlantic located 1,200 miles from the nearest continent. During the final years of his life, Napoleon could be found reflecting on the nature of his power. His thoughts were recorded in a remarkable document, the *Mémorial de Sainte Hélène*, in which his attendant, the Count Emmanuel de Las Cases, kept a careful record of the emperor's words and deeds over a period of eighteen months. Napoleon was especially preoccupied with the question of whether his influence would last. Surveying the lonely expanse of ocean from his room, it was not his historic achievements on the battlefield that gave Napoleon hope. It was the new rules he left behind. "My true glory is not to have won forty battles," he said. "Waterloo will erase the memory of these victories. What nothing can erase, what will live forever, is my Civil Code."[15] Napoleon was referring to the legal code that his jurists drafted and which the emperor imposed throughout France and the conquered territories. The Code gathered together a disparate collection of feudal laws, fused them with the Roman legal order, and created a transparent and systematic body of law. Today the Napoleonic Code underpins the legal systems of dozens of countries, from Romania to Egypt to Chile. The great emperor realized that the ultimate power lies not in the flash of today's achievements, but in shaping the very rules that societies live by.

Thus another defining characteristic of social rules is they are designed to last. We create rules to project a desired pattern of social interaction into the future, whether that means receiving regular shipments of bread on Thursday mornings, or banning the use of ozone-depleting solvents in the dry cleaning industry. They *institutionalize* new practices. This is one reason why social rules are so important for sustainability. Stewardship of the earth requires looking not only beyond quarterly profit statements and election cycles, but beyond the life spans of individuals. Social rules are the devices we use to achieve this. These rules may assume the form of religious doctrine, such as the Catholic Church's Catechism 2415 that specifies "Man's dominion over inanimate and other living beings granted by the Creator is not absolute; it is limited by concern for the quality of life of his neighbor, including generations to come." In other instances this future orientation can be achieved through a legal tool like a conservation easement, a relatively recent invention in which a landowner agrees, in exchange for tax benefits, to place an irrevocable condition on a property to ensure sustainable use of its resources by all future owners.

The durable quality of social rules is important because we cannot count on eternal goodwill or the unwavering vigilance of volunteers to sustain a worthy cause. Human attention wanders, as does that of the media and the political establishment. Occasionally moments of great political enthusiasm well up and puncture the otherwise placid stillness of a society. But after the revolution is over, and the ticker tape has been swept from the streets, people return to everyday concerns, tending their gardens and stock portfolios. It is then that the rules left behind determine the true legacy of a movement for change.

This phenomenon was most famously observed by political scientist Anthony Downs in his essay "Up and Down with Ecology: The Issue-Attention Cycle." Writing in 1972, at the dawn of the modern environmental movement, he tried to predict whether the dramatic growth in environmental concern in the United States would have a real impact. Anthony Downs asked the same question that vexed Napoleon: Will this last? He observed that public attention to social problems tends to move in cycles, with a quick burst of enthusiasm gradually giving way to disinterest as citizens turn their attention elsewhere. Downs also argued, however, that during the "up" phase of popular interest, if the public's concerns are institutionalized—if they are embedded in laws, regulations, and associated implementing agencies—then it is possible to address these large-scale, long-term problems in a sustained fashion. Downs was largely proven right on both counts. There was a general decline in public interest in environmental issues in the United States in the early 1980s; this corresponded with an economic recession and shows up clearly in the public opinion data. But the new rules laid down in the 1970s, such as the Clean Water Act and the Endangered Species Act, ensured sustained progress on these issues despite the inevitable swings in the public mood.

If stability is the mechanism through which rules generate benefits for society, it is also the source of their most pernicious effects. We have all encountered situations in which decision makers cling stubbornly to rules even when these stand in the way of doing the right thing. Generals can be found fighting the last war. Businesses resist change because "we've always done it this way." And government bureaucracies are famous for showing greater fidelity to following the rules than to getting things done. Permanence has its pitfalls. Yet perils can be found at either end of the spectrum between stability and change. We will see in chapter 6 that crafting smarter rules to promote sustainability is an incredibly challenging task in countries plagued by political instability—which is to say, in most of the world. In the end, institutionalizing new practices requires a balancing act. Social rules must be sufficiently sticky to prevent whimsical reversals, but they must not foreclose the possibility of future revisions in response to new ideas and changing needs.

To answer this book's title, then, let us begin with the observation that many of those who rule the earth are dead and gone. They built structures—laws, policies, codes and contracts—that cast a shadow on the future. Some of these structures, such as the Bill of Rights, were inspired by profound insight into the public good. Others, like rules restricting beach access, were put in place to benefit one group at the expense of another. Some rules—such as US crop subsidies, or Brazilian policies granting land to those who "improve" it by removing the trees—once served a noble purpose but have outlived their usefulness; yet they cast a shadow still, shaping what we plant and how we treat the land. Social rules are the ghosts of political battles past and are the legacy of social structures that we pass on to the future.

THE ROAD AHEAD

In the rest of this book we will peer into the machinery that runs society, its thousands of invisible levers patterning our actions in ways that are sometimes noticeable (Wednesday is trash day), but often taken for granted—like the assumption that trash disposal is the sole responsibility of consumers and municipalities, rather than the companies that design products with excessive packaging. Our wide-ranging tour will encompass the nature of laws, the fate of kings, and the rule of the McDonald's french fry. Along the way we will see that social rules are not limited to governmental laws and regulations. These are important examples of social rules, and I will refer to many such examples throughout this book. But the reach of rules, and the scope of this book, extend far beyond the activities of park rangers and politicians, encompassing also the rules created by paper manufacturers, neighborhood associations, and sports arenas.

The task before us is to build a platform, piece by piece, that will offer a new vantage point for seeing the world differently. Our viewing platform will have the following pieces. We begin and end with social change. In the next chapter (3), and in the final three chapters of the book (9–11), we will consider how social change actually works, and the role that you can play in that process if you so choose. Standing between these are five chapters (4–8) that reveal the invisible architecture of social rules that pattern our behaviors in so many ways. I will draw on examples from many dozens of countries, reflecting my own research specialty, comparative politics, which explores and compares the inner workings of diverse societies around the globe. At the same time, my assumption is that many readers are from my home country, the United States. I hope that the non-US reader will forgive the occasional bias toward examples and debates of particular interest to American readers.

Allow me to provide a bit more detail of what's to come. Out of the gate, chapter 3 ("Feasible Worlds") tackles one of the most important and humbling questions we can ask ourselves: Can we really change the world for the better? It turns out that social scientists have quite a lot to say about the possibilities for change. We will consider research findings that explain why we do not live in the best of all *feasible* worlds and why there are so many opportunities to make people and our planet better off. In chapter 4 ("A Perilous Journey"), we begin our study of the earth's rulebook by taking a closer look at one of the most powerful social rules of all: property. To understand who rules the earth, we need to appreciate who owns it, and how the rules surrounding ownership are made. To accomplish this, we will follow the cerulean warbler, a highly endangered migratory bird, as it traverses the Western Hemisphere searching for suitable forest habitat in which to rest. At each stop along the cerulean's journey, we will see how property rules affect its prospects for survival. In chapter 5 ("The Big Trade"), we consider the deeper, and often counterintuitive, relationship between rules and property, with a clear-eyed look at debates surrounding the use of market forces to combat pollution. My goal is to empower readers to participate in these debates without the ideological baggage that weighs down so much of the public discourse. In chapter 6 ("A Planet of Nations"), we will head into the corridors of government power, witnessing how different countries around the globe are grappling with environmental problems through rule-making systems that help or hinder sustainability.

Environmental problems move effortlessly across borders, thumbing their nose at our attempts to organize political life into cities, states, and nations. In chapter 7 ("Scaling Up") and 8 ("Scaling Down"), we will see how the distribution of rulemaking power across different levels of governance is shifting due to two major trends: the formation of the European Union and the unprecedented move by dozens of countries to decentralize environmental rulemaking power to local levels. After this wide-ranging tour around the globe, we return to the question of what it takes to bring about meaningful change. In chapter 9 ("Keep the Change"), we will see that the challenge is to not only break the patterns that cause us to get stuck in ruts (such as oil dependency), but also to establish "good" ruts, putting into motion self-reinforcing trends and new assumptions of normality. In chapter 10 ("Super Rules"), we will consider a special category of rules that decide how other rules are made, dealing with questions like who participates and which principles guide the creation of policy. Anyone hoping to make a lasting impact on the planet would do well to pay careful attention to super rules; the polluters certainly are. The final chapter ("Paper, Plastic, or Politics?") offers practical suggestions for those interested in taking part in rewriting the rules that govern the earth. I offer general principles of action that distill lessons from the research covered in the book, which action-oriented readers can pair with their own research into local political contexts.

3

Feasible Worlds

What would the world be like if high-speed trains arrived every ten minutes to whisk you away to the city of your choice? It would be a lot like Japan. What if our computers and coffee makers were not dumped in toxic landfills at the end of their lifecycle, but were instead reused as raw materials for new consumer products? Just ask Western Europeans. What if instead of crafting environmental rules in secret, governments were required to share all of the information shaping their decisions with any citizen who demanded it? The answer can be found in the United States.

The differences among the "worlds" experienced by citizens of Japan, Europe, and the United States stem in large part from variation in the rules underpinning them. In Japan, a national system of bullet trains (*shinkansen*) came about not because of an inevitable march of technological progress. It was the result of national and local rules that transformed a disjointed collection of railways into an integrated national system—a system that has not had a single fatal accident since its inauguration in 1964. In Europe, new rules make corporations responsible for collecting and recycling the electronic goods they sell to consumers. Because they must safely dispose of any toxic substances in their products, these companies have a strong incentive to remove heavy metals and other poisons from the manufacturing process. In the United States, the Freedom of Information Act empowers citizens to demand that government agencies send them all pertinent documents describing the rationale behind their decisions—a degree of transparency that is unheard of in Japan or Europe.[1]

Of course, these states of the world did not always exist. They were brought into being through deliberate acts of social change in which old rules were tossed and new ones put in place. Yet many people find the thought of social change too daunting. It seems unrealistic, out of reach. Compared to the dizzying pace of change in technology and popular culture, it appears that progress on big social problems like poverty alleviation, human rights, and environmental sustainability moves at glacial speed. Trends in music and media are here and gone faster than we can figure out how to operate our newest electronic device. We switch schools, jobs, homes, and hairstyles. But politics? New approaches to urban governance? Major shifts in the way that societies use energy? These forces seem so remote that many of us retreat to the comfort of the little things that are patently under our control. We dutifully carry our recycling to the curb, send a check to the Red Cross, or perhaps volunteer at the local school. We grab the laundry detergent advertising 50 percent less packaging. We "do our part," leaving the bigger changes to the seemingly impersonal forces of history and to people of great power and influence in distant locations.

This reluctance to get involved in promoting social change stems from a perfectly reasonable desire to be "realistic" in our efforts. Consider the following. Have you ever encountered a truly inspiring vision for a better world, only to have your bubble burst by the thought that it's just not realistic? This hopeful vision might arrive in the form of a captivating film, or a charismatic public speaker who advocates a future characterized by peace, prosperity, and sustainability. You ride that high for a while, perhaps vowing to yourself that you'll commit to the cause from that day forward. Then as the days and weeks pass, the initial surge of emotion fizzles and reality creeps back in (often with help from discouraging comments by others). However compelling the vision might be, you tell yourself that it's just not practical in today's world.

Hold that thought for a moment, and let's consider another.

Have you ever come up with a novel idea that you think might provide real value to others? Perhaps you dream up a way to revitalize an urban center. Or it could be a proposal to reduce the "carbon footprint" of your community through energy efficiency measures. Maybe you even have a bold vision for reducing world hunger. After enjoying the thrill of thinking through the novelty and implications of your scheme, the thought process then slams to a halt with the following question: *If my idea is so great, wouldn't someone already have tried it?*

The twin notions that the good ideas have already been tried, and that meaningful social change is out of reach, are not only deeply disempowering, they are also inaccurate. These habits of mind rest on the assumption that, while we may not live in the best of all imaginable worlds, for all practical purposes we live in the best feasible world. It's not everything we might hope for, but we figure it's probably the best we can expect to see in our lifetimes. In this chapter we will see that this perspective seriously underestimates the possibilities for significant shifts in how societies go about their business. I will not argue that any outcome is possible, or that every problem is solvable. I will argue, however, that we do not live in the best of all feasible worlds. Notice I did not say *imaginable* worlds. I am not referring to the far-out fantasies of utopians and science fiction novelists. I am talking about substantial improvements in human and environmental conditions that are economically, technologically, and politically feasible in the near term.

If these alternative outcomes are so feasible, you might ask, then why have they not already come about? This is the crux of the issue, and in this chapter we will see that many perfectly feasible scenarios are like boulders perched precariously at a cliff's edge, hindered by what social scientists call collective action problems—hurdles large and small that stand between the world we have and the world we can get. I am not going to simply tell you that you can make the world a better place. Rather, I will prove it. Fortunately, I have half a century of research findings to lean on when making the case. While researchers use solemn terms like "problems" and "failures" to describe the barriers to change, the pervasiveness of these impediments is actually great news. It means that countless changes are indeed possible because our world is teeming with potential but unrealized outcomes. All that is required is that somebody (or to be more precise, a group of somebodies working together) identifies and removes the hurdles standing in the way of win-win situations and other opportunities to improve upon the status quo. Many of these barriers are less formidable than you might imagine. Often it is a matter of reforming institutions—the rules that we rely on to coordinate human activities. In chapter 9, we will take a closer look at how social change works, with an emphasis on removing dysfunctional rules and creating new ones better suited to society's needs. The point of the present chapter is to establish that significant shifts in the way we treat our planet are well within our grasp. To begin, let's consider the story of how a small group of business entrepreneurs turned an "unrealistic" idea into a reality by creating new rules that are transforming the world's largest industry into a more sustainable enterprise.

RE-BUILDING THE WORLD

The American Society of Testing and Materials (now ASTM International) occupies a modern white building near the banks of the Schuylkill River that runs through West Conshohocken, Pennsylvania. It seems like an unlikely place for ruling the earth. Political leaders do not come to visit. You will not find swarms of reporters and news vans with towering satellite dishes gathered outside. Frankly, most people have never heard of the place. But what goes on inside the walls of ASTM is truly remarkable. ASTM is a sort of rulemaking factory. It is not a government agency, but an influential private sector rulemaking body with technical committees staffed by no fewer than thirty thousand experts. Electrical engineers, materials scientists, firefighters, food chemists, architects, aeronautics experts, and others work year-round like elves in Santa's toyshop, crafting hundreds of new industry standards on the proper design and use of everything from jet engines to bicycle reflectors.

ASTM's magazine, *Standardization News*, is not the kind of thing you will find featured on late-night talk shows. But in it you will find a world of rules, large and small, that pattern our daily existence. You can learn about rules like paint standard D562, which specifies that a Stormer-Type Krebs Viscometer is the best way for paint manufacturers to test the consistency of their products. It turns out that paint consistency is of keen interest to autobody shops and portrait artists alike. To appreciate the stakes at play with mundane-sounding D562, consider that this particular rule was created by a committee of 600 experts from 36 countries, chaired by the conservation administrator of the National Gallery of Art in Washington, DC. Or take standard F2508, which was issued by ASTM's Committee on Pedestrian/Walkway Safety and Footwear. F2508 lays out the proper procedure for determining the potential for a walkway to become slippery, using something called a tribometer. (This particular rule was crafted by the committee's aptly named Subcommittee on Traction.)

To govern such minute details of daily life might seem far removed from the question of who rules the earth. Indeed, you have to feel a certain sympathy for the ASTM expert trying to explain the importance of his or her work to a date over a candlelight dinner. But another ASTM committee issues standards on what it means to sustainably harvest forests. Others have created standards to determine if fish were killed by a chemical spill, and whether a company has done an adequate job of cleaning up a toxic waste dump. As these standards stream out of ASTM like widgets on an assembly line, they frequently become the norm in the industries concerned. Government agencies incorporate thousands of ASTM standards into local building codes and national health and safety regulations.

Like most people, David Gottfried had never heard of ASTM when he decided to enter the business of sustainable buildings. Trained as a commercial real estate developer, Gottfried is one of the founders of the Green Building Council, a consortium of industry leaders who are changing how we think about the built environment. The way buildings are designed, constructed, and operated has an enormous impact on our air, our water, and the livability of our communities. Buildings use over a quarter of the planet's commercially harvested wood and almost half of the world's steel and mined sand and stone. Constructing and operating buildings accounts for fully 40 percent of global energy demand.

Social rules were the furthest thing from Gottfried's mind when he began exploring what it would take to create a market for more environmentally friendly buildings. Traditionally, real estate developers have operated according to two simple rules: Make money and don't break the law. "We met the building code requirements for water and energy efficiency and not a bit more. We picked building materials according to price and market trends, never giving a thought to where they come from or where they would end up."[2] At the outset, Gottfried's primary motivation was to increase profits. As he tells the story in his autobiography *Greed to Green*, in the early 1990s Gottfried was working for Katz Construction, a real estate development company that was looking to diversify its investment portfolio during the economic recession. Sustainable buildings seemed to offer an unexploited niche. As a newcomer to questions of sustainability, he attended conferences where he sought the advice of leading environmental architects and engineers, hoping to quickly assimilate the state of the art in green design. At a conference hosted by the American Institute of Architects in Boston in 1992, Gottfried was enthralled by speakers like William McDonough who argued that Earth-friendly building designs were already technically feasible and often cost-effective.

If green buildings were so feasible, why weren't they being built? As a businessman, Gottfried understood that while these ideas might have great merit in theory, they hadn't penetrated the mindset of his industry. But instead of dismissing the idea as unrealistic, Gottfried used his knowledge of the industry to explore what it would take to create a lucrative market for green buildings. He met with construction companies, suppliers, and other influential players in the commercial building industry to learn why they chose some products over others, and why proven technologies like energy-efficient windows and recycled carpeting were not more widely used. As Gottfried explained to me during an interview in 2009, he concluded that if a transformation were to take place in the building industry, it would require changing the rules that guide its decisions. At the most basic level, there weren't even any agreed-upon standards for what constituted a green building or for assessing the environmental impacts of building materials. There was no way to translate consumer demand for green buildings into market incentives for producers, because the buyers and tenants couldn't distinguish the environmental impact of one building from the next.

Gottfried's investigation into the rules underlying the building industry soon led him to ASTM. In 1990, ASTM had created an Environmental Assessment and Risk Management Committee. The committee was chaired by Mike Italiano, an attorney who had been writing rules for ASTM since the early 1970s and who was deeply committed to environmental stewardship. Committee member Rob Josephs introduced Gottfried to Italiano, and the two men decided to create a new Green Building Subcommittee with Gottfried at the helm.

Despite their high hopes for the new initiative, it quickly ran into trouble. ASTM committees work by consensus, and while the Green Building Subcommittee had the benefit of a large and diverse group of experts, it also included industry trade group representatives who were loath to change. The trade groups consistently vetoed efforts to create standards for green buildings, and after three years there was little to show for the group's efforts. At Italiano's suggestion, he and Gottfried bolted from the ASTM and created a new organization, the Green Building Council. The council would operate in a similar manner to ASTM but would only allow representatives of individual companies, rather than trade groups, to take part. Corporate participation was crucial, both for cultivating private sector expertise and for ensuring the legitimacy of the enterprise within the building industry. But as Italiano explained to me, while some individual companies were ready for change, "Trade associations don't have brands to protect, so they're immune from public pressure." Officially launched in 1993, the Green Building Council brought together an eclectic group of builders, suppliers, architects, engineers, financiers, electric utilities, insurance companies, and environmental organizations that could think holistically about how to improve the environmental performance of buildings.

In 1998, five years after the launch of the Green Building Council, this small collection of industry mavericks launched a new rulemaking initiative that would prove to be the Council's most powerful innovation. Leadership in Energy and Environmental Design (LEED) was created to tackle a particularly thorny problem facing proponents of change in the building industry. As we saw earlier, the people who purchase buildings, and those who manage and live in them, have no way to reliably assess a building's impact on the environment. LEED would provide owners and occupants clear information about what they're getting for their money, akin to the nutrition information on a cereal box.[3] LEED-certified buildings would dramatically reduce water and energy use. They would recycle demolition waste, incorporate recycled materials into furniture, and landscape with native plants friendly to wildlife. LEED buildings could offer healthier air for building occupants by using materials with little or no "off-gassing"—lingering vapors from toxic chemicals used in the manufacturing process (think "new carpet smell"). Using a transparent scoring system, LEED would rank buildings according to their degree of greenness, providing developers and buyers with an incentive to exceed the minimum standards required by law. They brought in Rob Watson of the Natural Resources Defense Council to head up the effort, and drew on the expertise of people like Malcolm Lewis, president of the engineering firm CTG Energetics, who had long experience with environmental design and became chair of LEED's scientific advisory committee.

The strategy paid off. The world's first two LEED buildings were certified on March 30, 2000. These first movers served a specialized clientele willing to pay top dollar to go green: the headquarters of the Chesapeake Bay Foundation in Annapolis, Maryland, and the Kanalama Hotel, a sleek 5-star resort in Dambulla, Sri Lanka. The next group of customers included organizations whose missions emphasize public service and sustainability, such as natural history museums, environmental groups, universities, and government agencies. From there the idea spread like wildfire. In the middle of a severe economic recession, the US market for green building grew 50 percent between 2008 and 2010, to over $55 billion. During this period green buildings made up a third of all new nonresidential construction. By 2011, there were over 30,000 projects registered with LEED in dozens of countries, including over 1.6 billion square feet of commercial building space. Construction industry estimates placed the value of the green building market at about $100 billion in 2013, and projected a further doubling of the market by 2016.[4] Over 200,000 people have completed training courses to become LEED-credentialed professionals in fields like construction, interior design, and property management. Although LEED is no panacea (for example, building maintenance and operation receive short shrift relative to front-end design), it has transformed the construction industry, raising environmental standards as developers reach for the prestigious silver, gold, and platinum LEED categories.[5] As often occurs with rules created by the private sector, LEED has spilled over into the governmental realm. Dozens of American states, cities, and federal agencies have put in place rules requiring that new government buildings be LEED certified or otherwise meet its requirements. All new US embassy construction around the globe must now meet LEED standards.

THE ART OF THE POSSIBLE

LEED is only one among countless examples of how changes in social rules are changing the way people interact with their environment. These shifts are coming not only from private-sector initiatives like LEED. Ordinary citizens have demonstrated time and again that when we step outside the isolated world of "lifestyle change" environmentalism, and band together to revise the rules we live by, major advances are possible. Portland, Oregon, has transformed itself into a bike-friendly city—a quadruple win for improving health, reducing traffic, cleaning the city's air, and cutting global carbon emissions. (A quintuple win if you count the fact that bicycling is also great fun.) Between 1991 and 2010, Portland's dedicated bikeways increased from 50 miles to over 300 miles, while bicycle ridership tripled. This was spawned by the efforts of bike enthusiasts in the early 1970s, who helped change Oregon state law so that a portion of the state's highway construction budget is dedicated to bike paths. Down in Texas, meanwhile, the city of Austin has gone green. Voters have approved a series of measures to protect open space, which helps to conserve water in this drought-prone part of the world while providing recreation opportunities and protecting wildlife. City and state officials, working with groups like the Hill Country Conservancy and Save Barton Creek Association, have designated a third of Austin's land as a Drinking Water Protection Zone. The city has created several thousand acres of nature preserves in addition to the 30,000-acre Balcones Canyonlands Preserve, located just outside the urban periphery and co-managed by the City of Austin and Travis County.

Efforts like these come not a moment too soon, because the scale of the environmental problems we face today is truly formidable. Too many of our cities are jammed with traffic, too few children have access to clean air and green spaces, and too many dangerous chemicals find their way into our homes and consumer products. More than one out of every three freshwater fish species in the United States are threatened with extinction.[6] Worldwide, an estimated 500,000 people die from particulate air pollution each year—more than perish in all wars combined.[7] From 1990 to 2010, heat-trapping carbon dioxide emissions increased globally by 45 percent.[8] Fully 75 percent of the world's coral reefs are classified as threatened due to pollution, overfishing, ocean acidification, thermal stress, and coastal development.[9] Globally an estimated 2.5 billion people lack access to such basic sanitation services as pit latrines, instead using buckets or open sewers.[10]

There are many additional statistics we could consult to appreciate how serious the stakes are. But here I have a confession to make. By my nature, I can only tolerate so much bad news before my mind flips a switch and asks, *So what can be done*? I know people who can comfortably talk for hours about how the world is doomed. But I can only take so much before I grow impatient and need to know what's working well and why. It is not a matter of being an optimist or a pessimist, of debating whether the glass is half empty or half full. It's that I can't drink the empty part. I need options, and I suspect that many other people feel the same way. It is common for environmental orators to stun their audiences with chilling accounts of ecological disaster, but to then offer no practical solutions. Often these speakers are eminent scientists who, through decades of painstaking research, have catalogued a long list of environmental ills and can skillfully connect the dots and convey the scale of the problem. The problems are indeed very serious, and as a rhetorical strategy, an emphasis on pending catastrophe certainly grabs our attention. But it does not sufficiently motivate action.[11] Indeed it may have the opposite effect, as a sense of despair creates psychological detachment, turning people away from politics and civic engagement. People need to know what can actually be done to address these problems. Fortunately here is where social scientists can step in and offer some useful insights. So what does the research tell us about the possibilities for real change?

BARRIERS AND BREAKTHROUGHS

If LEED certification was such a good idea, why wasn't something like this already in place before reformers like David Gottfried arrived on the scene? If cities can reduce carbon emissions while saving on their electric bills, why are more cities not already doing so? Mind you, it's not difficult to see why some ideas have never come into being. Technological designs that ignore economic considerations, political proposals that require heroic assumptions about human behavior—the failure of these shortsighted plans is easy to comprehend. But ideas that would make the world better off while generating profit? Proposals that are technologically feasible and enjoy widespread public support? You have to wonder how it's possible that a free society filled with creative individuals could fail to take advantage of these opportunities.

It turns out that there are several types of hurdles that prevent us from living in the best of all feasible worlds. And this is great news—because these hurdles, while real, are not insurmountable.

SHARING INFORMATION

One reason why we do not live in the best of all feasible worlds is that decision makers—be they companies, consumers, mayors, parents, or presidents—often lack the information they need to make the best decision. When I say "best decision," I do not mean the choice most likely to help society as a whole; I mean the option that would best advance the decision maker's own interests.

The notion that people don't know how to best fulfill their own needs may seem strange, even patronizing.[12] But consider, by way of example, the process of buying a book. Let's say you decide that you would like to buy a nice work of fiction to relax with on vacation. If you walk into a bookstore (or browse online, the same logic applies), you immediately face a dilemma. Somewhere on one of the store shelves is the book that would give you greater fulfillment than any other book in the store. There is a book that would provide more laughs, more entertainment, more suspense, and would speak to your personal interests in more compelling ways than any other book. If only you could find it. So why is it that you don't find the book best suited to your needs? Consider what this would require. Hours of research scouring through book reviews and customer comments on the Internet. A team of bookstore employees tasked with assisting you after conducting detailed interviews to learn about your interests. Or perhaps you might read sample passages from several dozen books in the store to see for yourself which ones best match your mood and sensibilities.

Needless to say, gathering this information would be much too burdensome. And this is the essential point: Information carries a cost. Investing more of your time and energy might very well lead to a book that is better than the others, but the cost of doing so nibbles away at the benefit of reading the book in the first place. You could read several terrific novels in the time it would take to find the best one. As a result, you don't search for the best book in the entire store. You search for a good book. Maybe even a very good book. Perhaps you rely on a "suggested read" from an online site that uses a very simple (and low cost) software algorithm to divine your tastes. But you won't find the best choice.

It turns out there is an entire field of research, called information economics, devoted to the study of this phenomenon. One of the pioneers in this field was Herbert Simon, a brilliant and adept thinker who moved nimbly across the boundaries of political science, psychology, economics, and artificial intelligence. Beginning in the 1940s, Simon wrote numerous books and articles about our limited capacity to process information, research that earned him the Nobel Prize in Economics in 1978. "Because of the limits on their computing speeds and power," he wrote, "intelligent systems [notably the human brain] must use approximate methods. Optimality is beyond their capabilities; their rationality is bounded." As a result, "People satisfice—look for good-enough solutions—instead of hopelessly searching for the best."[13]

"Satisficing" applies to much more than the books we read on vacation; it also applies to the choices made by corporations. Herbert Simon demonstrated that, contrary to popular wisdom, firms are not actually profit maximizers. This may sound like a radical proposition, but with a bit of reflection we can see why this is the case. When hiring an employee, a firm faces the same trade-off that you do when searching for a book. A company could engage in countless hours of interviews with the candidates' former supervisors and coworkers. They could always search farther and wider to bring in a larger pool. But the cost is prohibitive, and so they hire what they hope will be a good graphic artist, or janitor, or sales representative, but they do not search for the best one. And here is where information economics relates to questions of the environment, for the same logic applies to a firm deciding what kind of windows to put in its new building. There may be energy-efficient windows that would lower the firm's heating bill while reducing fossil fuel use. But the company simply doesn't know about them.

The implications of information economics are greater still, because the same dilemma shapes a company's decisions about where to invest. There may be opportunities to earn money while doing good for society, but businesses—and for that matter shoppers, homeowners, cities, and countries—do not pursue these options because they are simply unaware of them. The information is too hard to find. As a consequence, we do not live in the best of all feasible worlds. Our firms and government agencies have not tried all or even most of the feasible ideas.

The high cost of information is the problem that LEED and the Green Building Council were designed to address. According to David Gottfried, green product standards for the building industry did not exist because the information just wasn't there. "Our concern about the environmental performance of building products and systems was entirely new," he explains. "We began to ask manufacturers for more information about their products, like the off-gassing profile over time as shown in an air chamber test.... In many cases, the company didn't have access to the information itself, and the cost of obtaining it could add up to hundreds of thousands of dollars."[14] By doing the research and creating a transparent rating system, the Green Building Council provided the missing information, making it easier for builders and manufacturers to be green and for consumers to notice and reward Earth-friendly business practices through their purchasing decisions.

LOWERING BARRIERS TO COOPERATION

Even if every firm, agency, and individual had access to all the information they needed, and even if all of these actors were pursuing their interests in rational and efficient ways, we would still live in a society that is not nearly as good as it feasibly could be. This is because people and groups often do not cooperate with one another *even when it is in their best interest to do so.* There are many circumstances in which participants would be better off if only they could find a way to join forces—forming an alliance, exchanging resources, trading ideas—and yet they don't get together. They may in fact know precisely what form of cooperation is needed, and fully appreciate the benefits, and *still* they do not do so.

What might explain this bizarre behavior? At first it seems highly irrational, but upon closer inspection it makes perfect sense. Failure to cooperate for the common good follows from what social scientists call collective action problems. This puzzle was first brought to light in 1965 by Mancur Olson, an economist at the University of Maryland, in his book *The Logic of Collective Action*. Based on his PhD research at Harvard, this little book turned conventional social science on its head. In the world of academia, authors routinely cite one another's work (as I do in the notes throughout this book) in order to build on the findings of a larger community, collectively gathering facts and insights like so many ants carrying twigs to the mound. If a book or article is cited in a couple dozen publications (which can be tracked using Google Scholar), this suggests some influence; a couple hundred citations would be deeply gratifying to most researchers, seeing that others have found value in the fruits of their labor. Mancur Olson's book has been cited in over 20,000 subsequent publications.

What's all the fuss about? Olson argued that cooperation (or "collective action") doesn't happen the way we might think. "Even when there is unanimous agreement in a group about the common good and the methods of achieving it," he wrote, "rational, self-interested individuals will not act to achieve their common or group interests" unless there is some sort of special incentive to get them to do so.[15] This is because cooperation is costly. It takes time to organize and attend all those meetings, to coordinate schedules, and return all those phone calls. Many organizations demand that members volunteer or pay dues to help advance their common agenda. Given the price tag, when people are considering whether to join a cause, they face a dilemma. Even if they can clearly appreciate the benefits of cooperation, they figure they can do one better: Reap the rewards of the group efforts without contributing to them.

At this point you might object that Olson paints a very cynical portrait of human behavior. But anyone who has tried to organize a group of volunteers, coordinate a student group, or stage a community event is familiar with this phenomenon. It's not that some of us are public spirited while others are selfish. (This may indeed be the case, but it is a separate point.) It is just inherently difficult to get people to contribute to things that jointly benefit a large group.

This finding is significant for two reasons. First, it contradicts what we were taught in our high school social studies courses. In a democracy, we are told, people with a shared interest or grievance will form a group and mobilize to advance their aims. They make common cause and press their demands on elected officials—right? Not so, says Olson. Incentives for free riding lead people to shun cooperative efforts that would produce shared benefits. But there is a second reason why Olson's insight into group behavior is so important. He presents us with an intriguing question: Given the inherent difficulties, why does cooperation ever occur? After all, we're surrounded with organizations and collaborations of all shapes and sizes. What's wrong with these people? Haven't they read Olson's book?

Olson argues that every organization—be it a corporation, a river conservation group, or an army—must find ways to overcome the free-rider problem by providing some additional benefits enjoyed exclusively by those who help the group. Within a company, this is straightforward: Employees contribute to collective goods like the profitability of the firm and shareholder value. The moment that a worker stops doing her part—spending the week on the ski slopes rather than finalizing the deal—her job is in jeopardy. Her access to the group benefits, in the form of a paycheck, is made contingent on her contribution to the group effort.

That's fine and well for a private firm, but how can you overcome the free-rider problem in a voluntary organization? Many professional associations solve this problem by providing their members with additional benefits that can be withheld from nonparticipants, like training workshops, access to research and data, and invitations to important networking events. The school Parent Teacher Association (PTA) is similarly situated: It provides collective goods (such as proceeds from a fundraiser) that benefit all of the school's families whether they contribute or not. But if PTA organizers combine their meetings with a dinner potluck or other social event, this strategy offers benefits to participants—and only to participants—in the form of food, camaraderie, and entertainment. Organizers have to find creative ways to sweeten the deal.[16]

Collective action problems are another reason why we do not live in the best of all feasible worlds. They refute the logic underlying the question,

If my idea is so great, wouldn't someone already have tried it? The relevant players may not have put a good idea into practice simply because of the short-term cost of cooperation. It is tempting to conclude that those who won't contribute to the effort without some extra incentive must not perceive the benefits of the group activity. If you have to provide food to get parents to show up at the PTA meeting, do they really care about the cause? Inevitably there will be some who care more about schmoozing over cupcakes than saving the school music program. But recall that free-riding behavior tempts even those who care deeply about the larger outcome and understand that it will take a group effort. A members-only benefit provides the little push required to move people from the sidelines to work together to achieve something they could not accomplish on their own.

We can draw an analogy here from chemistry and the concept of activation energy (Figure 3.1). If you want to unleash the energy from a chemical reaction, often you must first apply a small amount of energy on the front end to get the reaction going. Viewed in this light, the *If my idea is so great...* question is a bit like standing over a twenty-gallon drum of gasoline with a lit match and asking, "If this were so explosive, wouldn't it have already blown up?"

PEERING INTO THE GARBAGE CAN

Yet another reason why we don't live in the best of all feasible worlds is that the organizations we rely on to make the world a better place are frequently ill-suited to the task. And that is more great news. It's not as if a constellation of supremely efficient organizations have tried and failed to address

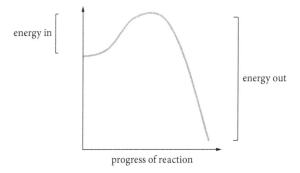

FIGURE 3.1 Activation energy

our most pressing problems. If you want to make a difference, there are many substantial and perfectly feasible improvements to be made.

What might lead organizations—businesses, government agencies, schools, environmental groups—to act in ways that do not advance their interests? This raises a larger question, namely, why do organizations do what they do? We might expect that organizational performance is shaped by external pressures, their life and death determined by the cruel but efficient logic of the survival of the fittest. This would also imply that those surviving for long periods must be doing something right. To a limited extent, this is true of firms in a competitive market economy, of nonprofit organizations vying for scarce donations, or of political parties courting voter fidelity. But there are other dynamics that shape what organizations do and why. In 1972, Michael Cohen, James March, and Johan Olsen wrote an influential research article titled "A Garbage Can Model of Organizational Choice," published in the journal *Administrative Science Quarterly*.[17] Organizations, these authors argued, often do not make rational decisions, in the sense of efficiently pursuing goals through an assessment of the costs and benefits of alternative strategies. This problem goes beyond the challenge of accessing information, discussed earlier. And it is distinct from the free-rider problems that inhibit the formation of groups in the first place. Cohen and his colleagues argued that even when organizations are formed and fully operational, with budgets, staff, and strategic plans, they often do not choose the appropriate means to achieve their collective ends.

These researchers observed that decision making within an organization typically involves so many competing priorities that it looks less like a rational pursuit of well-defined goals and more like a garbage can filled with a mishmash of unrelated items. As members come and go over the years, they dump into the garbage can their guiding philosophies, program priorities, and work routines, and these continue to exercise an influence long after their creators are gone. Policies and procedures are created to address the pressing problems of the day. Specific types of equipment are purchased, employees with the relevant skills are hired, and offices are opened in particular places for particular reasons. But as the years go by, the target population for the organization's services changes, as does the field of organizations it must compete with. And then they change again, and yet again. These shifting requirements, and the organizational responses they inspire, build up as layers of rules, routines, and administrative structures affecting how an organization operates. They influence the types of information it sees and ignores, and the skills it possesses. Any organization that has been around for more than a few years—be it DuPont, the Sierra Club, or the US Department of Agriculture—resembles a fossil dig more than a clean slate.

As a result, organizations do not design solutions to deal with problems as they arise. Rather, they have a stock of solutions and organizational capacities built up from past commitments, and are on the lookout for new problems to tackle with their existing solutions. Often organizational goals are ambiguous or contradictory. Consider government bureaucracies, which are famous for getting mired in "red tape"—the reams of rules they must follow to get anything done. In many government agencies, you cannot so much as order a replacement part for a vehicle without navigating a labyrinth of rules and procedures. Yet all of this red tape was created to address a specific problem: the rampant corruption and political patronage that characterized American government in the early 20th century, when city bosses funneled public resources to political cronies with a flagrancy that makes today's political favoritism look squeaky clean. Progressive Era reformers like Teddy Roosevelt put in place new rules to stamp out corruption by increasing oversight and standardizing how decisions are made. By giving agency staff less discretion in their decisions, reformers reduced opportunities for politically motivated decisions. The end result, unfortunately, is that US government agencies work reasonably well at promoting transparency and fairness, but can be maddeningly inflexible in response to new problems and new opportunities to fulfill their mission.

In recent decades a new reform movement has made significant strides in modifying the rules governing the behavior of bureaucracies, enabling them to be more flexible and responsive to public needs. This movement, called New Public Management, was catalyzed by the publication of a landmark study by David Osborne and Ted Gaebler called *Reinventing Government*, which documents the new strategies adopted by reformers ranging from city parks directors to military base commanders. These innovators have changed the rules governing their internal operations to foster a culture of customer service, providing incentives to their staff to innovate and reduce costs, and introducing more flexible budget systems that empower agencies to be more responsive to new challenges as they arise.[18]

So what does this mean for those of us who would like to have a positive impact on the world? It may be that the greatest hurdle to changing an outcome you care about lies within the very organization responsible for leading the charge. Sometimes change requires modifying the way an existing organization does business. But given the weight of organizational history, reformers hoping to change organizational behavior are, according to March and Olsen, "often institutional gardeners more than institutional engineers," moving through a thick tangle of past commitments and selectively weeding and seeding.[19] In many cases it makes more sense to create an entirely new organization or to shift responsibilities from one to the next. A company created with great vision and verve in the early 20th century to drill for oil may be poorly suited, in its expertise and management structure, to exploit new opportunities such as solar energy. An environmental organization designed to bring together small, influential groups of policymakers and scientists may be incapable of helping students and ordinary citizens bring about change in their communities. The good ideas have not all been tried because the organizations we create to advance our shared interests are predisposed to ignore entire categories of feasible changes.

CONFRONTING POWER

Reinhold Niebuhr was among the most original and influential thinkers of the past hundred years. A theologian and political philosopher born in Wright City, Missouri, in 1892, Niebuhr was a rare breed of intellectual embraced by both liberals and conservatives, although many of his followers today are unaware of their shared lineage. For conservatives, Niebuhr is considered a founder of the "realist" approach to foreign policy, which accepts at face value the idea that all nations seek to maximize power, rather than basing policy on what Niebuhr called a "romantic overestimate of human virtue and moral capacity."[20] Political progressives know Niebuhr as one of the architects of nonviolent civil disobedience. Martin Luther King, Jr. frequently cited Niebuhr as one of his most important influences.

Niebuhr's key insight was that social progress requires more than education and appeals to moral sensibilities. Doing good in the world requires dislodging power. His insights are pertinent to our discussion because another hurdle standing between the world we want and the world we've got concerns power, and in particular the disproportionate influence of relatively small numbers of actors who have a stake in the status quo. There are plenty of solutions to social and environmental problems that are technically and economically feasible but are blocked by these small groups. To promote social progress often requires that citizens exercise their democratic rights and confront, coopt, outmaneuver, or otherwise mitigate the influence of those who stand in the way.

What can academic research teach us about power as an impediment to change? Like all social organizations, academic disciplines are structured to privilege certain types of answers over others. We "discipline" our own, and for good reason. The shared standards of evidence, foundational texts, and common training enable research communities to flourish and create a shared body of knowledge. But in deciding which questions are most relevant, and sanctioning this method of study over that one, the disciplining process also creates blind spots. The field of economics, which inspired many of the contributions discussed in the previous sections, has some blind spots of its own. Few economists devote attention to things like power, social mobilization, and exploitation of the weak by the strong. The discipline tends to portray public problem solving as a managerial exercise of weighing the costs and benefits of alternative courses of action. Yet no serious analysis of the underlying causes of today's environmental problems can ignore questions of power, and here is where Niebuhr's insights offer guidance.

In his book *Moral Man and Immoral Society*, published in 1932, Niebuhr argued that groups of people act with less ethical consideration toward others than do individuals. "In every human group," Niebuhr wrote, "there is less reason to guide and to check impulse, less capacity for self-transcendence, less ability to comprehend the needs of others and therefore more unrestrained egoism than the individuals, who compose the group, reveal in their personal relationships." Building on this logic, Niebuhr chided educators and evangelists "who imagine that the egoism of individuals is being progressively checked by the development of rationality or the growth of a religiously inspired goodwill and that nothing but the continuance of this process is necessary to establish social harmony.... They do not recognise that when collective power, whether in the form of imperialism or class domination, exploits weakness, it can never be dislodged unless power is raised against it."[21]

Niebuhr's critique challenges the assumption that environmental education, awareness-raising, and other forms of information sharing are sufficient to move society off its current path of environmental destruction. (Indeed, Niebuhr's analysis raises some troubling questions about the very aims of a liberal arts education, which is based on the premise that critical inquiry, dialogue, and exposure to the world of ideas promote democratic habits and social progress.) Education is, of course, an extremely important component of movements for social change. And we have already seen the value of generating new knowledge and sharing information. In fact, the preferences and behaviors of powerful groups are not as fixed as Niebuhr portrays them; they can be influenced by information about new ways to pursue their self-interested goals. The point is that education and information-sharing strategies are rarely enough. Niebuhr argues that it is naive to imagine that "with a little more time, a little more adequate moral and social pedagogy and a generally higher development of human intelligence, our social problems will approach solution." Niebuhr admits that democratic dialogue and mutual accommodation can solve some problems. "But will a disinherited group, such as the Negroes for instance, ever win full justice in society in this fashion? Will not even its most minimum demands seem exorbitant to the dominant whites, among whom only a very small minority will regard the inter-racial problem from the perspective of objective justice? Or how are the industrial workers to [fare] when the owners possess so much power that they can win the debate with the workers, no matter how unconvincing their arguments?"[22]

Writing at a time when few intellectuals concerned themselves with sustainability, Niebuhr did not address environmental questions per se. But his understanding of power bears directly on the causes of the environmental crisis and the potential for change. When an elementary school decides to start an organic garden, this provides a meaningful way to connect children with their environment and helps build a constituency over the long-term. But it does not in itself offer a serious solution to the environmental problems confronting us today, which are of a scope and scale that require modifying the very rules that we live by. (New standards promoting nutritious, pesticide-free produce in school cafeterias would be a great start.) Changing the rules, in turn, requires confronting power.

WHAT IS POLITICALLY FEASIBLE?

To say that a more sustainable world is perfectly feasible, if we could overcome political obstacles, raises the question of whether it's feasible at all. If one were to argue, for example, that a carbon-neutral economy would be quite feasible if only we could overcome opposition from every industry that relies on fossil fuels (which would include the entire economy), this would make a mockery of the word "feasible." To reiterate, I am not suggesting that we can overcome every constraint, and this is as true when dealing with politics and power as it is with other types of hurdles. But success in confronting powerful interests that stand in the way of progress is often more attainable than we assume. For David Gottfried and Mike Italiano, this was a matter of moving green building standards to a new rulemaking organization where industry trade groups had less sway. For Dr. June Irwin, the country doctor turned community activist whose story we followed in chapter 1, the power shift entailed the election of a new mayor who was sympathetic to her concerns about pesticides and children's health.

The perception that power is unassailable, and things never change, works to the distinct advantage of those who benefit from the status quo. What better weapon than an appearance of invincibility, dissuading dissenters from mobilizing for change? This sort of self-censorship is explored by John Gaventa, a sociologist at St. Francis Xavier University, who conducted his PhD research in the coal-mining towns of Appalachia and published the results in his book *Power and Powerlessness*.[23] Gaventa builds on a research tradition exploring the "three faces of power."[24] The first face of power can be seen when one group directly influences another through violence or other forms of coercion. The second face of power manifests itself when a group controls the rules of the game that determine who gets to participate in decision making, which arguments and interests are considered legitimate, and how binding rules are made and revised. (We will consider this dimension of power in greater depth in chapter 10, on the subject of "super rules.") The third face of power is when a group shapes the very wants, desires, and aspirations of others. Those of us who assume that power is unassailable are controlled and manipulated with great efficiency because we impose constraints on ourselves, relieving those in power of the burden of responding to a coordinated challenge.

In the long run, the possession of power by any one person or group is temporary. Knowing this, powerful groups create rules to lock in their advantages. Often power obstacles persist because no one has bothered to organize for change, for reasons discussed earlier in this chapter. People are often unaware that alternative paths are available and lack information about who controls the decision-making process. Free-rider behavior can make it difficult to launch a coordinated response, and requires that a few enterprising people think of creative ways to motivate participation. Recall that not every group faces this problem to the same degree. A for-profit business can readily marshal resources to sustain a lobbying presence in a state or national capital. They often outnumber public interest groups not only in the number of bodies that they can place in relevant settings (courtrooms, committee hearings, state capitals) but in their ability to marshal research, fund public relations campaigns, and support political candidates to their liking. In 2013, all of the top twenty spenders on lobbying in Washington, DC, were corporations or industry associations. In just this one year, these 20 groups spent over $205 million to ensure that the rules favor their interests.[25]

Yet when people organize for change, they can and do have significant impacts, even in the face of powerful opposition. We will see in chapter 9 that when a rulemaking process is monopolized by a small group of powerful actors, the arrangement is often quite fragile, characterized by long periods of stability punctuated by rapid change when a key actor is replaced or when larger-scale developments—from elections to economic crises, natural disasters, administrative reshuffling, or popular uprisings—result in a change in how decisions are made.

CREATING VALUE

The preceding examples suggest that we are surrounded by unexploited opportunities to make the world a better place. This is intuitive to the business entrepreneur, who is constantly on the lookout for opportunities to create value for customers by lowering costs or enhancing quality through the creation of new goods and services. (If we lived in the best of all feasible worlds, the potential for business expansion would be small indeed!) But value creation is not just something that happens in the economy. You can apply entrepreneurial skill not merely to produce profit, but to enhance social well-being.[26] (We will see in chapter 5 why these two goals are not one and the same; markets are hard-wired to ignore many of the goods and services that societies demand, such as clean air and water.) So what exactly is value creation?

Let's take an example. You enter a room and find two boys arguing over an orange. Each claims he saw it first, and the two are engaged in a nasty tug-of-war. There is no other food to be found in the vicinity. How might you resolve this situation? When I use this example in the classroom, students almost invariably suggest the solution that may have come to your mind: Slice the orange in two and give a portion to each. Well it turns out that the boys wanted the orange for very different purposes. One of them was hungry and wanted to eat the fruit, while the other wanted the orange peel for a cake recipe. In light of this new information, we can see a better way to resolve the conflict. Apportioning the peel to one and the fruit to the other creates greater value than our earlier, more hasty settlement. To put a finer point on this idea, let's assume that each boy would have been willing to pay a dollar for the orange. Under the "slice it" scenario, each child receives 50 cents of value—that's what he would have been willing to pay for half—and the total value in this society of two people adds up to one dollar. Under the "peel it" scenario, each boy receives one dollar of value. (For simplicity, we will assume that our young combatants have no interest in the other part of the fruit.) Now the total value in this little society equals two dollars.

I first encountered this example in the work of Roger Fisher and William Ury, experts on negotiation and conflict resolution at Harvard Law School. In their book *Getting to Yes*, Fisher and Ury extol the virtues of value creation during negotiations as a strategy for identifying win-win solutions.[27] But value creation carries broader implications for the way we think about the world and the possibility of changing it for the better. We were raised to believe that money doesn't grow on trees, but here we see this isn't quite true. Value (which we often measure with money) can be created and destroyed. Consider the case of a relative who goes shopping for your birthday present and spends $75 on a hideous lamp shade that you discretely slip into the recycling bin a few months later. The economy has just lost $75 in value relative to a cash gift.[28] Just as value can be destroyed, innovative approaches to social problem solving can create value where there was none before.

I gained an appreciation for value creation when I accepted a position at Harvey Mudd College in the fall of 2003. Within a week of my arrival on campus, I was asked to direct the college's new Center for Environmental Studies. At the time I was designing new courses while juggling several research projects; taking on a major administrative responsibility was unthinkable. But when a senior colleague asks an untenured professor to do something, the transaction could not be fairly characterized as a request. "Just try to have something," advised a sympathetic colleague, "even one thing for the Center to show during your first year."

Shortly thereafter, the director of a local conservation group, whom I had invited to speak in one of my courses, asked if I could do him a favor. He was organizing a high-profile conference and did not have a place to hold the event. Renting a venue large enough to accommodate the expected audience would cost his organization thousands of dollars. For me, it was a simple matter of making a phone call to identify an empty lecture hall on the scheduled evening, which the college made available for free. In exchange, I asked that he highlight the Center for Environmental Studies as a co-sponsor of the event in his publicity materials, giving me the "one thing" to show during my first year. This provided a huge benefit to me, at no cost to his organization.

Value creation carries important implications for the way we think about politics and social change. Politics is commonly portrayed as a zero-sum game in which a gain for one group means a loss for the other. This impression is reinforced by our most visible political events, notably national elections, which are structured to ensure that a win for one party comes at the expense of another. During presidential debates, the goal is not to come up with good ideas that will benefit all—the aim is to "win." Candidates vehemently defend their positions and are not permitted to change their minds, for this would be considered a sign of weakness. We will not see a politician in a televised debate extend an eager hand to the opposition and say, "I think I've figured out a way to combine the best of both our ideas!"

This picture of politics as irreconcilable conflicts among warring factions has a long pedigree. For Karl Marx, people's interests stem from their position in the economy, as laborers or owners of capital, and are therefore inherently at odds; the essence of politics, in the Marxist model, consists of grabbing resources from others, whether through subjugation, exploitation, or revolutionary action. This perspective is popular among those who study power, considered in the previous section, and many academic research traditions portray politics in us-versus-them terms today. And to be sure, zero-sum interactions constitute an important part of political reality; a win for candidate A is, after all, a loss for candidate B. Slave owners opposed emancipation for the same reason that coal companies oppose a carbon tax today: They have a lot to lose. But zero-sum political struggles are not the whole picture. I noted earlier that disciplines create their own blind spots. The field of economics, which downplays power and conflict, deserves credit for the concept of value creation, which has broadened the vision of those of us trained to see the world as one grand tug-of-war. Politics is also an arena for creating value.

Eugene Bardach, an expert in policy analysis at the University of California at Berkeley, shows that opportunities abound for creating value in the public sector.[29] One strategy is *rummaging*, the discovery of novel uses for readily available resources. Motor voter initiatives, for example, allow people to register to vote at department of motor vehicle offices when applying for their driver's license. This new rule was first adopted in 1975 by Michigan Secretary of State Richard Austin and is now required by federal law in all fifty states. Another strategy for value creation discussed by Bardach is *complementarity*, in which distinct activities are joined together to enhance their impact, such as the use of public works construction to reduce unemployment.

Even in foreign policy—an arena often portrayed using the zero-sum metaphor of a high-stakes chess game—policymakers have shown that smart rules can enhance social value. Fisher and Ury point to the example of Arab-Israeli negotiations over the fate of the Sinai Peninsula following the Six-Day War of 1967. The Israeli military, which had learned of a pending attack by Egypt, launched preemptive air strikes that destroyed the Egyptian air force and went on to occupy the Sinai, a large chunk of Egyptian territory. In negotiations mediated by the Carter administration, both parties demanded complete control of the land. But in a manner reminiscent of the divided orange example discussed earlier, negotiators discovered that while Egypt truly wanted its territory back, Israel was more concerned that the Peninsula would be used to stage future military strikes. Carter brokered a deal that returned the land to the Egyptians with the condition that they establish large demilitarized zones where they cannot station troops. Appearances to the contrary, politics is a sphere filled with possibilities for win-win solutions.

OF RUIN AND POSSIBILITY

In sum, the reality that we experience today is not a fair measure of what is realistic. The good ideas have not all been tried, and the possibilities for change are far greater than most people imagine. After all, if today we find that the workings of our economy and society are tilted in the wrong direction, this is because of changes made in the past: structures built, laws passed, contracts and treaties signed, city codes and design standards put into practice. So the question before us is not whether change is possible. Change is ubiquitous. The question, rather, is who is participating in the process.

Our next task is to take a closer look at this institutional landscape we have inherited, its countless rules pushing us along pathways of action, some wise and others foolish. To understand who rules the earth, we will begin by considering one of the most powerful social rules of all: property rights.

PART II

Who Owns the Earth?

4

A Perilous Journey

PROPERTIES OF NATURE

For sheer splendor, few natural wonders can outperform the seasonal migration of birds across the planet. Each year about 100 billion birds traverse the globe in search of greener pastures during the winter and return to their mating grounds when temperatures warm. Record-holders like the Arctic Tern travel thousands of miles to reach their destination. But even the less athletic members of the avian family feel the urge. As winter approaches, groups of California mountain quail hobble along by foot 15 miles down to the safety of the valleys below, and make their way steadily back up the mountain slope come spring. On a planet dominated by human artifacts and controlled environments, birds remind us of the ancient things—cycles of nature stretching back well before the first humans walked the landscape. This is equally true of their morning song. If you stood at the North Pole with a microphone powerful enough to detect it, you would hear an enormous wave of music slowly circling the earth each day, as a chorus of millions of birds awaken, their song tracking the leading edge of sunlight moving westward around the earth's perimeter.

When we think of the study of the earth's natural wonders, what comes to mind are images from the natural sciences—lab coats and test tubes, gloves and galoshes, fish nets and soil samples. And when we wish to learn more about our environment, we consult these same sources of expertise. Take birding guides, for example. If you visit the southeastern United States in the spring, you may be fortunate enough to catch a glimpse of an adorable little blue bird called the cerulean warbler, literally the "sky blue singer." Consulting a trusted birding guide such as Peterson's, you will learn certain things. The cerulean is 4½ inches in length. It has a thick pointed beak suited to its preferred diet of insects. It breeds in North America, where it builds cup-shaped nests high in the trees to hatch its chicks. But there is other information about the cerulean warbler, equally important to its survival, that you won't find listed in your field guide: Who owns it?

This question weighed on the mind of Oliver Wendell Holmes, the famous Supreme Court justice, in a landmark ruling in 1920. Holmes and his colleagues on the high court were considering whether the US government had the right to regulate the hunting of migratory birds. By the late 19th century, North American migratory birds were being wiped out by the millinery trade, which used the feathers of exotic birds to decorate fashionable ladies' hats. Several hundred thousand birds were killed for this purpose each year, then shipped to fashion centers like New York and London from as far away as Bogotá, Colombia. In what was possibly the first global citizens' campaign for the environment, conservationists in North America and Europe lobbied to stop the practice. In response, the United States signed a treaty in 1918 with Britain (on behalf of Canada, which was still under British control) to prohibit the hunting or capture of all migratory birds.

The laws governing property in 19th-century America, however, specified that the states, not the federal government, owned the wildlife within their borders.[1] When federal officials tried to implement the treaty, the state of Missouri would have none of it and sued the US government, arguing that state ownership applied equally to migratory birds. Holmes was unconvinced. "Wild birds are not in the possession of anyone," he wrote in the majority opinion, "and possession is the beginning of ownership. The whole foundation of the State's rights is the presence within their jurisdiction of birds that yesterday had not arrived, tomorrow may be in another State, and, in a week, a thousand miles away."[2] Holmes was more right than he knew; almost two-thirds of "our" birds in North America spend most of the year in Central and South America and the Caribbean. The Supreme Court decision represented the first step toward a proactive federal role in protecting American wildlife. A century later, this rule would save the bald eagle, the national emblem of the United States, by allowing the government to regulate pesticides that were destroying this majestic bird's eggs. The court decision did not, however, address the matter of who controls the land and water that birds require for survival—with consequences that we will consider shortly.

The question of who owns the birds is shaped by one of the most powerful rules we use to govern the conduct of human affairs: property rights. Property rights are social rules that specify the relationship between people and things. Because these "things" include natural resources—birds and valleys, mountains and oil, freshwater and farmland—it is no exaggeration to say that the rules we devise for property determine whether sustainability is possible. Property rights specify who gets to make the decisions concerning the physical environment, what we may extract from it, and what obligations, if any, we have toward the planet and one another when using these resources.

Property is not a focus of research in the natural sciences, but has been studied and debated intensely by scholars of law, politics, and economics for over 300 years. So like birds traversing state borders, we need to breach some disciplinary boundaries to get our heads around the question of who rules the earth, combining the biologist's understanding of the contours and status of the natural world with the social analyst's insight into how human institutions operate. To do so, we will first explore what exactly property is (and is not), revealing that there are many possible arrangements of ownership rights in a given setting. These rules change from one place to the next and they shift over time. Viewing the world through the prism of property rights, we will then sample the rules governing our planet by tracking the journey of the cerulean warbler from its wintering grounds in the Peruvian Andes to its spring destination in North America.

PROPERTY AND THE PEACH

At my home in southern California, I have a variety of fruit trees that I nurture with almost fanatical devotion throughout the year, mulching and pruning and watering the oranges, lemons, and avocados that grow in abundance in the warm climate of the San Gabriel Valley. I own these trees and the land under them. But what exactly does ownership mean? Behind the idea of property, and its twin concept ownership, is the notion that it is for a particular person (or group), rather than anyone else on the planet, to decide how a given resource is to be used. Writing in 1766, Sir William Blackstone, a towering figure in the development of contemporary law, described property as "that sole and despotic dominion which one man claims and exercises over the external things of the world, in total exclusion of the right of any other individual in the universe."[3]

Armed with these weighty words, I exercise that power as I care for "my" trees, and above all my coveted peach tree. For most of the year it stands as a scraggly and uninspiring feature of the landscape. But as July approaches I go into action, driven by thoughts of the sweet harvest just around the corner. As the owner of that tree, I take pains to cover it with a net as the fruit ripens, to prevent local house sparrows from devouring the crop. Spreading the net requires an elaborate balancing act involving a ladder and a broom (and often a fair amount of cursing) to avoid tearing its lattices on the gnarled branch tips. With ownership I not only bear these costs but also reap the rewards, and so I jealously guard the crop, literally the fruits of my labor, come harvest time. I enjoy treating my wife and son to a perfect fat peach plucked from its hiding place in the upper reaches. As for my neighbors and friends, I decide who gets the fruit, which ones, how many, and when. I can be miserly or welcome all comers—the point is that as the owner of this resource, I am in control. A defining characteristic of *private* property rights, such as those governing my peach tree, is they are transferable. I can sell the peaches if I wish, and I often barter in exchange for apricots and tomatoes. I could charge for the rights to harvest the peaches if I were so inclined. Such is the level of control one has as the owner of things, whether rivers, forests, or factories. This is why answering *Who rules the earth?* requires that we understand who owns it.

Ownership differs in crucial respects from mere possession. Any thug can use his labor to possess something. A property right, in contrast, is a rule that is socially sanctioned and backed by the force of law. This point was emphasized in 1776 by Adam Smith, the eminent Scottish political economist, in his book *The Wealth of Nations*. Adam Smith is often portrayed as an advocate of unfettered capitalism, but a closer reading of his work shows that he was a strong proponent of government regulation. "It is only under the shelter of the civil magistrate," he wrote, "that the owner of that valuable property, which is acquired by the labour of many years, or perhaps of many successive generations, can sleep a single night in security."[4] Knowing that my ownership is secure, I have incentives to care for my peach tree because I can reap the benefits. Smith praised the salutary effect of property rights on the "small proprietor" who "knows every part of his little territory, views it with all the affection which property, especially small property, naturally inspires, and who upon that account takes pleasure, not only in cultivating, but in adorning it."[5] As Carol Rose argues in her book *Property and Persuasion*, property is designed to *do* something—it advances socially sanctioned goals like the creation of value through investment and trade.[6]

The nature of property rights, however, is considerably more complex than a simple tale of ownership and the productive investment it inspires. As the owner, I also choose whether to spray my peach tree with toxic pesticides or tolerate the occasional bug. What if I decide to use pesticides and thereby harm my neighbor's property, reducing the number of butterflies that visit her garden? If I leave damaged peaches to rot on the ground, this causes the local fly population to increase (as I was sternly informed by an elderly neighbor one day), interfering with the ability of others to dine outdoors. Do I have the right to cut down the trees on my land, and to run my air conditioner day and night—actions that pump carbon dioxide into the atmosphere and thus contribute to rising sea levels, which scientists tell us will flood coastal properties around the globe?[7] Does my right to cut down a tree depend on whether it provides habitat for birds? Does it matter which birds—abundant or endangered—use the tree? What if I planted it myself? May I grow my trees so tall that they block my neighbor's view of the San Gabriel Mountains? How about running my sprinklers all day—does that right come with the property, and should such a right vary from one region to the next based on the availability of water? Do I have the right to store rusted-out automobile chassis on my driveway? How about running a night club out of my garage, advertised with a flashing neon sign?

Clearly property rights entail tradeoffs that are more subtle than the political rhetoric that pits private property against government regulation. There are important decisions to be made, and these are decisions that we as citizens must track carefully if we wish to promote sustainability in our communities. As Harvard Law School professor Joseph Singer points out, property rights *are* a form of regulation. They are social rules that determine who gets what, when, where, and why, and are enforced by governments that must adjudicate among competing demands. When property claims clash—my tree versus my neighbor's view—the government must resolve these conflicts by referring to broader social priorities that have been encoded in law. The question, argues Singer, is not the appropriate balance between private property and social regulation, but "What kind of property regime should we construct?"[8] What should be the relative weight of goals like public health, unrestrained scientific inquiry, free speech, wildlife conservation, industrial development, poverty alleviation, and foreign investment? The answers to these questions vary from one society to the next and evolve over time. Property rules reflect social priorities and encode particular configurations of power. Property rights are political creations.

It is comforting to think of wild birds soaring free above the world, unconstrained by these sorts of questions and the labyrinth of rules that underpin human civilization. The very word *wild* signals, after all, the absence of human influence. But every bird must land, and when it does it is subject to the power of our earthly rules. We shall see that the annual migration of the cerulean warbler, and the hurdles and havens it encounters while trying to reach its destination, reveals as much about ourselves as it does the fate of the millions of species with whom we share the planet.

A BIRD'S EYE VIEW

Every spring upward of 2,000 bird watchers take part in the North American Breeding Bird Survey (BBS). Along with its counterpart event, the Christmas Bird Count, the BBS is the world's largest experiment in citizen science. Coordinated by the US Geological Survey and the Canadian Wildlife Service, volunteers set out in the early hours and walk, bicycle, or drive along specified routes scattered across North America. Each route is divided into a series of stops, where the volunteers spend three minutes recording by sight or sound all birds in the area. They write their observations on standardized forms and send these to the government agencies, who then plunk the information into a database that provides a comprehensive picture of the status of our birds from one year to the next.

These data reveal a troubling pattern. Since the survey began in 1964, over a third of North American seabird species have seen their numbers drop. Birds inhabiting American deserts and grasslands have suffered a similar fate. There is good news to report as well. Thanks to rules protecting wetland habitat in the United States, bird populations in peats, swamps, and other wetlands are increasing. But on the whole, of the 400 bird species counted in the BBS, almost two-thirds are in trouble due to declining numbers, external threats (especially habitat destruction), or small population distributions.[9] These data are consistent with the findings of scientists at the international level. Globally, 1,226 (1 in 8) bird species on our planet are threatened with extinction.[10]

Few birds are in straits as dire as the cerulean warbler. Mind you, BBS volunteers are experienced birders with an average of ten years of participation in the survey; this group doesn't miss much. And every year for the past four decades—as volunteers show up at sunrise with their clipboards, binoculars, and coffee-filled thermoses, and crane their necks to search for the elusive cerulean in the upper branches—they have recorded an average of three percent fewer than the previous year.[11] The cerulean warbler is the most rapidly declining bird in the eastern United States.

Why is this little blue bird at risk of extinction? There are enormous gaps in our knowledge of migratory birds, particularly concerning their fate in wintering grounds in the tropics where scientists are few and data are scarce. (Ornithologists call this data deficit "white map syndrome," signaling the blank spaces on population distribution maps in the tropics.) We don't even know for certain the flight patterns of many of the better-known birds. But scientists strongly suspect that the main culprit in the decline of the cerulean is habitat loss—the felling and burning of the steep-sloped broadleaf forests preferred by these birds. Who owns these lands, and what are the rules shaping their decisions? To answer this question we will take what is known about the cerulean's migratory route and habitat preferences, and on this basis identify some plausible landing spots.[12] Overlaying this biological information with social science research on the property arrangements along its route, we can construct a composite picture of the landscape of social rules that await the cerulean on its odyssey, and which ultimately determine whether it arrives at its destination. Our journey begins in the tropical forests of southern Peru.

Callanga Valley, Peru

In March of every year, cerulean warblers throughout the South American tropics feel an age-old biological itch and prepare to fly north. Ignoring the adage that birds of a feather flock together, during the winter months the ceruleans are party crashers, one or two birds mixed in with a flock of other species like brightly colored tropical tanagers. If we were to select a representative bird to follow on this trip, a reasonable starting point would be the lush mountain slopes of Manu National Park in Peru (Figure 4.1). Conservation scientists have designated Peru a "megadiversity" country, one of a handful

FIGURE 4.1 Property rights along the migratory route of the endangered cerulean warbler

1. Callanga Hacienda, Peru
2. Cerulean Warbler Reserve, Santander, Colombia
3. Sierra Nevada de Santa Marta, Colombia
4. Barro Colorado Island, Panama
5. Atlantic Coast, Costa Rica
6. Lancandon Rainforest, Mexico
7. Santa Catarina Ixtepeji, Mexico
8. Gulf of Mexico
9. Boone County, West Virginia

of species-rich nations that are blessed with a disproportionate inheritance of the world's biological wealth like some favored heir in the disbursement of Earth's estate.[13] Manu is the crown jewel of the Peruvian national park system, spreading over 15,000 square kilometers from the lowland Amazon jungle to the Andes Mountains that form a spine running up the middle of the country. (If we liken an aerial view of mountainous Peru to a stegosaurus with its head cocked to the right, Manu sits on the right rear haunch.) With 925 recorded bird species, Manu National Park provides habitat protection for one in every nine bird species on the planet.

Manu is also the site of a rich human history. The evolution of property rights in this part of the world has shaped and reshaped the landscape of Manu with as much force as the ecological relationships that transform its forests and waters through the millennia. Consider the small village of Callanga, located just within the southwestern border of Manu National Park, and situated in the cerulean warbler's preferred habitat between 1,500 and 5,000 feet in altitude.[14] Callanga is a place where the weight of past decisions is felt in the present. The vine-covered remains of ancient Incan walls and roads are in a state of decay from the relentless onslaught of centuries of tropical heat and rain. But the people here continue to speak Quechua, the language of the Incan Empire, as do many communities throughout the Andes today. The Callanga Valley was once a vibrant center of trade for the Incan emperors who owned these forests, ceruleans and all. Trade in property followed a pattern that anthropologists call the vertical archipelago. Goods manufactured by Incan settlements in the mountains were exchanged for raw materials from the unconquered tribes of the Amazon jungles below.[15] Our understanding of property rights in the Incan period comes from anthropologists like Catherine Julien, who spent years studying ancient artifacts called *quipos*, mysterious knotted and colored threads that Incan administrators used as a decimal system to keep track of their vast holdings. Carefully decoding the meaning of the quipos, she discovered that after a successful military campaign, Incan rulers would sometimes confiscate their subjects' property outright. More commonly, however, they demanded that conquered settlements like Callanga provide labor in service of the empire. Different towns specialized in roles such as salt miners, soldiers, or "feather workers."[16] It appears that our cerulean warbler's ancestors may well have been plucked, lending a splash of sky blue to the elaborate headdresses of Incan royalty, although their habit of hanging out in the upper reaches of trees would make them a tougher target than most.

As is the case throughout the cerulean warbler's migratory route, the property rights governing the forests of the Callanga Valley have shifted over time to serve different purposes, carving out new roles, defining new rights and responsibilities, and privileging some uses—and some users—over others. The rules governing cerulean habitat in Peru underwent a radical transformation with the arrival of the Spaniards. The conquistadors made no bones about their intentions. Upon their arrival on the shores of Latin America, captains of Spanish vessels were required by the crown to read aloud a statement declaring ownership of all the lands. This declaration was recorded in 1566 by Friar Diego de Landa and rediscovered by the renowned Mayan scholar William Gates, who provided the first English translation in 1937. I discovered Gates's book on an obscure shelf in the basement of City Lights bookstore in San Francisco in 2009. The declaration reads:

> I, [name], servant of the high and mighty kings of Castile and León, the conquerors of the barbarian peoples, being their messenger and captain, notify and inform you: That God, our Lord, One and Eternal, created the heaven and the earth.... All these people God gave in charge to one who was called Saint Peter, that he might be Lord and superior of all the people of the world, that all should obey him and he should be the head of the entire human lineage, wherever men might be and live, and under whatever law, sect or belief, giving to him the entire world for his service and jurisdiction.

The logic of the argument continues, noting that the Pope "made donation of these islands and mainlands of the ocean, to the Catholic Kings of Castile, who were then Don Fernando and Doña Isabel...so that his Majesty is king and lord of these islands and mainlands, by virtue of said donation." The declaration states that if you convert to Christianity and welcome the invaders, all is well. It continues:

> If you do not do this, and maliciously set delays, I assure you that with God's aid, I shall enter with power among you, and shall make war on you on all sides and in every way I can, and subject you to the yoke and obedience of the Church and of his Majesty; and I shall take your wives and children and make them slaves, and shall sell them as such and dispose of them as his Majesty shall command; and I shall take your property and shall do you all the harm I can....I protest that the deaths and harm which shall thereby come, will be your fault, and not that of his Majesty, nor ours, nor of these gentlemen who came with me.[17]

This declaration marked the arrival of a new property regime in Latin America, and soon the people and trees and wildlife of Callanga were firmly under

Spanish control. For as Gates wrote in his book, Philip II of Spain was "seeking not merely world supremacy but world ownership."[18] So how did the crown manage its newly acquired property and how did this affect cerulean warblers? The control of land is intimately tied to the control of people—if you own one you can control the other—and the Spanish sought dominance over both. One option was to enslave the entire continent. Christopher Columbus, who went on slaving expeditions throughout the Lesser Antilles, assured Ferdinand and Isabella that Indians were superior to Africans as slaves and could be had in "infinite" numbers.[19] But the Spanish crown eventually banned the practice of owning humans outright. In its place, they created a new forced labor system called the *encomienda*.[20] The encomienda system likely encouraged deforestation throughout cerulean habitat because Spanish owners of encomienda lands could not transfer the property to their descendants; the land would revert to the Crown upon the death of the second generation. This created incentives to sell the trees in exchange for gold, silver, and other forms of property that carried rights of inheritance. These metals were also a common form of tribute from conquered communities. From 1533 to 1560, the Spanish shipped home 1.6 million pounds of silver and gold from Peru[21]—equivalent to the weight of four Boeing 747s. Clearly the rules governing property on the continent had shifted.

Over the coming centuries, cerulean warblers passing through the Callanga Valley were affected by successive waves of property rules serving new political and economic agendas. First the encomienda system was replaced with haciendas—independent landed estates that included full rights of inheritance. The new system led to an unprecedented concentration of land ownership in the hands of Europeans over the coming centuries, as a result of purchases and outright theft from Indian communities.[22] Like so many social rules, the hacienda system persisted long after its creators were gone. As recently as 1965, Callanga was still part of a hacienda. The local people had worked as laborers on the hacienda and lived in the shadow of its political influence for as long as anyone could remember. But that year the last owner of the hacienda died, prompting a local rebellion. The people of Callanga occupied the property and divided it up, controlling their land—and with it, the fate of cerulean warblers—for the first time since the conquistadors arrived in the mid-16th century.

Peasant occupations of this sort were a common occurrence throughout Peru in the 1960s, as the pressures and resentment stemming from centuries of European land confiscation were on the verge of exploding. These events attracted the attention of the Kennedy administration, which advocated land reform programs that would break up large property holdings throughout Latin America and distribute the land to peasants for cultivation. The idea was to preempt the appeal of communism, which demands a more radical restructuring of the rules to invest all property rights in the government. But in 1968, the Peruvian military staged a coup and promoted a program of agrarian reform far more ambitious than American officials had envisioned. The military government distributed over 7 million hectares to some 300,000 peasant families in the Andean region by the mid-1970s.[23] Any individually owned commercial estate over a certain size (typically 37 to 136 acres for irrigated farmland in the Andes) was susceptible to expropriation.[24] Similar land reforms took place up and down the cerulean migratory flyway, changing the way forests are owned and managed today in Bolivia, Nicaragua, Guatemala, Colombia, and Mexico. A key component of these land reforms was that anyone working land owned by absentee landowners should have a right to take over the property. These new property rules encouraged deforestation by requiring prospective owners to make "improvements"—notably cutting down trees and setting fires to prime the soil for farming or pasture—as a prerequisite for acquiring rights to the land. In this way, rules created for the purpose of reducing poverty and increasing equality throughout the region set in motion a decades-long process of forest destruction. It was smack dab in the era of land reform that Peru's military government created Manu National Park in 1973, as part of a broader effort to reorder the rules governing natural resources.

Junín Department, Peru

As I write these words in mid-March, ceruleans are beginning their northward migration with the goal of arriving in their mating grounds in the United States and Canada in a month's time. It is the tail end of the rainy season in the Peruvian jungles and tropical downpours pose a serious risk, even for much larger birds. For a cerulean warbler, which weighs less than a cherry tomato, rough weather is often fatal. Birds will delay migration to avoid it. When the skies clear, ceruleans prefer to travel at night, using the stars to fix their course. Taking off from the slopes of Callanga and heading northward over Manu National Park, it must be a beautiful sight—the moonlight reflected in the snow-capped peaks of Nevado Verónica, which rises to almost 20,000 feet in elevation on the left. To the right, an infinite sea of treetops stretches across the Amazon basin, punctuated by the explosive orange blooms of flowering coral trees.

Continuing along the foothills of the eastern slope of the Andes, cerulean warblers leave behind the relative safety of Manu National Park, moving across the park's northern boundary and into a complex mosaic of property rules. Here there's both good news and bad. On the positive side of the ledger, the Peruvian government has created a new Forest and Wildlife Law that is among the most forward thinking in the world. The law includes new rules to promote sustainable harvesting practices and to ensure that the timber industry benefits small and medium-scale producers while protecting wildlife. The bad news is that the new regulations are up against deeply ingrained patterns of social interaction that hasten the destruction of the forests.

Governments own and manage over 80 percent of all forests worldwide, the remainder falling under private ownership or local community property.[25] As in most tropical countries, the government of Peru has struggled to govern its forests effectively. In 2005, researchers estimated that up to 90 percent of the timber extracted from the Peruvian Amazon was illegally harvested or traded.[26] The Forest and Wildlife Law is designed to bring the situation under control through a system of contracts. The government, as owner of the forests, leases the harvesting rights to a community or private company for a one-time cut or in the form of a logging concession lasting up to 40 years. In exchange for the right to harvest, users are responsible for complying with plans specifying the volume and location of trees to be felled.[27]

As a result of the new rules, today migrating cerulean warblers can take a rest in large patches of healthy forest. Based on a review of satellite data, a research team led by Paulo Oliveira of the Carnegie Institution reports that very little destructive logging occurs within forests covered by these new contracts.[28] Outside of these areas, however, the illegal felling of trees continues unabated, and high-quality habitat is ever harder to find. The situation unfolding in the Peruvian Amazon demonstrates a broader principle: To understand the impact of a new rule, we need to consider the larger system of rules into which it is introduced. These prior commitments—and the relationships, habits, and expectations they produce—make up an elaborate web that can quickly ensnare new attempts to regulate human affairs.

In the Peruvian forests this web of commitments includes unwritten rules inherited from the era of the Rubber Boom of 1860–1920, when the milky sap from Amazonian rubber trees (*Hevea brasiliensis*) supplied most of the world's high-quality rubber. All coordinated human activity requires rules, and this is especially true of something as complex as extracting latex from remote and inhospitable jungles and transforming it into the rubber tires coursing over the streets of London, Shanghai, and New York. The rubber barons who descended on the Amazon to exploit this lucrative market knew this all too well, and created chains of obligations and dependencies to gain access to the trees, machinery, and labor (including, in some cases, forced labor) they needed to extract rubber. After a brief resurgence during World War II, the Amazonian rubber economy eventually collapsed due to competition from British plantations in Asia. But the rubber barons' system of rules remains in place today, facilitating illegal harvests of tropical timber outside of the areas permitted under Peru's new forestry law.

Miguel Pinedo-Vasquez of Columbia University and Robin Sears of the international School for Field Studies provide a fascinating portrait of how the system works. From the time a tree is felled with a chainsaw until the wood is milled and transported, the process is guided by a hold-over from the Rubber Boom era known as *habilitación*, or enabling. Habilitación involves the extension of credit down a long line of intermediaries connecting people of money and influence to those who can get hold of falsified harvesting permits, to still others who have the muscle and know-how to extract timber. The latter group includes loggers, boat captains, and local *materos*, or tree hunters, who have grown up hunting and fishing in the area and can direct the logging team to valuable local tree species. A single mahogany tree can fetch over $10,000 on the US market. As credit extends down the system, and the trees move back up, participants take uneven cuts of the profits, but all share a stake in maintaining a system of old rules that is undermining the new ones. Newcomers such as foreign timber companies quickly become entangled in this web of relationships and "are forced to adopt the system or fail economically, politically and socially."[29]

Cerulean warblers feel the effects of these rules as they pass over a patchwork of land at varying levels of degradation. In the near future, migrating birds hoping to find a safe haven in the branches of Peru's forests will face yet another threat: rules promoting the extraction of fossil fuels. Following the rules encoded in civil law systems throughout Latin America and the world (recall Napoleon's cherished Civil Code, discussed in chapter 1), the Peruvian government owns the subsoil of the entire country. As fate would have it, the tropical forests of South America sit on top of enormous reserves of oil and gas. In neighboring Ecuador and Bolivia, oil and gas development has had ruinous environmental consequences, prompting widespread protests as forests are cleared for access roads, pipelines, and drilling platforms. In 2003, following a political shift in Peru to favor unregulated markets, the government changed the rules to reduce the royalties that oil companies must pay the government. This has fostered a boom in fossil fuel development in the cerulean's flight path. According to a team of researchers led by ecologist Matt Finer, almost three-quarters of the Peruvian Amazon is slated for exploration by multinational oil companies.[30]

Cordillera Occidental, Colombian Andes

By early April, cerulean warblers are beginning to cross into Colombia. You might imagine that a border crossing matters little to a bird traveling 3,000 feet up in the air. But this political transition is not merely some pen-and-paper construct, lines on a map with little relevance to the wilder creatures of the world. As in Peru, the choices that Colombians make about property have a direct bearing on whether migratory birds and other wildlife have food and shelter. The stakes are especially high in Colombia, which is the biological equivalent of a superpower. Colombia harbors more species of birds than any other country in the world. But as cerulean warblers make their way along the valleys that crawl up the western half of the country, high-quality habitat is increasingly hard to find. In this region, known as the Cordillera Occidental, most of the original tree cover has been removed. The reason can be found in your coffee mug—or, more precisely, in the property rules that Colombians have devised to help supply the 1.6 billion cups of coffee consumed globally every day. Our beloved caffeinated crop thrives on moist forested slopes at mid-elevations—precisely the preferred habitat of our cerulean.

The forests of Colombia's coffee-growing regions have been knocked down by a one-two punch of property decisions: the expansion of land devoted to coffee in the late 19th and early 20th century, followed by the promotion, beginning in the 1970s, of "sun-grown" coffee—a cheery euphemism for cutting down trees to make room for more of the red-berried bushes. In the late 19th century, Colombia's political leaders undertook a massive program of property rights reform designed to promote smallholder agriculture. The government gave up its title to millions of acres of land, handing over the property rights to farmers.[31] At the same time, growing demand for coffee in North America and Europe provided an opportunity for these new landowners to generate income through exports. The chain of events linking property rules to coffee to forest destruction has been documented by Andrés Guhl of the University of the Andes. Guhl shows that the expansion of coffee farms occurred precisely in those parts of the country where public land was turned into private property for farmers. The new property regime, combined with the new market, led to a rapid transformation of the landscape. Between 1892 and 1925, the number of coffee trees increased tenfold to 350 million. By 1997, coffee production supported more than a third of Colombia's agricultural workforce.[32]

The second blow of the one-two property punch was an aggressive effort to replace traditional shade-grown coffee with sun-grown plantations as part of the commercial intensification of coffee production that began in the 1970s.[33] Ceruleans have been spotted in significant numbers in the shade-grown coffee plantations of Colombia. But bear in mind, even shade-grown coffee typically involves cutting down the jungle and replacing its rich mix of plant life with a small number of economically useful species including the bushy green coffee bean tree. (It is not, as the "shade-grown" label might suggest, coffee beans grown in a relatively intact forest.[34]) Highly managed plantation forests are rarely as useful to wildlife as the real thing.[35] Still, shade-grown is far better than no trees at all, which is the method prescribed by sun-grown coffee.

Recognizing that ceruleans are in trouble, conservation organizations throughout the Americas are forging alliances in an effort to rescue their forest habitat before it's too late. Like their forebears who put an end to the hat-feather trade a hundred years ago, these groups are acutely aware of the role that property and economic initiative can play in helping or harming the earth. This time they are harnessing market power to protect Colombia's forests by touting the benefits of shade-grown coffee. The cerulean warbler is considered by many to be "the 'signature bird' of the shade coffee movement," according to the group Rainforest Alliance.[36] You can even buy Cerulean Warbler Conservation Coffee, marketed by Thanksgiving Coffee Company of Fort Bragg, California. Another property-based strategy for protecting the cerulean is to buy its habitat outright. In 2006, a Colombian nonprofit group called ProAves partnered with the American Bird Conservancy to create the world's first protected area devoted primarily to this little bird—the Cerulean Warbler Reserve in Santander, Colombia. But at only

545 acres, this privately owned reserve is far too small to slow the cerulean's decline. As they flap along over millions of acres of Colombian real estate, few ceruleans are likely to ever find it. But a parallel development in the country could provide a vastly more powerful tool for conservation. A radical shift in property rights is underway to recognize the long-standing demands of indigenous peoples. The nature of this shift becomes clear as cerulean warblers begin to arrive at the northern reaches of Colombia, where an ancient culture is taking advantage of the new rules governing the land.

Sierra Nevada de Santa Marta, Colombia

As the Andes mountain chain stretches to the north, it branches into a Y shape, with the base of the letter reaching down into Peru and the arms extending upward across the western portion of Colombia. At the northernmost tip of Colombia, hundreds of thousands of migratory birds make one last stop to rest their wings and build up their calorie count before heading out over the Caribbean sea. Many of these, including ceruleans, use as their launching pad a mountain of almost unfathomable size known as the Sierra Nevada de Santa Marta. Located only twenty-nine miles from the Caribbean, and topping out at 19,000 feet, it is the world's largest coastal mountain. In 2010, a team of geologists led by Camilo Montes of the Smithsonian Tropical Research Institute reported that the mountain has undertaken a migration of its own. The team examined traces of ancient magnetic fields recorded in the mineral content of the mountain, portions of which are a billion years old. They discovered that over the past 170 million years, the mountain has traveled 1,300 miles across South America, riding a tectonic plate from northern Peru to its current position on the Colombian coast, where it rotated clockwise a quarter turn to assume its current position overlooking the sea.[37] For birds, this migrating stone massif is prime real estate. Santa Marta harbors more endemic species—birds found nowhere else in the world—than any other continental land mass on the planet.[38]

The people who live here are no less remarkable than the landscape. The Kogi, Arhuaco, and Wiwa people survived the Spanish conquest and still live on the mountain that they know as the Heart of the World.[39] Draped in white canvas clothing that accentuates their flowing black hair, the people of the Santa Marta command a stark presence. The impression is accentuated by the white conical hats that Arhuaco high priests wear in a symbolic representation of their mountain's snow-covered peaks. Although the three groups have distinct languages, they share a cosmology and associated set of rules known as the Law of Se, or The Original Law, that emphasize harmony between human actions and the broader universe. The worldview of these ancient peoples, and the way it is implemented in their land management practices, has a lot to do with whether North Americans find birds in their backyards come spring.

Their approach to the land is guided by another social rule known as the Law of the Mother, a complex code enforced by the priests, called Mamas or Mamos, who are tasked with maintaining the well-being of the universe.[40] "The Great Mother gave us what we needed to live," a Kogi priest explains in the BBC documentary *The Elder Brothers*, which offers outsiders a rare glimpse into Kogi life, "and her teaching has not been forgotten, right up to this day. We all still live by it."[41] The Austrian-born anthropologist Gerardo Reichel-Dolmatoff, who spent his life studying the indigenous peoples of Colombia, provided some of the first published accounts of how the Kogi craft rules to govern the land: "In the course of centuries of being forced higher and higher into the mountains by encroaching settlers, the Indians' ecological awareness has been sharpened to a point where a precise knowledge of soil characteristics, temperature, plant cover, rainfall, drainage, slope exposure, and winds has begun to form a coherent body of procedures and expectancies."[42] The power to make these rules distinguishes the priests from the others. "Even the highest-ranking Kogi shares in the subsistence level, wears the same threadbare clothes, and lives in the same small hut as his lower-ranking compatriot," observed Reichel-Dolmatoff. "The difference consists in traditional power, in authority, and in the ability to establish rules of correct procedure."[43]

As we will see in chapter 8, "local" decisions are never truly local; all local rules are embedded within larger sets of national and international institutions, and this is true of rulemaking by the indigenous peoples of northern Colombia. This region is part of a bold property experiment in Colombia called a *resguardo*, or indigenous reserve. Resguardos assign basic land rights to local inhabitants and back these up, at least in theory, with the police power of the national government. Over the past two decades, Colombia has been at the forefront of an international movement to formally recognize indigenous peoples' property rights on millions of acres of land for the first time since their dispossession by European settlers.[44] This change was prompted by a political mobilization by indigenous peoples that took Latin America by storm beginning in the 1980s. In 1982, the

First Indigenous Congress was held in Bogotá, bringing together indigenous leaders from dozens of countries, including over 2,000 representatives from across Colombia. These groups soon formed transnational alliances, connecting formerly isolated tribes with large and politically influential indigenous organizations from the far corners of the earth.[45] In the Americas, the movement spread up and down the cerulean flyway and all the way to Canada, where new rules recognize Aboriginal peoples' property rights over thousands of square miles of land in the northern territories.

In Colombia, the impact of this shift in property rights has been truly extraordinary. Today there are 300 indigenous resguardos covering 76 million acres and inhabited by over a million people. Indigenous lands cover fully 29 percent of the national territory. Self-governance in resguardos is guaranteed by the Colombian Constitution of 1991, which specifies that these property rights may not be transferred out of indigenous hands. Collective territories of Afro-descendant communities occupy an additional 14 million acres. Now *that's* something a cerulean can find.

With their new legal status as owners of a resguardo, and financial support from The Nature Conservancy and other international groups, the indigenous leaders of Sierra Nevada de Santa Marta have formed a nonprofit organization that is purchasing land from non-indigenous peasants on the mountain. According to a statement by the group, "In each case, the acquired territory becomes the collective property of the indigenous people of the Sierra Nevada and can never again be sold. In each case, we let the forest grow back which, along with the spiritual work of our Mamas (Mamos) helps the animals that fled those areas return there."[46]

Despite these gains, significant challenges face the people of this mountain and the wildlife they protect, for migratory birds are not the only ones that find this region to be of strategic importance. Like coffee, crops of coca and marijuana thrive at the mid-elevations preferred by ceruleans. And like our birds, Colombian drug dealers have a keen interest in access to the Caribbean, which provides a link to customers in the United States. Many of those involved in the drug trade are hardened leaders of insurgent guerilla organizations and right-wing paramilitary groups from Colombia's brutal civil war, which has persisted at varied levels of intensity for over half a century. Conflict between these dangerous groups and indigenous peoples have produced tragic consequences. Approximately fifty Wiwa leaders have been murdered by paramilitary organizations operating in the area, according to estimates from The Inter-American Commission on Human

Rights. But indigenous leaders have continued undeterred, acquiring and regenerating thousands of acres of forest up and down the mountain slopes, including trees that the cerulean depends on for its survival.

It is remarkable to consider how birds offer a thread of connection among the people of the world. As a soccer mom in Ohio walks into her backyard with a cup of coffee and looks up to the trees, the type and abundance of birds there is affected by the political mobilization of indigenous groups throughout the Americas. Likewise the bird song enjoyed by an Arhuaco priest on his evening stroll is influenced by land use decisions in Ohio, and whether that soccer mom shows up at city planning meetings to demand the protection of open space. In its mix of hope and tragedy, the story of the Sierra Nevada de Santa Marta suggests that the well-being of people and nature are entwined in precisely the manner described by their indigenous cosmology.

Central America's Atlantic Coast

Most of what we know about cerulean migration comes from the Cerulean Warbler Technical Group, an international team of scientists who have been piecing together bits of evidence to form a coherent picture of the bird's whereabouts during its marathon flight. They are building on an early hypothesis about the cerulean's route devised by Ted Parker, who until his tragic death in a plane crash in 1993, was one of the world's foremost ornithologists. Parker was renowned for his innovative approach to identifying birds in the wild. Traveling the globe with a powerful microphone connected to a tape recorder, Parker combined technology with an almost superhuman memory capacity. He is believed to have memorized the songs of all 4,000 birds in the New World. In the early 1990s, Parker was part of a team of scientists brought together by Conservation International as part of its Rapid Assessment Program. The group paired Parker with experts of similar stature for the identification of plants and mammals and flew these elite teams to remote areas around the globe. There they would use their encyclopedic knowledge of nature to conduct lightning-speed surveys of species diversity (a process that would normally take years of research), with the goal of identifying high-priority conservation areas before they are destroyed. During a visit to Peru's Manu National Park, in a mere two weeks Parker found twenty bird species that leading ornithologists working in the area for a decade had never detected.[47]

During an expedition in 1992 to the Maya Mountains of southern Belize, Parker unexpectedly encountered large numbers of cerulean warblers, about 10–20 per day, foraging in the trees in a remote spot where he and his team had been deposited by helicopter with help from the Belizian Air Force. Parker was startled to find so many ceruleans hanging out in Central America. Prior to this discovery, it was assumed that ceruleans cruised northward nonstop from South America to the Gulf Coast of the United States. He published a paper hypothesizing that instead they hop across the Caribbean from South America to Central America, rest up, and then head out across the sea again to North America. Today the Cerulean Warbler Technical Group, under the leadership of Paul Hamel of the US Forest Service, is trying to figure out the rest of the puzzle. Hamel's team believes that from the northern crest of South America, some ceruleans catch the spring winds for a straight trip up across the sea toward Cuba and North America, while others make the shorter overseas flight to Central America (the Parker hypothesis), and still others hug the land, heading west across the Isthmus of Panama and continuing up Central America's Atlantic coast. Following the land route, this stretch of the voyage reinforces the point that property rights, and social rules generally, are neither preordained nor unchanging; they are human creations designed to propagate the political priorities of a given time and place.

Ceruleans make occasional stopovers at lower elevations, and have been sighted in the steamy jungles of Barro Colorado Island, a 3,700-acre land mass in the Panama Canal that was spared flooding when construction crews dug a waterway across the country a century ago. Until relatively recently, the trees of Barro Colorado were owned by the United States. President Teddy Roosevelt forced South American leaders to accept his plan to build a canal across the narrow isthmus, which would lower the cost of shipping goods between the Pacific and Atlantic oceans. At the time, Panama was part of Colombia, which tried to resist US demands. Roosevelt responded by supporting a military coup to install a government that broke from the Colombian federation and supported his proposal.[48] Under the new rules, the United States was granted its sought-after prize: exclusive property rights over a swath of rainforest five miles wide and fifty miles long, traversing the isthmus. Although Teddy Roosevelt is considered a hero of the American conservation movement, having established the national parks and national forest systems in the United States, the forests of Panama did not fare well under American stewardship. Between 1952 and

1983, forest cover throughout the canal basin shrunk from 85 percent to 30 percent.[49]

Today, the fate of the cerulean at this midpoint in its journey is once again in Panamanian hands. At midnight on December 31, 1999, with cameras rolling and the political establishment grinning, the canal and its patches of cerulean habitat reverted to Panamanian ownership, the result of a treaty signed with the Carter administration in 1977. As the canal undergoes a planned expansion, it remains to be seen whether Panama's political elites will make a priority of conserving the surrounding watershed.[50]

Continuing up the Atlantic coast of Central America and into Costa Rica, our blue bird is in luck. Here Costa Ricans have created new property rules with the explicit goal of protecting the country's forests. Through a government-run program, Costa Rican farmers conserve trees on their land and then sell these "environmental services" to those who benefit from the ecological functions that forests provide. These customers include local water users as well as people everywhere who rely on healthy forests to pull carbon dioxide out of the atmosphere and transform it into plant biomass—an important measure for combatting global warming. The program is funded by local water user fees, a national gasoline tax, and payments from corporations and governments around the globe who are compelled by law or conscience to reduce their carbon emissions. The program is also wildly popular in Costa Rica. By 2008, farmers had enrolled 1.5 million acres of land in the program, including dozens of properties throughout the cerulean's route in the Caribbean foothills, and received payments totaling $206 million.[51] A research team led by economist Rodrigo Arriagada evaluated the program's impact on the practices of landowners, and found that the forest cover on participating farms increased by about 11–17 percent as a result.[52]

Santa Catarina Ixtepeji, Mexico

By mid-April ceruleans are arriving in southern Mexico, where they fly inland to rest and refuel before the final part of their journey, which will take them over the Caribbean and into the United States. Those that alight in the trees of the Lacandon Rainforest near San Cristóbal de las Casas enter a situation of utter lawlessness, the result of an uneasy truce between the Mexican government and local Zapatista rebels. In January 1994, indigenous Mayan leaders formed the Zapatista Army of National Liberation, led by the enigmatic ruler Subcomandante Marcos, with his signature black ski mask and pipe. Although their tactics were more militant than those of indigenous groups in South America, their demands were essentially the same: protection of indigenous property rights. In particular, the Zapatistas opposed a move by the Mexican government to change the rules governing property to favor private enterprise. By openly challenging the legitimacy of the Mexican state and its ruling party, the uprising helped to usher in democratic reforms including the introduction of competitive national elections in 2000. In Chiapas, however, the truce between the central government and Zapatista leaders has produced regional autonomy alongside anarchy; there is little in the way of working property rules for the forests of Chiapas. The fate of these lands, and the people and wildlife that depend on them, is still anyone's guess.[53]

Birds traveling farther west to the Mexican state of Oaxaca encounter an altogether different reality. Here ceruleans benefit from a celebrated property experiment called community forestry, which is designed to simultaneously benefit local communities and the environment. Community forestry has been studied by economist Camille Antinori of the University of California at Berkeley and her colleague David Barton Bray, a cultural anthropologist at Florida International University. They report that Mexico has gone further than any other country in the world in vesting local communities with the power to make decisions affecting their forests.[54]

Community forestry in Mexico builds on an innovative property institution dating from the Mexican Revolution, called *ejidos*. Like the indigenous resguardos of Colombia, the ejido is based on age-old communal property rules used by indigenous people. An ejido is neither government-owned, nor private property—it is land owned in common by a community. Major decisions about how the land will be used, and who gets to use it, are made in local general assemblies. Typically some land is set aside for use by the community as a whole, while other parcels are allocated to individual families subject to approval from the assembly. Delivering on Emiliano Zapata's revolutionary promise of *tierra y libertad* (land and liberty), the ejido is as central to the Mexican political imagination as images of Zapata and Pancho Villa riding into villages with ammunition belts strapped across their chests. This popular system expanded throughout the 20th century, and today more than 30,000 ejidos govern over half the land in Mexico.[55]

Located in cerulean warbler habitat north of the city of Oaxaca, the ejido of Santa Catarina Ixtepeji demonstrates how Mexico's property rights regime empowers local communities. Until 1983, any decisions concerning the forests around this small community were made by a government-run timber operation. Around this time, communities throughout Mexico began demanding a greater say in the decisions affecting local forests. When the government tried to renew its harvesting contracts in Ixtepeji, the community staged a series of protests and eventually gained control over this portion of the cerulean warbler's migratory route. Today Ixtepeji is a showcase for community forestry, one of 2,000 Community Forest Enterprises that have sprung up across Mexico over the past four decades. Balancing conservation and development goals, the community operates a commercial timber mill that markets wood from its 37,000 acres of production forests, which also harbor profitable nontimber products such as white mushrooms and pine resin. Another 10,000 acres are in designated conservation areas to protect wildlife and water sources. Profits from these activities are invested locally to support services like schools, roads, and social security payments to the elderly. The rules governing forest use are made by the ejido's community assembly in consultation with national environmental authorities, while day-to-day operations are run by local ejido officials. Any community member who breaks the rules is subject to fines or exclusion from community enterprises.[56]

The Caribbean Sea

Shortly after sunset on the evening of their departure, ceruleans launch from their resting spots in southern Mexico and begin a nonstop flight to North America, where they hope to arrive by morning or early afternoon. Birds employ diverse strategies when making these long trips. Larger birds like peregrine falcons can save their energy by catching a warm air current or ocean breeze, which allows them to float upward and then soar in the direction of their destination before catching the next updraft. For small birds like ceruleans, it's a marathon flapping session, requiring a strategy that balances energy efficiency (which would suggest a moderate pace) with the need for speed to take advantage of favorable weather conditions.

As the birds head out over the sparkling blue Caribbean waters, the property rules governing the earth below change abruptly as the surface turns from forest to open sea. Here too, the rules have evolved over time to promote the varied aims of their creators, including the most infamous rulers of this portion of the planet: pirates. In the 17th and 18th centuries, pirate ships coursing throughout the Caribbean had quite specific rules to govern the disposition of their ill-gotten loot. The economist Peter Leeson, who studies the institutional dimensions of pirate life, finds that pirate vessels were not the lawless places of Hollywood lore; they were governed by constitutions and led by captains who were democratically elected by their crews. Writing in 1724, seamen Charles Johnson described the rules governing the division of spoils on the ship of the famous pirate Bartholomew Roberts. Pirates were expected to be forthcoming in revealing the amount of their stolen treasure: "If they defrauded the Company to the Value of a Dollar, in Plate, Jewels, or Money, Marooning was their Punishment," reported Johnson. "If the Robbery was only betwixt one another, they contented themselves with slitting the Ears and Nose of him that was Guilty, and set him on Shore, not in an uninhabited Place, but somewhere, where he was sure to encounter Hardships."[57]

Today property rights in the waters of the Caribbean are governed by an international treaty known as The Law of the Sea. This powerful collection of rules specifies who owns the oceans. Until fairly recently, each coastal country owned a stretch of ocean extending outward from its shore for three nautical miles, equivalent to just under 3½ miles. (The three-mile limit derived from a social rule called the "cannon shot rule," which held that ownership extends as far as a country can defend its territory. At the time the principle was codified in the late 18th century, three nautical miles corresponded to the upper range of the latest high-powered canons.) Beyond this, the seas were governed according to a rule known as freedom of the high seas—a centuries-old code of conduct specifying that no country has the right to restrict passage across the open ocean.

As technologies changed and political aspirations expanded, the old system of rules began to crumble.[58] In 1945, President Harry Truman issued a new rule—the Truman Proclamation—unilaterally declaring American ownership of oil and gas resources in the continental shelf beyond the traditional three-mile limit. The rest of the world quickly followed suit, as countries laid claim to oceanic property anywhere between twelve and 200 miles from their shores. While this chaotic mix of conflicting property rules was developing, a more systematic effort at reform was underway, inspired by yet another change in who rules the earth. In the 1950s and 1960s, dozens of former colonies gained political independence, particularly throughout Asia, Africa, and the Middle East. (Most of Latin America gained independence from Spain and Portugal in the early 1800s.) These new nations were eager to assert their autonomy in an international system of law originally designed, and still dominated, by the former colonial powers. Wealthy nations, for their part, sought greater clarity on the rules of the game governing maritime commerce. Still others cited the environmental imperative of regulating use of the world's oceans. In an impassioned plea before the UN General Assembly in 1967, the ambassador from Malta, Arvid Pardo, reminded delegates, "The dark oceans were the womb of life; from the protecting oceans life emerged. We still bear in our bodies—in our blood, in the salty bitterness of our tears—the marks of this remote past."[59] In a grueling and complex negotiation spanning several decades, the final settlement rewrote the rules governing the Caribbean waters that stretch out below the cerulean as it continues on its nighttime journey.

Today every coastal country holds property rights over the sea and underlying seabed according to the following rules: From the shore to twelve nautical miles—the so-called territorial waters—the coastal nation calls the shots and owns all the resources including the airspace above.[60] Out to 200 nautical miles, the coastal nation still has exclusive rights to the resources, but other nations are allowed to pass military vessels through the area. Beyond that, the freedom of the high seas is preserved. The open sea belongs to everyone and no one, subject only to rules agreed upon in international treaties.

To appreciate what all of this means for the cerulean warbler, recall why the United States and others were so eager to change the rules in the first place. The Law of the Sea facilitated the rapid expansion of offshore oil development along the coastline stretching from eastern Texas to Louisiana, Mississippi, and Alabama. This corresponds precisely with the most important migratory bird passage in North America.

As ceruleans approach the American coast, they form part of a massive front of birds of all shapes, colors, and sizes moving across the Caribbean. Birds that migrate at night use light from stars for navigation. Offshore oil platforms disrupt this age-old adaptation through a practice known as flaring, in which these 50,000-ton feats of modern engineering send flames soaring eighty feet into the air to burn off the natural gas that is extracted along with oil. This disorients migratory birds, which can be seen flying frantically in circles around the platforms until they eventually drop from exhaustion and drown just miles from their landed destination. Fortunately the American government created new rules to ban flaring on offshore oil platforms by 2015.

The impact of the American rules governing oil extraction are even greater for seabirds, as became painfully clear during the BP oil spill of 2010.

As with any social rule, ownership of offshore oil deposits implies both rights and responsibilities. The appropriate mix is decided by the country that holds the property right and reflects that country's political priorities. Prior to the spill, the rules governing oil development in the United States favored first and foremost the rapid development of the resource. Oil companies enjoyed a cozy relationship with regulators at the federal Minerals Management Service, which was tasked with the contradictory goals of approving safety permits and aggressively promoting oil exploration. Agency officials received large cash bonuses for expediting the approval of permits; frequently they allowed the oil companies to fill out the paperwork themselves, penciling it in for the officials to then trace in pen.[61]

BP's Deepwater Horizon rig was operating under these rules when the platform exploded in flames and collapsed into the sea, killing 11 workers and creating the largest oil spill in US history. This time seabirds and other marine life dodged a bullet. The oil plume was 5,000 feet underwater and composed of a light biodegradable crude, and so it did not impact wildlife on the scale of the 1989 Exxon Valdez spill in Alaska (which immediately killed a quarter million birds) despite leaking almost twenty times as much oil.[62] In response to the public outcry following Deepwater Horizon, the Obama administration changed the rules governing oil extraction. Today the agency officials responsible for promoting oil development are separate from those charged with oversight of safety and environmental protection.

Boone County, West Virginia

Throughout April, upward of a million migratory birds arrive on the Gulf Coast of the United States each day. Scientists specializing in "radar ornithology" have given us stunning images of a shape the size of a hurricane moving from the sea to the shore as masses of birds head inland across the cypress swamps and pine and live oak forests of the coast (Figure 4.2). Viewed from the ground, it is a sight to behold. There is even a greeting party, with celebrations throughout the American South including the North Alabama Birding Festival, the Great Louisiana BirdFest, FeatherFest in Galveston, Texas, and the Mississippi Flyway Birding Festival. While barbecue and binoculars are at the ready, ceruleans continue northward, making their way toward the Appalachian Mountains where they hope to create a new generation of cerulean chicks in the forested slopes of Tennessee, Kentucky, West Virginia, and Ohio.

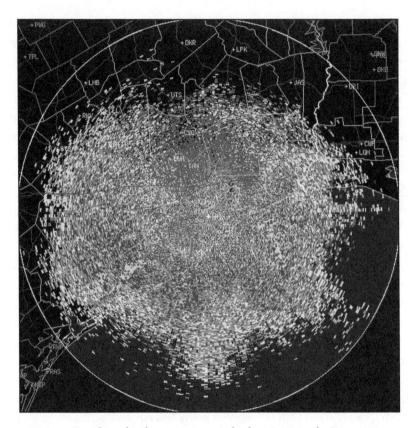

FIGURE 4.2 Doppler radar showing migratory birds arriving on the Texas coast

National Center for Atmospheric Research, courtesy of Sidney Gauthreaux, Clemson University.

The Appalachian mountain chain stretches from Central Alabama up through the northeast, defining the landscape in much of the eastern United States and reaching all the way to Newfoundland in Canada. It is one of the planet's most ancient mountain systems, more than ten times as old as the Andes. The Appalachians are home to so many wildlife species that biologists count this stretch of Earth among the top conservation priorities in the world.[63] But here at the end of their journey, cerulean warblers encounter something altogether different from the lush mountain forests that their ancestors returned to for millions of years. This is coal country. Today in places like Boone County, West Virginia, located in the heart of cerulean breeding habitat, our bird finds a dead and barren landscape interspersed with patches of trees (often along roadways, for visual effect) like some bad haircut. In an effort to find ever-cheaper ways to feed the country's appetite for coal, mining companies are literally destroying the state of West Virginia. The ecological destruction is the result of a practice known as mountaintop removal, an extreme version of strip mining in which the top of the mountain is blown off to expose coal seams. After the blasts, enormous machines called draglines, reaching twenty-two stories in height, shove the "overburden"—the term used to describe the earth that does not contain coal—into adjacent streams and valleys. Hundreds of mountains and over 1,000 miles of streams in the Appalachians have been destroyed as a result.[64]

You might expect that the United States—an advanced industrial nation and early pioneer in the development of environmental policy—would have rules to prevent this practice. It does. Under the Clean Water Act, the protection of the nation's waterways is clearly spelled out as a public priority of the highest order. The Endangered Species Act also forbids any actions that harm threatened wildlife and contains clear provisions for the protection of habitat. Government regulators have simply failed to enforce the rules. Under pressure from the fossil fuel industry and their political allies, wildlife officials have refused to list the cerulean warbler as a threatened or endangered species despite clear evidence from the Breeding Bird Survey showing that its numbers have declined by three-fourths since the 1960s. In 2002, the government went so far as to write a new rule recategorizing mining waste as "fill" so that it would be subject to less stringent regulatory standards under the Clean Water Act. But the battle is far from over. American environmental laws contain special rules that allow citizens to sue to sue their government if it fails to faithfully implement the laws on the books. Lawsuits filed by groups like the Ohio Valley Environmental Coalition are contesting mining permits, while groups like the Audubon Society and Defenders of Wildlife are pushing for federal wildlife protection.

The rules governing property in West Virginia affect more than birds. West Virginia has one of the highest concentrations of poverty in the United States. For years, struggling families have been selling their land rights to corporations that have little interest in the future of the land or its people and wildlife. This land grab has been facilitated by rules governing subsurface property rights in the United States, which work differently from those in Latin America and most of the world. Elsewhere oil and mineral rights are owned by the government; this is why, in the ceruleans' winter home, it is Peruvian lawmakers who will decide whether the forests are cleared to make way for oil and gas exploration. In the United States, the fossil fuels below the earth's surface belong to the property owner above.

If you want to control a resource, ownership is not enough; you also need the ability to access and use it.[65] This is precisely why the conquistadors sought to control not only land, but the indigenous labor needed to extract the gold and silver it contained. In Appalachia in the early 1900s, small landowners sold their mineral rights for a pittance to mining companies that had the capital to exploit the underlying coal. The result of all this buying and selling is that in West Virginia today, land is concentrated in the hands of a few. This is documented in a meticulous seven-volume study called *Who Owns Appalachia?*, in which sociologist John Gaventa and his colleagues reviewed the court records for 20 million acres of land throughout the region. The results were astonishing. These researchers discovered that 44 percent of the land is owned by 1 percent of the owners. Fully 72 percent of those with land title are absentee owners, principally multinational corporations. As you might imagine, the people of West Virginia are bitterly divided over the practice of mountaintop removal. What some see as the economic lifeline of isolated rural communities, others see as the cause of their demise.[66] Coal company executives are not waiting for the outcome of the public debate; they have spent millions of dollars to unseat lawmakers and judges not to their liking.[67]

The situation in West Virginia is bleak, but it is not hopeless. In recent years, people have put in place creative new property rules to protect those forests not yet destroyed. Others are beginning the process of restoring the one million acres that have been razed to the ground by coal companies. The first of these developments involves the spread of a new legal tool called conservation easements. To understand how conservation easements work, consider that property rights can be joined together or broken apart. In a metaphor that every law student knows by heart, property rights may be understood as a bundle of twigs, each representing one right with respect to a given resource. For a parcel of land, one twig represents the right to build on the land. Another covers access to minerals. Yet another represents the right to harvest the trees or draw water from the aquifer below. Many such distinctions are possible; the component rights may be bought and sold separately and are often subject to different sets of rules. The essential point is that a "landowner" may own a mere couple of twigs or a fat bundle.

Conservation easements are legal agreements in which the landowner chooses to forfeit the right to develop part of the land—forgoing the right to cut down any trees on the upper third, for example, or allowing public passage on a popular trail that traverses a meadow. In return, the landowner gets a reduction in property taxes, which is the government's way of rewarding those who make a sacrifice to promote the public good. Importantly, a conservation easement and its restrictions on land use are permanent—as in, forever. When the landowner sells the land, the easement travels with it to the new owner.

Laws permitting conservation easements spread from state to state throughout the 1980s and 1990s, resulting in the rapid growth of nonprofit groups called land trusts, which specialize in brokering these and related land conservation deals. By 2010, there were over 1,700 land trusts across the United States protecting a total of 47 million acres of land in perpetuity.[68] Groups like the West Virginia Land Trust and The Nature Conservancy are making it easier for ceruleans to find a healthy forest in which to raise a family.

Even bigger changes are underway in the region, offering a glimmer of hope for the cerulean. The destruction of the Appalachian forests was facilitated in part by a government rule originally intended to protect the environment. Under the Surface Mining Control and Reclamation Act of 1977, companies that mine the earth are required to then grade the land, bulldozing it back in place and packing it down tight with heavy equipment. This rule was designed to minimize the risk of landslides on denuded slopes. But its implementation has prevented the growth of young trees, which require looser soil in which to spread their roots. In recent years the rule has been modified, like so many along the cerulean flyway, this time thanks to a group of reformers who launched something called the Appalachian Regional Reforestation Initiative. Bringing together scientists, citizens, regulators, and industry representatives, this group created new guidelines that require a four-foot layer of loose soil at the top of reclaimed areas. Not only is this better for forests and wildlife, but it reduces the cost to industry, requiring one pass with a tractor rather than dozens, saving on labor and equipment costs. Over 85 million trees have been planted on 125,000 acres of land since 2005 as a result of this effort.[69] The group has teamed up with scientists hoping to restore the American chestnut tree, which made up a quarter of the hardwood forests in the eastern United States until it was decimated by chestnut blight in the 1950s. Using a resistant variety of the tree developed by researchers, abandoned coal mines will be covered with millions of trees to restore the once majestic chestnut forests.

A BIRD IN THE HAND

Faced with continuing threats throughout their migratory range, the little dots of cerulean blue that inhabit the upper branches of trees throughout the Americas continue their decline like lights blinking out in slow sequence on a Christmas tree. To understand why, and what it will take to reverse the situation, requires that we understand the institution of property. Returning to our original question, then, who owns the cerulean warbler? According to the Migratory Bird Treaty, no one, strictly speaking, is allowed to own or capture these birds. But the treaty says nothing of the land that the birds require for their survival. Like the freedom-loving West Virginia communities who depend on coal companies and their destructive practices for their survival, you have to wonder: When someone else controls your land, your resources, and your livelihood, when their decisions dictate whether you live in comfort or deprivation—how free are you really?

Far from being free as a bird, cerulean warblers—and indeed all the world's species, including our own—depend for their survival on the rules that we create. The variety of property arrangements found along the cerulean's route demonstrates that property involves choices. Rules of ownership determine not only whether we leave our children with a healthy planet or a collection of bleak and polluted landscapes. They also establish who reaps the benefits and who suffers the consequences of our decisions. These choices are too important to leave to others; the options and tradeoffs should be weighed and decided upon by the citizens affected by them. Toward this end, in the next chapter we will descend from the heights of transcontinental migration to consider some down-to-earth decisions facing us about who should own the air we breathe and whether we might use property rights and markets to help tackle some of our toughest environmental problems.

5

The Big Trade

As we continue our exploration of who rules the earth, we find that the economy, once you look inside it, relies on a vast system of rules and regulations, its cogs and wheels spinning day and night to enable the countless transactions that make up a modern economy. The relation between markets and rules is a fascinating one, far more complex than is suggested by the usual debates over government regulation versus free enterprise. Markets rely on rules. But increasingly, the reverse is also true: Some of our most innovative environmental policies and regulations have embedded within them market incentives designed to promote pro-environment behavior. To appreciate the stakes, let's begin by considering what is arguably the greatest environmental tragedy—and biggest environmental success story—of all time.

GETTING THE LEAD OUT

The removal of tetraethyl lead from gasoline has had a profound impact on human health and well-being worldwide. The change began in the United States in the late 1970s, soon spread to Europe, and over the next two decades diffused throughout the entire world. This shift was prompted by an innovative set of rules that actually assigned property rights to poison—and in the process created incentives for widespread changes in corporate behavior.

Under the Clean Air Act of 1970, the US Environmental Protection Agency had the legal authority to regulate tetraethyl lead, which had been added to gasoline since the 1920s to boost engine performance. The original decision to add "ethyl" to the chemical mixture sloshing around in our gas tanks took place despite dire warnings from health experts. Foremost among these was Alice Hamilton, Harvard's first female professor and the country's leading expert on the health impacts of lead, which she knew intimately from her studies of worker exposure in the largely unregulated "dangerous trades" of the time. In 1925, the US Surgeon General convened a special meeting to decide whether ethyl production could proceed despite the known health risks. Hamilton argued that it would be reckless to deliberately disperse throughout the air a substance whose toxic effects (notably damage to the human nervous system) were well known for centuries. "You may control conditions within a factory," Hamilton argued, "but how are you going to control the whole country?... [If] this is a probable danger, shall we not say that it is going to be an extremely widespread one, an extraordinarily widespread one?"[1]

Hamilton was right. The effects became immediately clear as the first ethyl gasoline was produced in the chemical factories. Workers at Dupont's Deepwater plant nicknamed their workplace the "House of Butterflies," a reference to the hallucinations afflicting workers exposed to lead fumes. At a Standard Oil processing plant in Elizabeth, New Jersey, five of the plant's forty-nine workers died from lead poisoning, typically following acute bouts of insanity, and another thirty suffered severe illness.[2] But the road from knowledge to action has never been a straight one. Benjamin Franklin, writing to a friend in 1786 of this "mischievous Effect from Lead," commented wryly, "you will observe with Concern how long a useful Truth may be known and exist, before it is generally receiv'd and practis'd on."[3] One hundred fifty years later, Franklin's observation still applied; Alice Hamilton's objections were ultimately ignored, and soon ethyl gas became a staple at service stations everywhere. Manufacturers of leaded gasoline launched an aggressive public relations campaign to make sure the new product was widely used (Figure 5.1).

The result was the largest environmental health disaster in US history. By the late 1970s, nine out of every ten kids ages one to five had blood lead levels considered unhealthy by the US Centers for Disease Control, primarily as a result of leaded gasoline.[4]

THEY didn't pass you when your car was bright and new—and you still don't like to be left behind. So just remember this: *The next best thing to a brand new car is your present car with Ethyl.*

If you buy a new high-compression car, you'll of course use Ethyl. But if you must make your old car do, give it Ethyl and feel lost youth and power come back as harmful knock and sluggishness disappear.

These days, when we have to do without so many things, we can at least make the most of our cars. And even if you don't measure the fun of driving in dollars and cents, you'll find that Ethyl makes real money savings in lessened repair bills. Ethyl Gasoline Corporation, New York.

Ethyl fluid contains lead. © E. G. C. 1933

BEWARE OF IMITATIONS

All Ethyl Gasoline is red, but not all red gasolines contain Ethyl fluid. The color is for identification only and adds nothing to performance. Look for this Ethyl emblem on the pump (or its globe).

The all-round quality of Ethyl is doubly tested: at the time of its mixing, and through inspection of samples taken from pumps. Ethyl's margin of anti-knock quality over regular gasoline is greater today than ever before.

NEXT TIME STOP AT THE ETHYL PUMP

FIGURE 5.1 Ethyl Corporation advertisement promoting the use of leaded gasoline

The decision to introduce lead into gasoline is a story that has been told many times. Less well known is how the problem was fixed through inventive rules that transformed an industry that had come to depend on leaded fuel. With clear evidence of harm, and a new clean air law that authorized swift action, the EPA could have simply mandated that every oil refinery switch production to unleaded gasoline by a given date. But EPA officials were aware that some operations (notably smaller refineries) faced more difficulty than others in making the change, which required new chemical processes to produce high octane gas without the use of lead. So instead of issuing a one-size-fits-all regulation, EPA officials decided to give companies a choice. First, the agency issued each oil refinery permits allowing the use of a specific quantity of lead. This enabled regulators to put a cap on the total amount of lead released into the environment. Then they did something unheard of—they allowed the companies to buy and sell the permits. Recall that the right to sell is one of the defining characteristics of a private property right; in effect, the EPA had created a new market for poison. The approach is not as devilish as it sounds. Oil refineries that were faster to make the switch wouldn't need permits and could sell them to the laggards for a profit. Rather than simply reach the mandated level of pollution control and then stop, as is the norm with traditional approaches to environmental policy, the tradable permits gave firms an incentive to keep going, reducing pollution and selling permits to their competitors. It didn't take long for the companies to catch on. In late 1983, 10 percent of all lead used in the industry was covered by permits traded among firms; by 1987, the figure reached over 50 percent. As promised, the EPA then gradually reduced the total number of permits so that even the slow-goers ultimately had to make the change.[5]

The introduction and eventual phase-out of leaded gasoline offer a clear illustration of how social rules shape our physical environment, and even the chemical composition of our bodies. Over ten years, the rate of elevated lead levels in the United States dropped from nine out of every ten children to one in ten—with the remainder resulting from continued exposure to contaminated soil and paint.[6] A corresponding physical change occurred across the planet's surface; in soils and sands and snow, we see a peak in lead content during the height of ethyl consumption in the 1970s, followed by a decline during the phase-out (Figure 5.2).[7] A similar trend can be seen in the tissues of plants and animals around the globe, which scientists use as indicators to track the environmental fate of lead. The rise and fall of lead can be traced in the green moss clinging to trees in Norway, in the muscles of reindeer

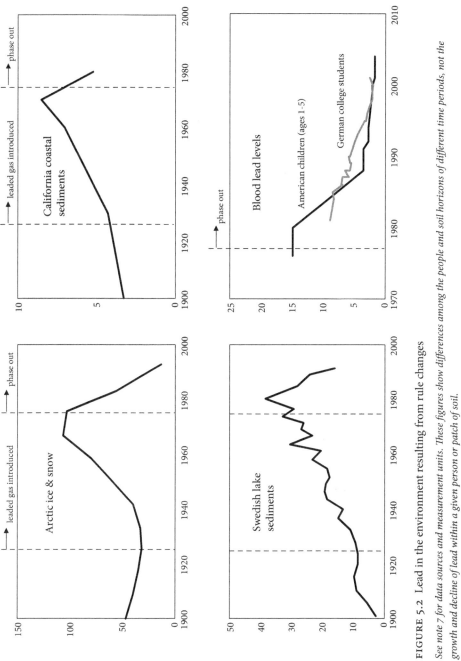

FIGURE 5.2 Lead in the environment resulting from rule changes

See note 7 for data sources and measurement units. These figures show differences among the people and measurement units. These figures show differences among the people and soil horizons of different time periods, not the growth and decline of lead within a given person or patch of soil.

peacefully eating meadow flowers in Sweden, and in the sands of Southern California beaches. The pattern appears in spruce needles and poplar leaves in Germany, in the livers of cod swimming in the Baltic Sea, in measurements of urban air quality in Italy, and in the vast ice sheets of Greenland.[8]

In addition to the benefits for human health and the environment, the US tradable permit program removed lead at a cost saving of hundreds of millions of dollars compared to conventional approaches to regulation.[9] Rather than forcing every company to adopt the same technology at the same time, permit trading built a measure of flexibility into the regulatory process while maintaining strict control over the total amount of lead allowed. You might expect that a rule capable of protecting the environment while reducing the associated cost to industry would enjoy broad political support. But while market-based regulation carries some appeal for moderates (fiscally responsible environmental protection—what's not to like?), it plays less well in polarized political settings in which participants cling stubbornly to ideologies. Opponents of environmental regulation are uninterested in making distinctions among different types of environmental policies, categorizing any government attempt to shape industry behavior as anathema to a healthy market economy. For many political progressives, incentives to increase profits are considered the cause of environmental problems, not the solution; efforts to deliberately blend the two goals are viewed with deep suspicion.

We will see that market-based regulations can, when used with care, offer significant social and environmental benefits. To make sense of this, we first need to understand the relationship between rules and markets more generally. This is a big question, surrounded on all sides by supercharged rhetoric, and little wonder: The entire Cold War turned on the question of the appropriate balance between market exchange and government intervention in a modern economy. Who should control property—governments, corporations, nonprofit organizations, neighborhoods, families, individuals, or something else? These questions continue to define the terms of political debate within and among countries. And these are just the kinds of questions that can benefit from careful research, which offers an antidote to more reflexive (and less reflective) forms of political discourse.

Hacking a path through the ideological thicket, we will consider two issues of great importance for the question of how property rules can promote or hinder the transition to sustainability. First, what role do social rules currently play in the operation of market economies? Second, is there a role for market incentives in environmental policy? This question turns the first one on its head, looking not at the rules hiding within markets, but at the potential for designing market incentives into the fabric of our rules. Make no mistake, market-based environmental regulation is all the rage. Untethered from its modest beginnings in the United States, large-scale experiments in market-based regulation are now underway in dozens of countries around the globe. These experiments carry important implications for environmental quality in our communities. Evaluating the record of past successes and failures can help us to make wise decisions about whether to support or oppose initiatives like environmental taxes and cap-and-trade programs, and whether these can help tackle the biggest, baddest environmental problem of them all: climate change.

THE RULE OF THE FRENCH FRY

We like to think of the private sector as a freewheeling source of fast-paced innovation, rather than a roost of rules and regulations. But in practice, all corporations operate through social rules. The most agile start-up firm would be completely lost without them. The level of coordination required for a company to hire staff, rent office space, secure loans, and sell goods and services across long distances requires a lot of rules. Coordinating activities among large numbers of companies—suppliers and insurers, equipment maintenance companies and import-export firms, truck drivers and container ships, security and information technology companies, realtors and accounting firms, and many others—would be impossible without reams of rules. The political scientist James Q. Wilson spent decades studying the performance of organizations in market and government settings, and concluded that there are no fewer rules in the private sector. He used the example of the McDonald's french fry:

> McDonald's is a bureaucracy that regulates virtually every detail of its employees' behavior by a complex and all-encompassing set of rules. Its operations manual is six hundred pages long and weighs four pounds. In it one learns that french fries are to be nine-thirty-seconds of an inch thick and that grill workers are to place hamburger patties on the grill from left to right, six to a row for six rows. They are then to flip the third row first, followed by the fourth, fifth, and sixth

rows, and finally the first and second. The amount of sauce placed on each bun is precisely specified. Every window must be washed every day. Workers must get down on their hands and knees and pick up litter as soon as it appears. These and countless other rules designed to reduce the workers to interchangeable automata were inculcated in franchise managers at Hamburger University located in a $40 million facility.[10]

The rules governing economic organization can foster innovation or they can crush it. 3M Corporation has a rule requiring that at least 25 percent of the revenue generated by its more than ninety product divisions must come from products invented within the previous five years. Contrast this with the rules governing economic organization in Shaker communities, whose Millennial Laws specified, "No new fashions, in manufacture, clothing, or wares of any kind, may be introduced...without the sanction of the Ministry, thro' the medium of the Elders of each family thereof."[11]

The rules that businesses follow also shape how they treat the environment. When McDonalds changed the rules governing how its food is wrapped, the company reduced its trash output by an astounding 30 percent.[12] Jewelry seller Tiffany & Co. has put in place rules to ensure that none of its jewels originate from countries where diamond sales prop up dictators engaged in murderous civil wars.[13] Environmentalists and industry groups have joined forces to create The Forest Stewardship Council, a rulemaking body that certifies sustainably harvested timber.[14] This growing collection of corporate rules favoring sustainability is matched by a mountain of rules promoting pollution and waste, often for reasons that have more to do with habit than with smart business practices.[15] The point is that corporations operate through rules.

Proponents of free markets would find little to object to in this portrayal. So long as businesses are free to select their own rules and regulations, the discipline of market competition will, at least in theory, reward those whose rules work well. If that means strict standardization of french fry width, or some more forgiving culinary code, power to them. But here is the key distinction: According to the free marketeers, rules per se are not the problem; it's government rules and regulations that get us into trouble. But it turns out that businesses can't function without extensive government regulation. The unregulated "free" market is a mythical concept, not unlike the unicorn. This counterintuitive finding comes from an unexpected quarter: the field of economics.

REGULATIONS ARE A CAPITALIST'S BEST FRIEND

Hernando De Soto, a Peruvian economist, published a remarkable book in 2000 titled *The Mystery of Capital: Why Capitalism Triumphs in the West and Fails Everywhere Else*. De Soto advanced a simple but powerful argument: The reason why most developing countries have failed to enjoy the fruits of economic growth is they lack the government rules that make markets work. He demonstrated the pathologies that plague rulemaking in his home country with an ingenious bit of field research. De Soto and his research team set up a small garment workshop in Lima, the Peruvian capital, and attempted to legally register their business with the government. Time and again they were told by officials to come back another day, provide one more piece of information, fill out yet another form, visit another office for approval, and jump through one bureaucratic hoop after the next to register their property. Instead of doing what any sane person would do—give up—they kept at it for six hours per day, every day, to see how long it would take to play by the official rules governing property rights in Peru. After 289 days of effort, the group was finally able to register their business. But that was just the first step. Permission to construct a small building for the business—with the stated goal of having just a *single* employee—took almost seven years, a process involving 207 administrative steps across 52 different government bureaus. All this to obtain something as simple as legal proof of one's property.

As you might imagine, most Peruvians do not stick it out. Since it is all but impossible to conform with the law, millions of people operate outside it. So why is this a problem? Without a paper trail and legal proof of their property, such as homes and storefronts, they cannot leverage those assets to do things like secure a loan to expand a small business. Unable to move their wealth around, their possessions amount to what De Soto calls "dead capital" that is unavailable for economic growth.

De Soto's research on the importance of government regulation for capitalism provides a window into a larger field of study known as institutional economics. Researchers in this field emphasize that markets for the buying and selling of goods and services are themselves institutions—the collections of rules that make coordinated social activity possible. Standardized weights and measures, legal recourse for the enforcement of contracts, police power to prevent theft or trespass, bankruptcy laws, patents, trademarks, insurance regulations, land use zoning, federally insured banking deposits, air traffic regulations to facilitate the movement of goods, and countless other rules are needed to make markets work. Douglass North, an economic historian at Washington University in St. Louis and pioneer in this field, shows that the historical development of modern market economies was made possible by an accompanying expansion in the superstructure of rules underlying the economy. In his book *Institutions, Institutional Change, and Economic Performance*, North demonstrates how these binding rules allowed market transactions to evolve from face-to-face interactions in the village square to complex multibillion dollar deals among strangers on different continents.[16] As economies developed, and the need for legally binding standards grew, increasingly the job fell to governments to establish the ground rules for market transactions. In short, modern capitalism would be impossible without extensive government regulation.

Today there is a substantial research literature showing that healthy economies require strong, competent government bureaucracies. Sociologist Peter Evans of the University of California at Berkeley and his colleagues have concluded that the quality of government institutions goes a long way toward explaining the pace of economic growth in East Asia and around the world.[17] In places like Singapore and South Korea, highly effective government agencies reward staff performance and work in close coordination with the private sector to foster investment and economic development. By creating an institutional environment conducive to growth, these countries have transformed themselves from relative backwaters to prosperous societies over the past half century—practically overnight, by historical standards. In stark contrast, government institutions in poor countries like Kenya and Guatemala are unstable, unprofessional, and often corrupt, focusing their energy on channeling resources to friends and family rather than serving the public. As Harvard economist Dani Rodrik put it, "markets need to be supported by non-market institutions in order to perform well." Rodrik has spent over twenty years studying the relationship between economic growth and political institutions around the globe. He has come to the conclusion that functioning government rules are the missing ingredient in poor countries: "A clearly delineated system of property rights; a regulatory apparatus curbing the worst forms of fraud, anticompetitive behavior, and moral hazard; a moderately cohesive society exhibiting trust and social cooperation; social and political institutions that mitigate risk and manage social conflicts; the rule of law and clean government—these are social arrangements that economists usually take for granted, but which are conspicuous by their absence in poor countries."[18] In their book *Why Nations Fail*, Daron Acemoglu and James Robinson reach a similar conclusion. They also find that countries with strong democratic institutions are the most likely to meet these conditions, which could account for the fact that worldwide, democratic countries also tend to be wealthier countries.[19]

WHEN MARKETS FAIL

Clearly markets require extensive collections of rules, including government regulations, in order to function. But there is another reason to pay close attention to the market-government dyad: Sometimes the market mechanism itself breaks down and actually fails to provide society with the goods and services that consumers demand. These "market failures," as they are known to economists, include spillover effects, or externalities. Externalities arise when the cost of producing a good is not borne by the producer but is instead foisted on someone else. Let's say the owners of a cell phone manufacturing plant decide that, rather than reduce or properly dispose of their pollution, they will dump it in a river, imposing a cost on those who use the waterway for fishing and recreation. This is not just bad manners. It is a breakdown in the operation of a market economy. To understand why, consider the important function that prices play in a market exchange. Let's say you go shopping for a cell phone, and discover two models that have essentially the same functions, but one phone is considerably less expensive than the other. The price difference tells you something: The manufacturer of the less expensive cell phone is more efficient, having figured out a way to provide the same product at a lower price. It isn't necessary to write a letter of appreciation to reward this behavior and encourage more of the same. With only your self-interest in mind, purchasing the less expensive phone automatically rewards the efficient producer. (The same thing occurs when, faced with a choice of two equally priced phones, you choose the better quality model.) In this way the impersonal mechanism of prices enables a continual ratcheting up of efficiency. This mechanism works up and down the chain of production for the cell phone. The cell phone manufacturer uses price information to decide which metals to buy, which engineers to hire, and which shipping companies to use when sending the goods to your store.

At this point some readers may object that gains in efficiency could be had as a result of exploitive labor practices or other unsavory business practices.

This raises larger questions about the extent to which the logic of the market should be balanced against other logics, such as morality, community, and civic responsibility. I am personally very sympathetic to this point of view, expressed by Lionel Jospin, the former prime minister of France, as the desire for a "market economy, not a market society." But I'm also aware that it is easy to adopt a rather smug attitude toward the wealth-creating capacity of markets. We owe so much of our modern lifestyles to the operation of market mechanisms that, whatever the shortcomings—and I will discuss a number of these below—it's important to not be dismissive of their role in producing prosperity and alleviating poverty. This is the line I wish to walk: taking very seriously the benefits of markets while being cognizant of their limitations.

The point about spillover effects, however, is a narrower one. These are cases where the price mechanism itself fails. If the market is to reward efficiency, prices must be a reasonably accurate proxy for the cost of production. When a company doesn't actually bear the full cost of producing something, but instead passes it along to someone else (dumping its toxic waste in the river), the price of its product is artificially low. The price no longer accurately reflects what it cost to produce the good. With the price distorted in this way, consumers purchase "too much" of the product—more is manufactured than would be the case in an efficient market.

The concept of externalities was first described by the British economist Arthur Pigou in 1920. A half century before the rise of the modern environmental movement, Pigou used the example of factory pollution, "for this smoke in large towns inflicts a heavy uncharged loss on the community, in injury to buildings and vegetables, expenses for washing clothes and cleaning rooms, expenses for the provision of extra artificial light, and in many other ways."[20] Since Pigou's time, economists have published literally thousands of research articles on the topic of externalities and the environment. But we needn't dive into the calculus-filled pages of economics research journals to understand the idea, which can be demonstrated with a simple example involving the price of gasoline. In the United States, the true cost of producing gas for automobiles includes the cost of stationing troops in the Middle East to ensure access to oil. Joseph Stiglitz and Linda Bilmes estimate that the cost of the Iraq war to the American people exceeds 3 trillion dollars.[21] That cost, however, is borne by taxpayers and distributed throughout the economy; it appears nowhere in the price we pay at the pump. If it did, gasoline would be so prohibitively expensive that people would flock to bicycles, buses, and alternative fuels faster than any environmental awareness campaign could inspire. A simple back-of-the-envelope calculation shows that if we spread this cost over ten years, and divide it by the 134 billion gallons of gas consumed annually in the United States, it would increase the cost of gas by over $2 per gallon.[22]

Or consider electricity. Over a third of the electricity used in the United States is generated by coal-fired power plants, which produce greenhouse gases that are changing our climate. As with any business, the owners and managers of these power plants pay close attention to the cost of coal, labor, equipment, and other inputs. But the cost of carbon dioxide emissions doesn't figure into their calculations. Instead it is passed along to others, including coastal states like Florida that will lose real estate over the coming decades, as sea levels rise by a meter or more due to global warming. Researchers at Florida State University estimate that changing sea levels could wipe out $6.7 billion in land values in Dade County alone.[23]

If these sorts of costs were folded into the price of coal—in other words, if coal had to compete on a level playing field—solar-generated electricity would look cheap by comparison. Consumers would bemoan the economic inefficiency of the fossil fuel industry, calling them starry-eyed idealists whose technology just can't compete in the real world of competitive energy markets. Yet time and again, cost comparisons between solar energy and fossil fuels conclude that solar can't compete with the price of fossil fuels. It's an unfair fight. The oil, gas, and coal companies are not paying for the true cost of production. When prices no longer reflect costs, the torrent of consumer dollars unleashed by market forces flows not toward the more efficient producer, but to firms that were more effective in lobbying for rules that allow them to pass their costs on to society. "The current rules of the game are steeply stacked against the new-energy entrepreneurs," argue Fred Krupp and Miriam Horn of the Environmental Defense Fund. "Even the best ideas will fail in a contest as rigged as the $5 trillion energy business is today.... [I]nnovators are up against the most powerful companies in the world, companies that have spent decades successfully pushing for subsidies, trade agreements, and regulatory structures that favor their business."[24]

To fix this market failure typically requires some form of government regulation. Indeed, much of environmental law and policy can be understood as rules crafted by governments to make companies pay for the full cost of their production processes rather than passing the cost on to the public—in effect internalizing the externality. When companies have to pay for their pollution, the prices of goods created through highly polluting practices rocket upward, consumers shift their purchasing preferences, and producers have to scramble to find cleaner, more socially responsible alternatives.

PROFITING FROM REGULATION

Diesel Power Magazine is not a place where you would expect to find editorials arguing for stricter environmental regulations. The diesel industry has been the focus of numerous air quality regulations and has fought against many of these tooth and claw. Yet in an article published in 2012, editor Jason Thomson said to his fellow diesel fans that it's time to recognize that strict regulation makes for innovation. "It's ironic," he explains, "that the emissions regulations many enthusiasts hate gave birth to the diesel performance we have today. Common-rail injection, VGT turbos, and 400hp clean-burning trucks (with a warranty) wouldn't have happened if manufacturers didn't need to meet EPA regulations."[25]

The innovation-from-regulation dynamic at play in the diesel industry is part of a wider phenomenon. In a research article titled "Regulation as the Mother of Invention," policy researchers Margaret Taylor and Edward Rubin teamed up with historian David Hounshell to explore the relation between technological innovation and government controls on sulfur dioxide emissions.[26] Sulfur dioxide is spewed out by coal-burning power plants when sulfur, which occurs naturally in coal, combines with oxygen during the combustion process. The sulfur pollutes surrounding areas, affecting human health and damaging ecosystems by increasing the acidity of the soil and water. ("Acid rain" is the result of sulfur dioxide interacting in the atmosphere with water and oxygen to produce sulfuric acid, which is then carried to the ground in rain, snow, fog, or dust.) From a technological standpoint, this is a solvable problem. Devices called scrubbers, which remove the sulfur after coal is burned, have been commercially available in rudimentary form since 1926, when they were first used in England. But these researchers point out that for forty years, scrubber technology had been neither adopted nor improved upon in the United States. It took government regulation to spawn innovation. Following the passage of strict federal standards in the 1960s and 1970s, the number of patents for new scrubber technologies took off (Figure 5.3).

To be sure, environmental regulation can impose substantial costs on industry. According to the White House Office of Management and Budget, major environmental rules promulgated by the EPA from 2000 to 2010 cost the US economy as much as 28 billion dollars.[27] But looking at the cost column of a balance sheet does not tell the whole story. Apple Computer has high costs—about 200 billion per year—but is one of the most profitable companies in the

The Big Trade 109

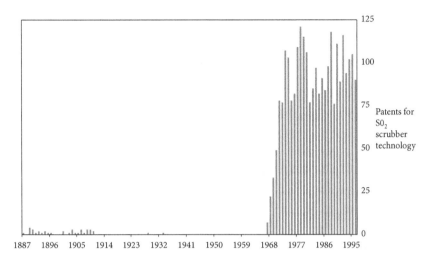

FIGURE 5.3 Innovation in response to clean air laws

From Margaret R. Taylor, Edward S. Rubin, and David A. Hounshell (2005) Regulation as the Mother of Innovation: The Case of SO2 Control, Law & Policy 27(2):348–78.

world. According to the same White House figures, the economic benefits of the EPA rules were at least three times greater than the costs.[28]

We see the same phenomenon at the level of individual companies, which often profit from regulation. Prompted by government requirements for technological change, investments in new industrial processes and new kinds of expertise translate into new products and services. This phenomenon has been studied by Michael Porter, director of the Institute for Strategy and Competitiveness at Harvard Business School. Michael Porter is not your stereotypical environmentalist; a world-renowned expert on business strategy, he is best known for writing books about how corporations and nations can more effectively make tons of money. But in the course of his research on business strategy and the environment, Porter noticed a peculiar pattern: Businesses seemed to be profiting from regulation. He also discovered that the stricter regulations were prompting more innovation than the wimpy ones.

The Dutch flower industry provides an illustration. For many years, the companies producing Holland's world-renowned tulips and other cut flowers were also contaminating the country's water and soil with fertilizers and pesticides. In 1991, the Dutch government adopted a policy designed to cut pesticide use in half by 2000—a goal they ultimately achieved. Facing increasingly strict regulation, greenhouse growers realized they had to innovate if they were going to maintain product quality with fewer pesticides. In response, they shifted to a cultivation method that circulates water in closed-loop systems and grows flowers in a rock wool substrate. The new system not only reduced the pollution released into the environment; it also increased profits by giving companies greater control over growing conditions.

Porter found a similar result in the American electronics industry. For many years companies cleaned circuit boards using chemicals like methyl chloroform that harm the ozone layer. As the environmental consequences of these practices became clear, the EPA banned ozone-depleting chemicals, requiring a complete phaseout by 2000. Industry scientists, who had to think up new ways of ensuring the quality of their products, soon discovered safer alternatives that did a better job of cleaning circuit boards while reducing costs by 30 to 80 percent.

In each case, government regulation was essential, goading industry into action. According to Porter and his colleague Claas van der Linde, tough rules are better for innovation than more tepid approaches. "Enact strict rather than lax regulation," they write. "Companies can handle lax regulation incrementally," meaning they'll patch together Band-aid solutions rather than rethink the way they create their products.[29] Porter is quick to point out that not all regulations are alike—some are more conducive to innovation than others. Rules that focus on improving environmental outcomes, rather than mandating one-size-fits-all technological standards, are more likely to promote these win-win outcomes for businesses and the environment. Porter's findings have inspired a deluge of research examining whether the numerous cases he reviewed were the exception or the norm.[30] As we might expect, this larger body of research finds that environmental regulation is not always beneficial to all firms, and some regulations are indeed more conducive to innovation and profit than others. The takeaway message is that if we wish to protect the environment while fostering a healthy economy, it's not about more or less regulation. It's about smart regulation.

INTRODUCING MARKETS WITHIN RULES

Up to this point we've been considering how rules, and government regulations in particular, are essential for making markets work and producing prosperity. It appears that the merchant needs the bureaucrat more than the former is apt to admit. But now let's turn the tables and consider the market-government relationship from the other direction: Are there cases where government regulators should make use of market incentives, incorporating the profit motive into the design of their rules?

In recent years, a series of controversial experiments in environmental rulemaking have attempted to do just that. At the most basic level, market-based environmental regulation uses financial incentives to induce people to change the way they interact with the planet. Deposit-refund schemes for car batteries encourage consumers to return them for safe disposal. Rules that require mining companies to post bonds for environmental damage before their work begins—rather than chasing them through the court system after the fact—give these companies an incentive to minimize damage in order to maximize the return from their bond. Pollution taxes create incentives for factories to reduce emissions, often beyond mandated government standards. Although taxes are a taboo subject in some political circles, economists consider taxes to be a supremely efficient regulatory tool because they are easy to administer and they let each firm decide how to address its pricey pollution problem. Cap-and-trade programs, like the one that helped wean the United States off of leaded gas, are yet another market-based approach, helping those who help the planet by allowing them to sell their unused permits. As these experimental programs unfold, important decisions are being made about who has the right to pollute, who owns the earth's natural resources, and who decides which technologies to adopt and which processes to pursue—in short, who rules the earth. Here again I will disregard rigid ideological positions in favor of an evidence-based inquiry into the potential benefits and pitfalls of market-based environmental policies.

CAP AND TRADE

Let's begin by returning to the origins of cap-and-trade programs in the United States, where their lineage can be traced to two distinct parental lines: economics professors and street-level government bureaucrats who faced a tough dilemma.

Economists debated the possibility of tradable pollution rights long before they were put into practice.[31] The shift in thinking began in 1960 when Ronald Coase, a British-born economist at the University of Virginia, published a paper proposing a radical new approach to regulation.[32] Coase was concerned about what government regulators might do to address the problem of externalities. Should a company that imposes a burden on the surrounding community be required to pay a tax, or might there be some other way to motivate firms to take responsibility for their actions? Like most economists, Coase was obsessed with the idea of maximizing social welfare. From this perspective, the desirability of an economic activity or regulatory action can be weighed by looking at the benefits it brings and the costs it imposes on society. Given several alternatives, economists will advocate the choice that maximizes benefits minus costs (the "net benefits," in economic parlance). This is what economists mean when they talk about economic efficiency. So how does this relate to the environment? Among the various ways to reduce pollution—including choices such as who should clean up, how much, how fast, and by what methods—some solutions are more efficient than others. From a business standpoint, more efficient environmental regulation achieves a given level of environmental quality at a lower cost. For the environmentalist, efficient regulation means more bang for the buck: greater environmental protection for a given level of expenditure.

The problem is that regulators rarely have the sort of detailed information needed to promote efficient regulation. This would require a nuanced understanding of compliance costs at thousands of businesses large and small. At one plant, waiting an extra year before installing a new pollution control device could translate into hundreds of thousands of dollars in savings, preventing interruptions during a crucial phase of expansion. For another business, which has been looking to experiment with new production processes, it may be possible to reduce pollution even faster than regulators envision. Government regulators don't have this information at hand, but the companies do. Knowing this, Coase made a bold proposition: What if we were to assign property rights for a given amount of pollution—through government-issued pollution permits, for example—and gave the power to the businesses themselves to buy and sell those rights?[33] Those who can afford to change would sell permits to those who can't; the latter would be eager to pay for the right to pollute. This bargaining process would reduce pollution at a lower cost than is possible with one-size-fits-all regulations. In theory, this would happen regardless of which firms receive the pollution rights at the outset, so long as they are allowed to trade those rights. Coase suggested that if we were to insert this market mechanism into our regulations, we could get efficient outcomes without requiring government regulators to be omniscient about the cost information for each firm.

All of this sounds pretty abstract, and indeed it was. Coase was writing well before the United States made a serious commitment to cleaning up the air, and his work garnered little attention outside of specialized groups of academics in law and economics (where this work helped to earn him the Nobel Prize in Economics). Building on Coase's insights, throughout the 1960s and 1970s, economists such as David Montgomery of CalTech and John Dales of the University of Toronto worked out the math, laying out a theoretical argument for how tradable pollution rights might work.[34] Sequestered in conversation with their professional peers, Coase and his colleagues had the luxury of ignoring considerations other than economic efficiency that matter in the real world and should inform wise public policy: fairness, equality, democratic participation, protection of vulnerable populations—you name it. But this ivory-tower exercise soon came face to face with reality, when officials from the newly formed Environmental Protection Agency took on the daunting task of implementing tough new regulations under the Clean Air Act.

As is often the case, change came about when practitioners took an idea developed by academics and combined it with their own in-depth knowledge of how things work in specific social settings. A central figure in this story is Paul DeFalco, Regional Administrator of the EPA's San Francisco office, who was charged with writing the rules to make the Clean Air Act a reality. DeFalco's efforts provide a classic illustration of how rulemaking determines the legacy of a political movement. The enormous social energies unleashed by the US environmental movement, which came to a head as millions of Americans turned out for the first Earth Day in 1970, inspired lawmakers to enact a wide range of new rules to improve the nation's environmental quality. When Congress passes a new piece of legislation, it sets out broad goals but leaves it to federal or state agencies to determine exactly what needs to be done, where, and how. After the clamor of protest marches fades into the history books, it is the comparatively quiet work of rulemakers like DeFalco—scribbling in the margins of draft reports, debating proposals in city planning commissions, and arguing points in courtrooms—that determines the lasting impact of the movement. This is not a time for us to turn our backs and let the lawyers and scientists "work out the details," although certainly we rely on their expertise. This is precisely where members of the public should pay careful attention to what comes next.

By the mid-1970s, rulemakers like DeFalco faced a dilemma. Under the Clean Air Act, each state had until 1975 to comply with federal air quality standards or face stiff penalties. But what if a factory wished to expand its operations, especially in areas like Los Angeles that were already out of compliance with the new national standards? The idea behind the Clean

Air Act was never to bring economic growth to a screeching halt. (It is safe to say this was not President Nixon's intent when he signed it into law.) The predicament facing DeFalco cut to the very heart of the question of how to promote economic prosperity without spoiling the environment.

DeFalco's answer, issued in 1976, was the Offset Interpretative Ruling for California. This rule allowed a company to expand its operations so long as it compensated for any increases in pollution by reducing emissions from its facilities elsewhere in the area by a greater than 1 to 1 ratio. The offset program provided a way to expand industrial activities while reducing pollution *below* the levels that existed prior to the new development. The new approach proved to be so popular that Congress incorporated offsets into the Clean Air Act amendments of 1977. But then the EPA went a step further. Agency officials created new rules that allowed pollution offsets not just within the same company, but across firms, with polluters buying and selling allowances from one another. In effect, they were experimenting with the very idea that Ronald Coase had proposed in the economics literature.

When regulators first floated the idea, both industry and environmental groups basically freaked out. One of the first experiments was in Virginia, where Hampton Roads Energy Company wished to build an oil refinery near a wildlife refuge.[35] The new plant would impair local air quality, so EPA administrators suggested to company president Jack Evans that he consider paying another refinery in the area to reduce its emissions. Evans figured the regulators had finally lost their minds. He was quoted in the *Oil and Gas Journal* as saying the EPA was "foolish" to think he would pay his competitors for the right to pollute. The industry journal ran an article blasting the new rules, arguing that they would slow innovation and increase unemployment. Environmental groups had a similar reaction. They were concerned that cap-and-trade would undermine the tough standards established in the Clean Air Act. The Natural Resources Defense Council sued the EPA in an unsuccessful attempt to halt the program.[36]

Despite the opposition, the agency proposed similar programs around the country. Soon firms started to sign on, apparently reaching the conclusion that saving money was not so foolhardy after all. Building on these early experiments, the EPA created new rules in 1979 allowing "bubbles" and "banking." The bubble approach built on DeFalco's earlier invention: A firm could treat multiple facilities within an area as if they were one source, in effect creating an imaginary bubble around the region. Instead of having to reach a certain amount of cleanup at each facility, the firm could decide where to reduce and by how much, so long as it achieved an acceptable aggregate pollution reduction within the bubble. The banking rule introduced a time dimension: Firms that cleaned up beyond mandated levels could bank their pollution—saving that "right to pollute" for future use. As with the other approaches, regulators decided how much pollution had to be reduced, but they provided leeway in how to get there.

THE "GRAND EXPERIMENT"

As regulators were experimenting with market-based regulations throughout the latter part of the 1970s, the first big test for the new approach came with the phaseout of leaded gas, which began tentatively in 1974 and ended with a complete ban in 1996. But it was with the Acid Rain Program, created under the Clean Air Act Amendments of 1990, that emissions trading came into its own. In the intervening years, champions of tradable permits were attracting attention at the highest political levels. Two key figures in this history are Robert Stavins, an economist at the Kennedy School of Government at Harvard, and Fred Krupp, president of the Environmental Defense Fund, a prominent environmental group. In the late 1980s, Stavins served as research director for an influential bipartisan report released by Senators Timothy Wirth and John Heinz titled *Project 88: Harnessing Market Forces to Protect Our Environment: Initiatives for the New President*. Using jargon-free prose easily digestible by policymakers, the report described the potential for expanding market-based solutions, suggesting they might be used to tackle problems like wetlands conservation, ozone depletion, climate change, and urban waste management. Meanwhile, Fred Krupp had witnessed the success of a water conservation and transfer program in Southern California, and he became convinced that market-based approaches could more effectively advance the aims of the environmental movement.[37] As a staunch supporter from within the movement, Krupp was challenging the other large environmental groups to keep an open mind.

They found sympathetic policymakers within the administration of newly elected president George H. W. Bush. The Bush administration was working with Congress to revise the Clean Air Act to include new rules requiring cuts in sulfur dioxide emissions. The new rules mandated that industry reduce its annual emissions by 10 million tons below 1980 levels. They also authorized the EPA to achieve these reductions through emissions trading. Compared to the hodgepodge of experiments in the 1970s, the Acid Rain Program was a more refined affair, using the benefit of hindsight to create clearer accounting standards and mechanisms of exchange. As with other cap-and-trade programs, the agency regulated the maximum amount of pollution by controlling the number of permits in circulation, which it reduced over time. The program proved to be even more popular with industry than the earlier experiment with leaded gas. Between 1995 and 2009, no fewer than 2,000 polluting facilities (mostly electricity generating plants and large industrial facilities) traded allowances for 255 million tons of sulfur dioxide emissions. That's a lot of pollution, equal in weight to almost 1,000 Empire State buildings. According to EPA figures, by 2011 coal-burning power plants in the United States reduced their sulfur dioxide emissions by two-thirds relative to 1990 levels.[38]

I had the opportunity to take a class from Robert Stavins in 1993, when the sulfur emissions trading program was first getting underway. By his own admission, Stavins is not the sort of warm and fuzzy professor who patiently nurtures students along. (As my academic advisor, Stavins brought me in for a brief meeting in which he said, "Think of course selection as investment, not consumption.") But Stavins has a first-rate intellect and little tolerance for claims that are not founded on solid evidence. Indeed, Stavins is among those who argue that the Michael Porter provides too glowing a picture of the profit-inducing potential of environmental regulation. Taking a look back at the sulfur dioxide program, however, Stavins concludes that the results provide the proof of concept for cap-and-trade. In an article titled "What Can We Learn from the Grand Policy Experiment?" published in the *Journal of Economic Perspectives*, Stavins reports that the program dramatically improved air quality in the United States at a cost savings of about $1 billion relative to traditional regulatory approaches.[39]

THE EMPTY JAR

The experience with cap-and-trade in the United States suggests that embedding market incentives within environmental regulations is potentially a very smart idea, improving the efficiency with which we protect ecosystems and human health. But recall that efficiency is all about the big picture—maximizing the aggregate well-being across an entire population, measured as total net benefits. It says nothing about the details of who reaps the rewards and who bears the costs. To reach an informed decision about the merits of market-based regulation, then, we need to consider a question that is second nature to any child but rarely enters the lexicon of the economics journals: Is it fair?

Let's take an example. Say an unscrupulous but enterprising young man steals $1,000 from his elderly neighbor, who was planning to just keep the money in a glass jar, and the youngster invests the funds in such a way that he doubles his money. His benefits are larger than his neighbor's loss, and so this act of larceny represents an increase in economic efficiency by any conventional understanding of that term. In theory, of course, he could return the original $1,000 to the jar such that he is better off without making his neighbor worse off. That is precisely what has economists and others so excited about regulations that promote more efficient outcomes—you get something for nothing. But this ethical gesture is a hypothetical, and a dubious one at that. The jar may remain empty while our pilfering protagonist enjoys the fruits of the transaction. So too in reality, efficiency gains often produce winners and losers.

Part of what made the early cap-and-trade programs politically palatable is that they didn't just reduce the cost of regulation to industry; they were designed in such a way that they actually reduced more pollution than would have occurred under traditional approaches. The money was put back in the proverbial jar. Under the Acid Rain Program, however, consider the implications of one plant reducing emissions while another buys up some permits and continues to pollute. Cap-and-trade only works if the pollutants are uniformly mixed and spread across the larger area in which trading occurs. Under those conditions, it doesn't matter which particular polluters clean up their act, so long as the total amount of pollution is reduced. But it turns out that sulfur dioxide has two quite distinct environmental impacts. As a cause of acid rain, it spreads throughout the atmosphere regionally, sometimes covering great distances. But sulfur dioxide also increases the risk of respiratory and cardiovascular disease for people living close to the source of emissions. As a consequence, companies that choose to buy pollution permits instead of upgrading their equipment could harm their local neighborhoods. Communities might be made worse off in these "hot spots" than if there had been an old-fashioned rule requiring that all firms adhere to the same standard.

Hot spots are a hot topic in debates surrounding market-based regulation, in part because industrial facilities are more likely to be located in low-income communities with high percentages of nonwhite residents. These concerns gained visibility in the 1980s when civil rights activists, under the banner of "environmental justice," began asking questions like which communities win, and which ones lose out, in environmental planning decisions. These activists pointed to areas like Louisiana's "Cancer Alley"—a stretch of land along the Mississippi River where over 150 industrial plants produce a quarter of the country's chemicals—as evidence that people of color and the working class bear the costs of industrial development. Environmental protection, they argued, should be understood not just in terms of average tons of pollution reduced or acres of green spaces preserved. We also need to ask who gains and who loses. By linking civil rights and environmental issues, groups such as Communities for a Better Environment and the United Church of Christ's Commission for Racial Justice redefined the American environmental movement, challenging the large environmental groups to diversify their base beyond middle-class Anglo constituencies. They also brought to light the long and underappreciated history of environmental advocacy by nonwhite communities, such as the movement to reduce pesticide use led by Cesar Chavez and the United Farm Workers in the 1960s and 1970s.[40]

I had a chance to witness these changes unfolding in 1991 at the first People's Earth Day, an environmental justice event held in Bayview Hunters Point, a predominantly African American neighborhood located in the industrial southeast corner of San Francisco. As I watched two elderly women from the community engrossed in a discussion about water quality with two young Greenpeace volunteers, it struck me that something new was afoot. That same year the First National People of Color Environmental Leadership Summit took place in Washington, DC. The movement spread fast, with new groups popping up and mainstream organizations embracing the environmental justice cause. Before long, policymakers in the Clinton administration were paying attention. The EPA created a new Office of Environmental Justice. On February 11, 1994, President Clinton issued Executive Order 12898, requiring all federal agencies to evaluate and redress any race or class bias in their environmental decisions.

The movement also attracted the attention of researchers in the social sciences. You see, nothing excites a statistically minded researcher more than a complete data set. And it just so happens that the US Census has collected data on the racial composition and income levels of communities throughout the United States for well over a century. Pairing these numbers with data on the location of polluting factories and how much they pollute, we can test whether environmental rules benefit some groups more than others and whether market-based approaches to sulfur dioxide reduction have created hot spots of pollution.

This question has been tackled most convincingly by Evan Ringquist, a political scientist at Indiana University.[41] Ringquist asked whether companies operating in poor and nonwhite communities tended to buy up more sulfur dioxide permits (allowing them to continue polluting at high levels) than those located in well-to-do Anglo communities. Combing through the pollution-trading records for 2,000 facilities, Ringquist discovered two things. First, his research confirmed a finding from previous studies showing that polluting facilities tend to be concentrated in urban working-class and nonwhite communities.[42] Environmental burdens in the United States are unfairly distributed. Of course, this finding suggests that the Acid Rain Program, and any innovation that promotes faster reductions as part of that program, could in theory disproportionately benefit these communities. But that again assumes uniform improvements across all polluting firms, which is precisely the sort of one-size-fits-all outcome that raises the cost of regulation, and which cap-and-trade is designed to avoid. So who bought and who sold permits under the Acid Rain Program? Which firms cleaned up more than they had to and made a handsome profit, and which ones paid out-of-pocket in order to continue bellowing sulfur dioxide into the air?

Ringquist found that polluting facilities in lower-income communities, and those in communities with higher percentages of Latinos and African Americans, were not in fact net importers of permits. Pollution was cleaned up just as quickly in these communities as it was in wealthier and Anglo populations. Given that there were no differences across social groups, and given that most polluting facilities are located in nonwhite working-class communities, we can conclude that the cap-and-trade program has disproportionately *benefited* lower-income communities and people of color.

THE MAD CAPPER

The potential for cap-and-trade rules to produce local hot spots of pollution remains a real concern—particularly as these market-friendly techniques become so popular that they take on a life of their own, torn from their conceptual moorings in economics and used for purposes unforeseen by their inventors. This Frankenstein-like scenario came to pass in 2004, when

President George W. Bush proposed a cap-and-trade program for mercury. Mercury is one of the most dangerous substances known, similar to lead in its deadly combination of high toxicity at low doses and long-term persistence in the environment. As with lead, the health effects of quicksilver have been common knowledge for centuries. In the 1800s, the British colloquialism "mad as a hatter" became a popular reference to the hat-making trade, where mercury used in the production of felt (where it helped to remove fur from animal skins) was known to cause erratic behavior among factory workers.

Mercury pollution gained further notoriety with the outbreak of Minamata Disease in Japan in the 1950s and 1960s. The chemical company Chisso Corporation had been dumping industrial waste containing methyl mercury in Minamata Bay in Western Japan. As the poison concentrated in the tissues of marine organisms, and made its way into seafood consumed by locals, the health effects were shocking. Gruesome photographs of children born with severely twisted limbs and contorted facial expressions spread throughout the international news media (and can still be found on the Internet), providing vivid symbols for Japan's nascent environmental movement in the 1970s. Similar outbreaks were reported in Niigata prefecture. The Japanese government estimates that by 2004, the human toll from these two incidents of mercury exposure reached 1,784 premature deaths.[43]

The mercury emitted by coal-fired power plants in the United States occurs at levels far below those that caused the more severe symptoms seen at Minamata or in the hat-making factories of Victorian England. But it's still a cause for alarm because mercury is so toxic that even minute amounts can cause brain damage. Mercury occurs naturally in trace amounts in coal. When the black rock is scraped from the earth, pulverized into dust, and burned inside power plants to turn their massive turbine blades, mercury is released into the surrounding community. Power plants are the primary source of mercury exposure in the United States, sending out forty-eight tons of mercury annually, where it accumulates in the fatty tissues of fish and shellfish and is transferred to humans.[44]

When the Bush administration proposed a cap-and-trade program for reducing mercury, the original proponents of cap-and-trade, such as Fred Krupp of the Environmental Defense Fund, vigorously opposed the proposal. Lisa Jackson, the head of New Jersey's state environmental agency, argued that cap-and-trade "simply does not work for emissions of a neurotoxin as dangerous as mercury."[45] Moreover, while mercury emissions do travel far and wide, they are most concentrated near the source. Communities located near firms buying mercury pollution permits would be exposed to greater danger than others. Led by New Jersey, a dozen states challenged the Bush proposal in the courts, with support from groups like the American Academy of Pediatrics. The Obama administration eventually reversed the Bush proposal and passed more uniform regulations in 2011. Despite opposition from politicians like Senator James Inhofe, who derided the measure as part of "EPA's job-killing regulatory agenda,"[46] the administration put in place new rules of the old variety, requiring that each and every power plant reduce its mercury emissions by 90 percent.

HOT PROPERTY

Unlike mercury, greenhouse gases like carbon dioxide would seem ideally suited to a cap-and-trade program. When carbon dioxide is released into the air from the burning of fossil fuels, it quickly spreads throughout the atmosphere. This normally benign gas has been slowly accumulating, molecule by molecule, over the past two centuries, changing the chemical composition of the air and trapping more of the sun's energy as it reflects off the earth's surface. From a strictly physical standpoint, it doesn't matter who reduces carbon emissions (or increases its uptake by growing trees) so long as, in the aggregate, we put less of the stuff in the air. That suggests that market-based regulations could fit the bill. Carbon dioxide also comes from countless sources: minivans and farm tractors, cement factories and oil refineries, massive forest fires in the Amazon and smoldering trash heaps on the outskirts of New Delhi. The changes required for a large-scale shift to a low-carbon economy are so broad that uniform technical standards can only accomplish so much.

It hasn't taken policymakers long to notice these advantages and to propose tackling global warming through market-based rules like cap-and-trade or carbon taxes. In 2005, the European Union launched its Emissions Trading System, coordinating a market for carbon permits involving 11,000 power stations and industrial plants in thirty countries. In 2009, India announced its own cap-and-trade program for carbon, worth an estimated $15 billion. New Zealand unfurled a carbon trading program the following year, while the city of Tokyo (with a population twice that of New Zealand) launched a municipal trading initiative covering its top 1,400 emitters of carbon. Soon policymakers began to knit these growing markets together, eager to tap larger pools of buyers and sellers. In 2012, Europe linked its Emissions Trading System with a cap-and-trade initiative in South Korea. Vietnam and Thailand announced plans to launch similar programs. Even China, the world's largest emitter of carbon dioxide, declared its intentions to join the European program. In his State of the Union address in 2013, President Obama threw his support behind the idea, stating, "I urge this Congress to pursue a bipartisan, market-based solution to climate change." His proposal had a chilly reception from lawmakers who questioned the science of climate change and argued that cap-and-trade was bad for the economy. (The environmental blog *Grist* noted the irony that communist China embraced this market-oriented approach before the United States.) With Congress unwilling to play a part, Obama issued new rules limiting carbon emissions from power plants, a first step in the slow process of turning America's titanic-sized economy away from the glacier ahead. But change was already underway at the state level. California successfully fended off a legal challenge from the George W. Bush administration and launched its own cap-and-trade program in November 2012 with the goal of cutting greenhouse gas emissions back to 1990 levels by the year 2020, and achieving an 80 percent reduction by 2050. Through an agreement known as the Western Climate Initiative, California coordinated its program with similar efforts across the border in the Canadian provinces of Ontario, British Columbia, Manitoba, and Quebec.[47]

Forests, too, are part of the growing global market for carbon. Forests contain about 45 percent of the carbon stored on the land. Capitalism is good at conveying the value of harvested wood, which can be bought and sold on the market, but not so good at valuing the carbon-capturing services that forests provide. Market-based regulations can provide landowners with incentives to keep trees on the land, rather than clearing them for crops and pasture. Under United Nations auspices, participants are now gearing up for a program that will allow industrialized countries to meet some of their carbon commitments by investing in forest conservation in developing countries—in effect paying farmers in poor countries for the atmospheric scrubbing services provided by their trees. Critics have been quick to deride this approach as allowing polluters to abdicate their responsibilities by paying off poor landowners. But the reality is much more complex; for rural families in developing countries, carbon capture is an opportunity to diversify the mix of products and services sold on their land. The UN initiative, known as REDD+, will result in payments of up to $30 billion per year to dozens of countries throughout the developing world. To put this figure in perspective, that's more money than all of the world's rich countries provided in aid to Africa in 2010.

THE BOTTOM LINE ON BOTTOM-LINE APPROACHES TO ENVIRONMENTAL PROTECTION

Major questions remain for the implementation of cap-and-trade and forest-based carbon markets.[48] Who will be given permits to pollute? Will these be auctioned off or handed out for free? If the ability to sell permits provides an incentive for clean-tech innovation, wouldn't it follow logically that the right to buy permits hinders innovation, allowing businesses to forestall needed changes? Will payments for forest conservation favor wealthy landowners over the rural poor, who have a harder time reaching international markets? When tropical forests are turned into commodities to be bought and sold, how will this affect non-market activities like firewood collection and the use of traditional medicines? Will all of these approaches add up to measures that move us far enough and fast enough to prevent the nightmarish scenarios forecast by climate scientists, who warn of droughts, floods, heat waves, and rising sea levels?[49]

As with any social rule, market-based or otherwise, the devil is in the details. Those details are being decided right now, in your name, with your resources, and with or without your active participation. So when someone proposes market-based regulations that will affect your community and your planet, how should you respond? As a voter, should you support proposed cap-and-trade legislation? As a consumer or shareholder, how should you react to a proposed carbon tax? If you want to reduce trash in your community, should you charge users for the size of their trash bins, subsidize the cost of composting bins, use a deposit-refund system to encourage proper disposal of toxic waste, or pursue some other strategy?

"It depends" would make for an awkward campaign slogan and an unsightly bumper sticker, but that's the one I'm going to paste on the end of this chapter. It depends on what? The desirability of market-based environmental regulations, and similar experiments taking place at the intersection of markets and governments, depends on the specific rules put in place and how these mesh with the interests, abilities, and vulnerabilities of those causing and experiencing environmental degradation in a particular place and time. That's a lot of qualifiers, but that is also why rough-cut ideologies (Regulation is bad! Cap-and-trade is selling our future!) make for poor guides to policy. We must investigate: What are the terms of the contract, the rules of the game, and the responsibilities of participants? What are the assumptions about who owns the property rights? When a proposal arises for private development on public lands, how do the rules ensure that it advances the public interest?

Regulations can, under certain circumstances, benefit from introducing market incentives to induce behavior change. But we must be smart in how we go about it. These approaches deserve neither outright rejection nor an uncritical embrace. Ronald Coase, the economist who first proposed tradable pollution rights, argued, "Satisfactory views on policy can only come from a patient study of how, in practice, the market, firms and governments handle the problem of harmful effects." Though he believed that his colleagues put too much faith in government regulation, "This belief… does not tell us where the boundary line should be drawn. This, it seems to me, has to come from a detailed investigation of the actual results of handling the problem in different ways."[50] Many of the successes of the US Clean Air Act, for example, came about not from market-based regulations but from one-size-fits-all mandates, like the requirement that all automobiles be equipped with a catalytic converter. Even when market-based tools are appropriate, they must be accompanied by no-exceptions-to-the-rule standards. We needn't be economists to make sense of these proposals. Every citizen has the right to demand that elected leaders provide clear explanations and solid rationales for their decisions. We should listen not just to the bankrolled experts carted out by government agencies and project sponsors at public hearings, but to independent researchers and experienced administrators, community leaders, and industry insiders who have worked with market-based solutions elsewhere and can help us to identify opportunities and avoid repeating past mistakes. Only then can we get beyond bumper sticker politics and move toward a careful assessment of the facts and their implications for the things we value.

Ronald Coase's call for pragmatism applies to a variety of important decisions at the intersection of markets and governments. Take foreign investment, and the question of whether and how poor countries should encourage it. In East Asia, regulators have courted investment by multinational corporations with a great deal of foresight and strategy, putting in place rules to promote the development of local industries. In Mexico, by contrast, a hands-off, "leave it to the market" approach to foreign investment resulted in the destruction of Mexico's fledgling electronics manufacturing industry.[51] Or consider privatization, which encourages private companies to take over functions like urban drinking water systems or waste treatment, services that were traditionally provided by governments. In Argentina, privatization of municipal water services appears to have significantly increased public access to safe water and is associated with reduced levels of childhood disease; in Colombia, privatization has had precisely the opposite effect. In other cases, privatization has been pursued with reckless abandon, involving the wholesale transfer of public assets into corporate hands, with disastrous results.[52] In each case, the essential point is to not simply leave decisions to the market or to get government out of the way, but to put in place rules that leverage the benefits and avoid the pitfalls of each.

Just as the rules governing the economy come in many flavors, over time governments themselves have evolved and diverged, disintegrated and consolidated, so that today the planet is ruled by a varied collection of sovereigns—presidents, prime ministers, kings, sultans, juntas, and ruling parties pursuing diverse agendas with uncertain accountability to the people living within their domains. To understand who rules the earth, the next step is to take a closer look at these things we call countries. Why do national governments make the choices they do? What leads them to plunder or protect the planet? We will begin by peering into a world that bears little resemblance to the relative peace and stability experienced by the citizens of wealthy nations. It is a tale of national governance gone wrong. The story begins with the armed hijacking of a taxicab carrying two young Peace Corps volunteers traveling along an isolated road in the jungles of West Africa in 1990. It's a story I know well—I was one of them.

6

A Planet of Nations

On December 24, 1989, a man named Charles Taylor marshaled a band of armed rebels in the northern part of Liberia, a small country on the coast of West Africa. Carpeted in green jungle crossed by the occasional red dirt road connecting remote ramshackle towns, Liberia had never managed to attract much attention from the outside world. It carried none of the economic clout or strategic importance of continental powers like Kenya and South Africa. To outsiders, Liberia figured as little more than a historical curiosity, the place where freed American slaves settled and founded Africa's first independent republic in 1847. Nor did Charles Taylor's activities attract much notice. Military coups are a common occurrence throughout Africa, as much a part of reality as the tropical downpours that bring life to a temporary standstill in thousands of villages across the landscape before people tentatively poke out their heads and resume their daily activities.

But this time something was different. Instead of racing to the capital and storming the presidential palace—as the incumbent dictator, Samuel Doe, had done a decade earlier—Taylor and his men were slow and deliberate in their progress, taking control of one town after the next. Rumors spread that the rebels were supported by Libya, a country that exercises much greater influence throughout the African continent than most people realize. Ultimately Charles

Taylor would orchestrate a catastrophic civil war in Liberia, a conflict that would engulf neighboring Sierra Leone and lead to one of the worst humanitarian crises of the past century.

At the time I was serving as a Peace Corps volunteer in Liberia, where my wife and I were assigned to work in President Doe's hometown of Zwedru, a remote place that could only be reached through days of travel along roads with mud pits the size of swimming pools or, alternatively, in a single-propeller plane that the tropical air currents would toss about like a toy in a bathtub. It was in Liberia that I first came to appreciate how national governance impacts the lives of billions of people every day. Milton Friedman, the famous libertarian economist, once argued, "Every act of government intervention limits the area of individual freedom." To those who worry that government rules are at best a nuisance, I suggest spending some time in a place that doesn't have them.

As the weeks and months went by, and the rebels slowly advanced, tensions grew. Any semblance of a free press had been crushed under Samuel Doe's dictatorship, and in the information vacuum that resulted, rumors quickly spun into wild stories. My wife was nearly trampled by a human stampede when someone ran into Zwedru's large outdoor market warning that Charles Taylor and his rebels were on the outskirts of town (which was not true). These anxieties were rooted in a collective memory of violent clashes among Liberia's ethnic groups that took place a few years earlier, when President Doe rigged an election. To say that the president displayed favoritism toward his own ethnic group, the Krahn, would be an understatement. The US Embassy in Monrovia estimated that President Doe had expropriated 40 percent of the country's national wealth for use by his extended family.

Bits of information reached us through BBC reports broadcast on ubiquitous handheld radios, one of the few reliable sources of news. From friends and neighbors we learned that a group of European missionaries living in a nearby village had mysteriously disappeared. A Peace Corps volunteer was assaulted by a soldier at a military checkpoint in the town of Robertsport. The soft-spoken physics instructor at the high school where I taught was strip-searched and beaten by border guards when trying to return from his home country of Ghana. A careful and understated man, he shared rumors that someone had been skinned alive at the border. Mind you, I was born with a resilient character, and became tougher in-country, surviving two bouts of cerebral malaria among numerous other challenges.

But what happened in Liberia over the following months and years is still beyond my capacity to process.

My own small run-in with the broader pattern of violence unfolding in the country began on a trip with my wife in a taxi returning from the coastal town of Greenville. As we bumped along the jungle road in the semi-slumber of a trip that must be endured rather than enjoyed, we suddenly heard shouting and noticed two soldiers running up to the road ahead with their guns in the air. One of the weapons was so big that it appeared to have been unhinged from a tripod by the particularly large soldier carrying it. I distinctly recall the terrified look in the cab driver's eyes in the rearview mirror, when he dispensed with the normal pleasantries afforded a paying customer and said simply, "We should stop." The soldiers broke into the car and crowded in with their weapons, smelling of palm wine and sweat. They rode with us for what felt like an eternity but was probably less than an hour. We had no idea of their intentions and made few demands, though my wife did tell the big fellow to move the barrel of his gun away from my head.

Heading down a branching series of side roads, eventually we reached a remote hamlet with three thatched huts clustered along the edge of an impressively tall jungle. As the soldiers ordered the driver to stop the car, my mind was only partially successful in suppressing thoughts of what could happen next. Fortunately, the soldiers disappeared into the forest without a word. Others were not so lucky. The following week, rebel soldiers commandeered another taxi occupied by foreigners; when government forces encountered the vehicle, they shot and killed everyone inside.

The situation continued to deteriorate, and eventually all Peace Corps volunteers were evacuated from the country with help from four US naval warships and 2,300 marines. My wife and I were dropped in London with plane tickets to anywhere and a voucher for two psychological counseling sessions, courtesy of the US government. The rest of the story came to me through American newspapers and occasional letters from Liberian friends. Army factions proliferated, and with them human rights abuses. Charles Taylor installed himself as president and his predecessor was cut to pieces by a mob in the streets of Monrovia. Thousands of children were abducted as soldiers, and Taylor's men implemented a policy of disfiguring civilians with machetes. I received word that one of my students was killed. An estimated 750,000 people fled their homes, and soon the fighting spread into Sierra Leone. Soldiers overran national parks where they hunted endangered elephants and other wildlife with automatic weapons.

The fighting mixed with ancient cultural traditions of animism and beliefs in supernatural powers, twisting these in macabre ways. Shadowy figures hired "heart men" to conduct human sacrifices in the hope of increasing their power. The most frightening picture I've ever seen was a photo in a *Time* magazine article showing a young soldier with a rifle moving through the streets of Monrovia wearing a woman's wig and a gruesome Halloween mask—attire that soldiers wore as fetishes to ward off the power of their opponents. For some reason the part of the picture that stands out most in my mind is his white tennis shoes, a reminder that once he was a teenager like any other. The war in Liberia and Sierra Leone continued for the better part of twenty years until Ellen Johnson Sirleaf was installed as the new president, offering a glimmer of hope for Liberia's future. Charles Taylor was tried and convicted for war crimes by a United Nations tribunal in 2012.

A year after my return to the United States, I witnessed the importance of national governance from a different perspective, when I took a job as a researcher at the Natural Resources Defense Council in San Francisco. In 1991, forestry experts from the group were preparing for their first visit to Russia, which was then part of the Soviet Union. My job was to learn everything I could about conservation policy in the vast country in advance of the trip. It was a fascinating experience for a young researcher. At a time when the Kremlin strictly controlled information, and there was no Internet to consult, I had to rummage for insights wherever I could find them, poring over CIA maps of the Russian Far East and leafing through Russian-language "Red Data" books on endangered species. On this basis I cobbled together a preliminary picture of likely threats and opportunities for forest conservation during Gorbachev's slow process of reform. Then something happened that was thrilling from the perspective of human rights, and yet deeply annoying for a researcher with a deadline. The Soviet Union simply ceased to exist. With a speed that no one predicted, populations rose up throughout the USSR and broke the country apart into a collection of newly independent states. Meanwhile, at my office in San Francisco, if a movie camera were to capture the scene from afar, it would show a single light in a high-rise tower late at night on December 30, 1991, as I went through my suddenly unfinished report, using my computer's search-and-replace function to substitute "Commonwealth of Independent States" for the now defunct term "Soviet Union."

It was at that time, and in light of my recent experience in Liberia, that I was struck by a certain irony: If we are going to do a better job of managing the earth's resources sustainably, we need institutions capable of governing over long time horizons. Yet most of the world's countries are the scene of ongoing political and economic upheaval. How can we achieve sustainability on a planet where most political systems are themselves unsustainable? How can we govern without functioning governments?

NATIONS MATTER

Questions like these require that we better understand the nature of the governments ruling the earth. When we take a closer look, it turns out that the news is not all bad. Sometimes it is downright inspired. There are plenty of examples today of countries, even poor and relatively unstable ones, that have put in place thoughtful and innovative rules to govern resources sustainably, to ensure that their citizens have clean air and water, and to improve the economic and cultural vitality of urban centers. To make sense of what leads governments to help or hurt the planet, we can avail ourselves of research from political scientists and others who spend their lives comparing outcomes in different parts of the world to understand why governments do the things they do.

Before proceeding, we need to put to rest a popular misconception about who rules the earth. It has become fashionable to claim that national governments are increasingly irrelevant for addressing the global-scale problems we face today. Some environmental commentators bristle at the concept of national sovereignty—the idea that nations still call the shots. Understandably, national sovereignty may seem a bit old fashioned. What about "thinking globally," that paradigm shift ushered in by the environmental movement and symbolized by the iconic image of Earth viewed from outer space?[1] We have all encountered this image countless times in popular media and nature documentaries, often accompanied by a soothing soundtrack and a narrator cooing something along the lines of "Viewed from outer space, our planet shows no divisive political borders."

Well our cameras have become a lot more powerful since 1972, when the Apollo 17 spacecraft first took that snapshot of our bright blue planet floating in space. It turns out that national borders are quite visible from space, and this is largely a result of the different rules at work in different nations. Consider the border between Brazil and Bolivia, shown in Figure 6.1. The windy Abuna river divides Bolivia's Pando region (lower right) from the state of Acre in Brazil. The curving white line in Acre is deforestation from the construction of highway BR-364, part of a plan by Brazil's military dictators in the early 1970s to shore up control over the remote reaches of the Amazon. Jagged lines branch off the highway in a classic herringbone pattern of deforestation. These are smaller access roads built by the cattle ranchers who streamed into Acre by the thousands in response to financial incentives from the government.[2] A team of researchers led by Stephen Perz of the University of Florida at Gainesville has conducted a more fine-grained analysis of the satellite data. They report that even within Acre, variation in forest cover from one place to the next is driven to a significant degree by the policies in place. Acre's forests are better protected in rubber-tapping regions governed by rules that strictly limit the amount of land that can be

FIGURE 6.1 National borders are visible from space

NASA/GSFC/METI/ERSDAC/JAROS, and U.S./Japan ASTER Science Team.

cleared.[3] Whether promoting highways, sustainable harvesting techniques, or the preservation of wilderness, nations and the rules they make are literally shaping the contours of the earth.

The idea that the nation-state will remain a central actor in the transition to sustainability swims against the main current of environmental thinking today.[4] Many commentators point to the transnational nature of problems like climate change as evidence that national governments are ill-equipped to address environmental issues. Add to the mix increasing economic interdependence, the power of multinational corporations, and the growth in the size and importance of the nonprofit sector worldwide, and nothing seems more outdated than the idea that crusty old governments hold the key to the planet's future. Yet the big levers required to shift economic development onto a more sustainable track—transportation infrastructure, energy incentives, agricultural policy, land use planning, and investment in maternal-child healthcare, to name a few—are controlled by national, and to a lesser extent provincial and local governments. Markets do not capture the value of many of the natural resources we rely on, and so more often than not, some form of government involvement is required to save a wetland or make urban air more breathable. Moreover, protecting our forests, air, and oceans often requires confronting powerful vested interests. No actor other than a government is capable of filling this role.

At the same time, there is no social actor better positioned to make things worse. Governments have a monopoly on the sanctioned use of coercive force and are often unaccountable to their citizens. Public bureaucracies in many parts of the world are rife with patronage and corruption. Everywhere governments can be found enforcing rules that are poorly conceived or out of date. It is the "can't live with it, can't live without it" quality of government that makes it so important to pay attention to the rules that governments make and whether these need to be reformed.

As is often the case with social rules, to appreciate the influence of national policy one must dig under the surface of things. What we see might be a farmer in Brazil setting fire to a patch of forest to make way for cattle. But underlying this seemingly local and personal decision is an elaborate system of national rules shaping the farmer's decisions. In 1991 Hans Binswanger, a development economist with long experience in Brazil, published an article cataloguing dozens of policies that push people to destroy the Amazon. These include tax exemptions for agriculture, which encourage people to turn forests into pasture. Today government subsidies for soybean production are hastening deforestation, which is further encouraged by rules called right to possession (*direito de posse*) that offer property rights to those who demonstrate possession of the land by cutting down the trees.[5]

Similarly, it might seem entirely natural in the daily life of a nation to find a group of citizens protesting against a company accused of polluting a local waterway. Yet the ability of citizens to organize, and the tactics they deploy to bring about change, are profoundly shaped by the rules made by national governments. Constitutional rules protect or deny their right to call attention to irresponsible behavior. Tax codes help or hurt their ability to form a nonprofit organization. A fair and functioning legal system is necessary to protect them from retribution. In practice, it doesn't take long for community organizations to realize that seemingly local, neighborhood-scale problems have roots in larger political decisions. The soup kitchen discovers that homelessness is accelerated by urban planning policies that ignore affordable housing. The health clinic finds that battered women are not reporting abusive spouses because they have inadequate protection under current law. The community garden flounders because of city health codes preventing people from selling produce grown on urban plots.

Grassroots activism and government policy are intertwined. Over the past twenty years, a research team led by Lester Salamon of Johns Hopkins University has been analyzing results from the most comprehensive survey of the nonprofit sector ever undertaken. Reviewing data from forty countries, the team has found that governments provide 36 percent of the funding for nonprofit organizations worldwide, more than double the amount provided by private foundations (with the remainder coming from fee-for-service arrangements).[6] In 2003, a separate research team led by Russell Dalton of the University of California at Irvine published findings from a study focusing on environmental nonprofits in particular. These researchers polled 248 organizations in 59 countries to find out what it is that environmental groups around the world actually do.[7] Almost half of these groups report interacting with national and local governments "very often" and another third do so "sometimes." John Clark, an expert on the voluntary sector in developing countries, concludes that nongovernmental organizations can "oppose the State, complement it, or reform it, but they cannot ignore it."[8] Let's now take a look at what they're up against when they do so. If governments are a big part of the answer to who rules the earth, then who rules the governments, and how do they govern?

CHARTING THE TERRITORY

There is hardly a patch of land on Earth that is not ruled by a nation, with the exception of the frozen expanses of Antarctica. Even there, a handful of countries are currently jostling for territorial control.[9] In the South China Sea, several countries are trying to claim the still un-countrified Spratly Islands, a rich source of oil and natural gas deposits. And then there's Bir Tawil, a bit of land wedged between Egypt and Sudan that is apparently so undesirable that each country is arguing that the other should claim it. These exceptions aside, we live on a planet of nations. The actions of national governments are sometimes quite visionary. In South Africa, the government's Working for Water program provides jobs for thousands of citizens who control invasive plants that threaten scarce water sources. In other cases they are maddeningly short-sighted—witness Chile's plan to flood thousands of acres of Patagonian wilderness for power generation, a scheme that was finally defeated in 2014 following environmental demonstrations involving hundreds of thousands of Chilean citizens. Every day national governments craft innumerable rules that will shape our planet, for better or for worse, long into the future. These actions receive less media attention than do global environmental summits and treaty-signing ceremonies adorned in diplomatic splendor—in part because there are so many countries (almost 200 in all), and they change governments so frequently, that it's hard to keep track. Despite the bewildering diversity of political arrangements ruling the earth, some generalizations about their rulemaking properties can be made.

Let's say you were to put on a backpack and traverse the world with a scorecard, jotting down each country's name and assessing how government rulemaking works in that place. For simplicity, the scorecard might have two categories that are especially relevant to the question of who rules the earth. The first category measures whether a country's rulers are accountable to their people. The second assesses how effective the government is at actually governing—fulfilling the basic functions expected of a modern nation-state, like ensuring security and promoting economic growth, social welfare, and environmental stewardship.

Researchers in the field of comparative politics have been tallying precisely these sorts of things for many years, allowing us to take count. First let's consider the source of authority for government rules, and in particular whether these originate from the people (democracies) or from a small group of self-appointed rulers (authoritarian regimes). Today, among countries with half a million people or more, there are about ninety-two democracies. These are places where environmental rules are made amid competitive elections, popular participation in politics, restraints on executive power, and the protection of civil liberties like the right to speak out and organize. Another twenty-three countries are ruled by authoritarian regimes, a motley assortment that includes monarchies (Saudi Arabia), military juntas (Madagascar), single-party systems (China), and personalistic rule by dictators (North Korea). The remainder of the world's political systems fall somewhere between the poles of democracy and autocratic rule, including an alarming number of essentially "stateless" countries with no government authority in charge.

Over the past two centuries, the general trend has been toward democracy (Figure 6.2). Reversals are so common, however, that few political scientists believe we're witnessing any sort of unstoppable march of history toward democratic forms of government.[10] In 2011, just under half of humanity lived in democracies,[11] partly due to the weight of autocratic China on the statistical scales. This should give us pause when thinking about the political challenge of moving toward a more sustainable world: Never in human history have most people been free to choose their political leaders. Still, the changes that have taken place in recent decades are extraordinary. As recently as 1914, most of the earth was ruled by kings, queens, and other members of hereditary monarchies whose rule-making authority derived from their DNA. In 1946, there were 71 independent countries; by 1979, following the end of colonial rule in Africa, Asia, and the Middle East, the number of sovereign nations had jumped to 149. In 1989, there were 48 democracies in the world; five years later, with the fall of the Soviet Union and democratic transitions in many other countries, the number reached 77. It is too early to tell whether popular uprisings throughout the Arab world will have a similar effect; this will depend on the aspirations of social movement leaders and whether they put in place rules that promote tolerance and political pluralism. Today the typical national setting in which the rules affecting the earth are made—if you were to blindfold yourself and toss a dart at a world map—would be a relatively new and partially democratic country, struggling to move past a legacy of authoritarian rule and often teetering on the brink of constitutional crisis.[12]

FIGURE 6.2 Growth in democracies

Data courtesy of Monty G. Marshall antd Benjamin R. Cole, Global Report 2011: Conflict, Governance, and State Fragility, *Center for Systemic Peace, Vienna, VA.*

SULTANS FOR SUSTAINABILITY?

The jury is still out on whether democracies or authoritarian regimes do a better job of caring for the planet. In her book *Mao's War on Nature*, Judith Shapiro of American University shows how authoritarian regimes can impose ruinous environmental rules on their people. In the mid-1950s, Ma Yinchu, an economist and president of Beijing University, tried to warn China's leaders that explosive population growth would soon pose a threat to national development. But Maoist China was deeply suspicious of intellectuals and had little tolerance for independent advice. Not only did the Communist Party ignore Ma Yinchu's warnings, but for his outspoken views they removed him from his university position and subjected his family to public humiliation. China's population rocketed upward until, two decades later, the regime responded to the crisis with coercive measures limiting each married couple to one child.

The link between anti-democratic and anti-environmental behavior stems from an additional source: oil. Fossil fuel dependency appears to not only hurt the environment—it's also bad for democracy. Michael Ross of the University of California at Los Angeles shows that regimes that are heavily dependent on income from oil and other natural resources tend to be less democratic. This finding holds not only in the oil-rich Middle East (the world's least democratic region), but around the globe. Part of the explanation lies in the fact that leaders who can fund government activities directly from the receipts of oil sales have no need for tax revenue. And regimes that don't need the people's money don't need their approval to go about their business. These same regimes, unsurprisingly, tend to oppose action to address global warming, which requires that we reduce dependency on oil and other fossil fuels.[13]

Yet the environmental record of authoritarian regimes is not all bad. In the mid-1980s, Indonesia's dictator, Suharto, implemented the world's most ambitious effort at pesticide reform, banning organophosphate pesticides and promoting ecologically-based Integrated Pest Management techniques.[14] Or consider the Himalayan kingdom of Bhutan, which today is a global leader in its commitment to conservation, having placed 30 percent of its lands in protected area status. And of course the green credentials of the world's democracies range widely—from Japan's leadership in energy efficiency to the sidelining of environmental issues in Brazil during the presidency of Lula da Silva, a hero of the country's pro-democracy movement.[15] After reviewing the historical and statistical evidence on democracy and environmental quality, Kathryn Hochstetler of the University of Waterloo finds that the advantage goes to democracies, but only slightly.[16]

The democratic or dictatorial qualities of a country are more likely to affect the "how" than the "whether" of environmental protection. Consider the story of Juan Nogales, a little-known figure who worked tirelessly to protect endangered species under Bolivia's dictators in the 1960s, long before environmental groups in wealthy countries took a serious interest in tropical conservation. This particular story was related to me by Armando Cardoso, an agricultural scientist and early pioneer of Bolivia's environmental movement. According to Cardoso, Juan Nogales was a man of no notable stature or influence in Bolivia. But he was a devoted conservationist who held a particular fascination with wild vicuñas, doe-eyed relatives of camels that look like a cross between a llama and Bambi. These graceful creatures have roamed the Andes for millions of years. But by the 1960s, vicuñas were on the brink of extinction due to hunting for their luxurious pelts, and the export of live animals to European zoos, where apparently vicuñas were in high demand. Nogales wanted to create a national park specifically dedicated to vicuña conservation. To play the role of park manager, however, he would need an official military rank; only then would he be acceptable to the dictators who ruled Bolivia one after the next during this period. Cardoso, who by all accounts was well connected in political circles in La Paz, used his influence to garner Nogales the unlikely title of Police Captain—bringing to mind a scene from Woody Allen's comedic satire *Love and Death*, in which the peace-loving protagonist makes his way across the field of battle clutching his butterfly net. But the result was a serious success: The vicuña population in Bolivia rebounded from roughly 1,000 animals in the late 1960s to over 30,000 in 2006, due to the new Ulla-Ulla National Fauna Reserve and a treaty signed by Bolivia and Peru, banning international trade in vicuñas. The treaty effort was spearheaded by Felipe Benavides, who pushed a conservation agenda within the dictatorship of neighboring Peru. These personalities, and their efforts to advance sustainability under the trying circumstances of military dictatorships, are virtually unknown today, even within South America. But the rules they created remain in place.[17]

THE CAPACITY TO ACT

The second measure on our scorecard of nations takes account of something that is pretty fundamental for a government: the ability to govern. This matters quite a lot for the future of our planet because a country must not only have the political will for sustainability, but a political way—the capacity to bring about substantial improvements in environmental quality and human welfare. Poor countries are obviously at a disadvantage here. If you take the size of a country's economy, divide it by the number of people living there, and array the results, they range from wealthy countries such as Switzerland ($49,960) and Singapore ($55,790) to middle-income nations like Mexico ($14,340) or Thailand ($8,190), to the very poorest such as Madagascar ($960) and Mozambique ($930).[18] Small economies, combined with inefficient tax collection systems, lead to emaciated government agencies that are highly dependent on foreign aid and whipsawed by shifts in foreign donor priorities. The overall size of the country matters too, affecting its ability to assemble a critical mass of administrative staff, equipment, and technical expertise. At one end of the spectrum, federal agencies in goliath-sized Brazil have an annual budget of over a billion dollars for environmental management activities. At the other end of the spectrum are "micro states" like the Pacific island nation of Tuvalu. These tiny countries have nameplates at the UN General Assembly just like any other sovereign state, but are so small that their governments lack the capacity to perform many basic rulemaking functions.[19] Still, these constraints are not absolute, and we see plenty of examples of small, poor countries proactively protecting their piece of the planet. On the tiny Caribbean island of Saint Lucia, a national campaign to protect the endangered Saint Lucia Parrot was so successful that it has become a model for conservation education campaigns throughout the world.[20] The Dominican Republic has put in place far-reaching conservation policies protecting its forests, sparing Dominicans from the land degradation that plagues neighboring Haiti.

Beyond poverty and size, a less widely appreciated constraint on government effectiveness is the impact of political and economic instability. Returning to the question raised at the outset of this chapter, how is it possible to promote sustainability within political systems that are themselves unsustainable?[21] We might imagine that events like the civil war in Liberia, or the breakup of the Soviet Union, are rare outliers in the experiences of nations and need not be tallied on our scorecard of national effectiveness. But it turns out that political instability is the norm. Consider the numbers. From 1946 to 2011, a total of 252 armed conflicts, mostly internal, took place in 153 countries. From 1970 to 2010, 1 in every 6 countries in the developing world and former Soviet Union experienced a complete collapse in central government authority for one or more years; most of these have occurred in the past two decades.[22] Between 1971 and 2009, a total of 111 successful military coups took place in developing and postcommunist countries.[23] Deeper changes are wrought when entire constitutional regimes are overthrown, such as a shift between democratic and authoritarian forms of government. From 1981 to 2010, 329 major changes in constitutional structure took place in 112 countries. The typical rate of turnover during this period was two major constitutional changes per country.[24] Between 1951 and 1990, the average lifespan of a democracy was eighteen years for countries with per capita income between $1,001 and $3,000 and six years for those under $1,000.[25]

How does all of this institutional upheaval affect the prospects for sustainability? It turns out that political turnover is a double-edged sword for the earth and its people. On the one hand, these shakeups often provide opportunities for real reform, a time when the legitimacy of the old regime and its rules are called into question and new leaders emerge with a vision for change. No one with a modicum of compassion for their fellow human beings would wish for a war or social crisis for the sake of advancing a particular policy agenda. But against a backdrop of crisis and upheaval, environmental reformers can be found moving quickly to put in place new ideas that had languished under the old regime. In 2009, for example, the governments of Liberia and Sierra Leone, emerging from 20 years of civil war, created a transboundary "peace park" in the lush Gola rainforest, giving protection to forest elephants, pygmy hippos, and ten species of primates.

The problem, of course, is that in countries characterized by chronic instability—which is to say, in most of the countries ruling the earth—these new rules might only last until the next crisis arrives. In these settings, savvy reformers often try to insulate their creations from future reversals by building strong social constituencies around the new rules—ensuring that local communities and other influential stakeholders benefit from the new arrangements, and involving multiple political parties and government agencies. In contrast, when environmental reformers put all of their institutional eggs in one basket—for example, relying on a frail new agency that is isolated from the rest of government and supported by a narrow political constituency—the new rules are unlikely to survive the next crisis or coup. Like building a sandcastle at the water's edge, it's just a matter of time.

SUSTAINING SUSTAINABILITY

The reason why social scientists use the vocabulary of "institutions" to describe social rules is we care about the enduring structures that funnel human energies this way and that. If a rule doesn't last, it's not an institution at all. A new forestry policy that is unlikely to survive beyond the next election may have symbolic importance, generating popular support for those in power. It may appease a constituency or attract some foreign funding. But ultimately this sort of transient proclamation matters little for the future of our planet. What, then, determines whether the rules made by governments have staying power? At the most basic level, we must ask whether a country has a bureaucracy that is staffed by reasonably competent and mission-oriented professionals who are in it for the long haul. In a place like

Hong Kong, a strong civil service tradition and successful anti-corruption campaigns have produced highly effective government agencies. In this setting, the challenge is to ensure that environmental considerations enter into the calculations of rulemakers who for decades have focused single-mindedly on the goal of rapid economic growth. But in countries with effective bureaucracies, at least you can count on the durability of new social rules. Elsewhere the challenge is of a different sort. In much of the world, government agencies are treated like warriors' loot, the lucrative spoils of political contests that confer upon victors the power and prestige that come with the ability to hand out jobs and valuable resources. The new president divides up and apportions agencies to political allies like a dinner host might carve up a roast turkey for a table of hungry guests.

There is not a government bureaucracy in the world that is completely immune from political influence. If there were, citizens would have no way to make the agency democratically accountable. But in many countries, political patronage—the practice of awarding agency positions to supporters of the ruling regime—is completely out of control. In patronage-based systems, political allegiance determines not only who is appointed to lead the agencies (a common practice in countries with effective governing systems), but also who fills crucial positions in management and operations, in effect robbing the agency of any enduring cadre of professionals. Government office buildings are filled from top to bottom with short-term political appointees who flow in and out with every change in the political tides.

Often these patronage-based systems are dripping with corruption. According to corruption researchers (yes, there really is such a research specialty), political corruption is the use of public office for personal gain and at the expense of the public good. Susan Rose-Ackerman, a legal scholar at Yale University, finds that natural resource management is particularly prone to corruption because harvesting and mining activities take place in remote locations far removed from media scrutiny.[26] Unless properly managed, natural resource exports also produce sudden influxes of cash due to wildly fluctuating prices on international markets. In a country that depends heavily on revenue from natural resource exports, when the international price of oil or copper or timber shoots upward, suddenly it's payday. This dynamic has not escaped the attention of those who see political office as a pathway to personal enrichment. On much of the African continent, the practice of using government posts for financial gain has reached such outlandish heights that researchers invented a new term—"kleptocracy"—to describe the high thievery that governs public affairs in this type of political system.[27]

For the purpose of our scorecard of nations, you'll be pleased to know that researchers have come up with measures of corruption that you can view on colorful interactive maps. Daniel Kaufmann and his colleagues at the World Bank report that corruption is widespread in developing and postcommunist countries, with the worst offenders including places like Angola, Haiti, and Uzbekistan. The most squeaky-clean political systems include those of Canada, Australia, and Sweden, among others. The United States does not quite make the top rank, while Italy is noteworthy as the most corrupt among wealthy Western countries, scoring below Malaysia and South Africa.[28]

In political systems plagued by corruption, environmental reformers must focus their efforts on improving transparency and professionalism. Reforming corrupt government practices is difficult to achieve (though not impossible) because corruption is itself an institution. As geographer Paul Robbins of the University of Arizona points out in his article titled "The Rotten Institution," corruption gives rise to its own rules, unwritten but clearly understood by participants.[29] Corrupt practices also create a self-reinforcing system. Those who take part grow accustomed to breaking the law, come to depend on the income, and become ensnared in relationships that prevent them from changing their ways. This was explained to me in vivid terms by a South American scientist some years ago, as he described his reluctance to accept a high-level post in his country's conservation agency: "If you stick your arm in a pile of crap," he explained, "it's unlikely to come out clean." Given the self-reinforcing dimensions of corruption, reformers must not only change the incentives facing participants—increasing the cost of corrupt practices and the benefits flowing to those who follow the rule of law; often they have to disentangle environmental rulemaking from institutional settings where corruption is deeply entrenched, by creating new agencies with new staff who are not caught up in the old system.

A STRANGE MEDLEY

To understand why it's often difficult for governments to govern, we must also appreciate the fact that laws do not mean the same thing everywhere. Every country has reams of laws on the books. And every country has rules governing the behavior of its people. But only in some parts of the world do these two sets of rules—the laws in force and the rules in play—correspond. Recall that nation-ness is a fairly new thing in the human experience, only a couple of centuries old. Nations are typically constructed on top of local rulemaking systems that predated the rise of the modern nation-state by hundreds, even thousands, of years. This is true not only in countries where national borders were imposed by colonial powers as they carved up the globe. It's equally true of the colonial countries themselves. In France, nationhood was achieved by cobbling together a diverse collection of regions through conquest (Alsace in 1648), marriage (Champagne in 1284), and outright purchase (Bourges in 1100), integrating these and other peoples over the course of several centuries into a semi-coherent whole. (Today regions like the Basque and Bearn still hold an uneasy relationship with all-powerful Paris.) This was typical of medieval Europe. The first modern nations were superimposed over a local mishmash of tribal customs, cultural mores, and religious and community institutions. These included German common law, commercial rules and customs agreed upon by merchants, dictates of the Roman Catholic Church, the revived Roman law that governed the universities, and rules laid down by the numerous courts created by manors, guilds, and royalty.[30]

In much of the world today the idea of a national government is younger still, and vies for influence with much older local institutions. Researchers call these local rules-in-use "informal" institutions because they are rarely written down and lack official state support. But their influence is just as real—and often more so—than the decrees and proclamations emanating from national capitals. The newness of national policy compared to local custom is described by the renowned anthropologist Clifford Geertz:

> One cannot write a history of "Morocco" or "Indonesia" (the first derives, in the sixteenth century, from a city name, the second, in the nineteenth, from a linguistic classification) that goes back much beyond the 1930s, not because the places didn't exist before then, or the names either, or even because they were not independent, but because they were not countries. Morocco was dynasties, tribes, cities, and sects, and later on *colons*. Indonesia was palaces, peasants, harbors, and hierarchies, and later on *indische heren*. They did not sum to colored polygons.[31]

Beneath the veneer of national unity, the rules created by older forms of human organization persist, often subverting the will of national policymakers. We saw in chapter 4 that Peruvian lawmakers passed an exemplary forestry law in 2001, yet deforestation in the Amazon continues with help from informal institutions dating from the era of rubber barons in the early 20th century. These local rules facilitate illegal timber extraction through elaborate chains of dependency among buyers and sellers of wood, labor, machines, and falsified harvesting permits.[32] It's not just that local customs can undermine well-intentioned national projects; too often national laws are imposed on local communities without allowing them to participate meaningfully in their creation. These new rules often have the effect of destroying age-old institutions designed to solve local problems. In other cases, outside interests circumvent national laws by manipulating local rules to their advantage. In his book *Shadows in the Forest*, political scientist Peter Dauvergne of the University of British Columbia documents how multinational timber companies have used local social networks throughout Southeast Asia to illegally access timber.[33]

The most effective rules are those that enjoy legitimacy among those governed by them. The legitimacy of national policy can't be taken for granted in places where, for decades, the sound of government vehicles rolling into town has inspired fear and hatred. In many cases, however, formal national rules and informal local rules complement one another, advancing a common purpose.[34] Strong communities are held together by what researchers call social capital—a sort of interpersonal glue that makes cooperation easier by discouraging cheating or free-riding behavior. These social bonds include mutual trust, expectations of reciprocal favors, and repeated face-to-face interactions.[35] In a small community—be it a village or a tight-knit neighborhood in a city—a person's reputation for fairness, benevolence, and reliability is a valuable resource, and this resource can be replenished or destroyed over time depending on the person's actions. When national environmental policies have the support of local communities, reformers can tap into local social capital, mobilizing community peer pressure to ensure that the rules on the books are enforced.

Consider the Sierra Gorda Biosphere Reserve, a tropical forest spanning a mountainous region in the middle of Mexico, where local communities have formal responsibility for managing the forest. To prevent illegal cutting, groups of crime-watching volunteers report infractions to local leaders who pass the information along to national authorities. In 1998, PBS aired a documentary on the Sierra Gorda featuring the work of community leader Pati Ruiz Corzo.[36] The film's subtitle, "How One Woman Has Created a Biosphere," rightly celebrates her efforts but reflects a widespread bias in media coverage of environmental issues. An emphasis on personalities makes for a good story, but the success of initiatives like the Sierra Gorda depends on rules that define the roles that local leaders can play. I had the opportunity to speak with Ruiz Corzo in 2002 as part of a research project on collaborations between national governments and local communities. She emphasized that her community couldn't possibly monitor and report illegal logging without the formal backing of the Mexican government; it would be far too dangerous. Moreover, Mexico has national policies that directly empower local communities to manage forests.[37] Particularly in cases where new national institutions are layered atop longstanding local traditions, it is the relationship between national and local rules that makes all the difference.

WHEN GOVERNMENTS DO RIGHT

In the spring of 2001, I was interviewing Nabiel Makarim, Indonesia's Minister of the Environment, in his hotel room in Washington, DC, and I was worried. For real insight into the workings of government, I find that interviews with top officials produce little; they have so much experience fending off questions from reporters that they repeat pre-prepared remarks rather than answering questions. For a penetrating analysis of what it takes to bring about changes in government policy, give me a seasoned second-tier official any day.

I was nonetheless determined to make the best of the opportunity as we drank strong coffee served in delicate china, and the minister chain smoked from his position on a yellow chaise lounge. After all, this interview was not supposed to take place. The previous day, I learned that Nabiel Makarim was flying to Washington to give a talk at the World Bank. At the time I was working as a research consultant at the World Bank, the lone political scientist amid a group of economists writing the World Development Report, which that year focused on the role of institutions in sustainable development. I asked my colleagues if they could arrange a meeting, and a senior bank official patiently explained that this was impossible given the minister's busy schedule. So I resorted to an information technology known to few young researchers today: the yellow pages. I flipped to "E" for embassies, ran my finger down to "I" for Indonesia, and soon spoke with the junior embassy official who was assigned the task of arranging the minister's travel logistics. He kindly agreed to schedule an interview in the minister's hotel room the next day.

In 1995, Nabiel Makarim had designed a novel approach to reducing industrial pollution in Indonesia, the world's fourth most populous country. The idea was to motivate environmentally responsible corporate behavior by informing local communities about the pollution released in their neighborhoods. As in Mexico, the idea was to mobilize local social capital to improve the effectiveness of national laws. Factory owners and operators in Indonesia must carefully nurture their reputations as upstanding members of the communities where they operate. Knowing this, environmental officials assembled data on almost 200 of the country's largest polluters. Instead of inundating citizens with pollution emissions data, the new program, called PROPER,[38] used a simple color scheme to publicize corporate compliance with the country's pollution laws. Each firm was given a rating of Gold (near zero emissions), Green (50 percent less pollution than allowed by law), Blue (in compliance), Red (out of compliance), or Black—a designation reserved for those placing local communities at serious risk. Every year the government shared the company color ratings with the national media. They even sent the ratings to banks and to the Jakarta Stock Exchange, affecting the firms' ability to raise capital. The results were impressive. In the first years of the program, corporate compliance with pollution standards increased from 33 percent to more than 50 percent without a single enforcement action by the government.[39] By 2011, over 1,000 companies were included in the PROPER rating system, which was replicated in a dozen additional countries.

Given his role in creating PROPER, I was interested to learn from Nabiel Makarim about relations between the central government and the citizens' groups that used its ratings to put pressure on polluting factories. But the conversation took an unexpected turn when I asked him to identify the biggest hurdle he encountered when launching the program. Departing from the usual press-ready script, he described some of the challenges of promoting institutional change in a young and fragile democracy. I came to appreciate that the creation of PROPER was an act of political midwifery guided by the minister's extraordinary knowledge of Indonesian political processes. He had to navigate the country's transition from dictatorship to democracy, deal with a simultaneous increase in the power of local governments, settle nerves amid a recent coup attempt, cope with a weak judicial system, broker agreements between citizens groups and government agencies, and build coalitions among government ministers who were jealously competing for influence. This sort of expertise does not appear in vivid color in a *National Geographic* photo spread. There we are more likely to find stories about scientists doggedly pursuing the mysteries of the earth's natural systems. But behind the scenes, the health of those natural systems depends as much on the political expertise amassed by people like Nabiel Makarim as it does on scientific expertise in areas like toxicology and air quality monitoring.[40]

There is a broader lesson to be distilled from the experiences with pollution control in Indonesia and forest conservation in Mexico. It has become fashionable to argue that national governments can do no right. Given the predominance of this point of view, it may come as a surprise to learn that policymakers can sometimes be found working hard, crafting good ideas, and putting in place new policies that work reasonably well.

Take Bangladesh's efforts to rein in population growth. From 1950 to 2011, the global population increased from 2.5 billion to 7 billion, as death rates declined due to medical advances without a corresponding drop in birth rates. The conventional wisdom was that poor countries with large rural populations could not be expected to slow population growth, because children provide inexpensive farm labor; only with increasing wealth and urbanization, it was argued, would parents decide that the benefits of having more children are outweighed by the costs. This model, known as demographic transition theory, provided a highly simplistic view of the causes of population growth. According to demographers John Bongaarts and Steven Sinding, fully 40 percent of pregnancies in the developing world are unintended. The problem is not just a lack of economic development, but the absence of birth control options for women. Shrugging their shoulders at the conventional wisdom, in the 1960s Bangladeshi officials began an aggressive public outreach effort focused on reproductive health. This included free contraceptives, home visits by female health workers, and education campaigns to counter social taboos and make women aware of family planning options. The campaign was accompanied by a broader government effort to expand educational opportunities for girls. In areas where the program was piloted, contraceptive use soared from 5 percent to 33 percent. Health officials soon realized they were onto something. They launched a nationwide program over the ensuing decades that cut Bangladesh's birth rate in half, from six to three births per woman in the late 1990s, compared to five births per woman in neighboring Pakistan.[41]

Or consider renewable energy. Denmark has emerged as a global leader in renewables as a result of rules laid down by the Danish government and active support from grassroots constituencies.[42] Since the 1970s, officials have consistently promoted wind power as a complement to traditional energy sources. The country has invested heavily in turbine research and uses taxes and subsidies to compensate wind generators for the environmental benefits of their power. Danish energy pricing policies spur the development of new technologies, while their land use plans explicitly encourage turbine use. The combined impact of these rules has been a vast expansion in wind power. By 2014, renewable energy made up 28 percent of all electricity generated in Denmark. Per capita carbon dioxide emissions in the country have declined by a third since the 1970s.[43]

Over in the Philippines, we find one of the world's most impressive programs to protect coastal resources. Responding to decades of overfishing, the national government now coordinates a system of strict no-take reserves that benefit adjacent fishing communities while enhancing tourism. The reserves are run by community organizations and local governments, which are empowered by national laws to play a proactive role in coastal management. From its modest beginnings in the 1970s, the program has grown to include more than 600 marine protected areas throughout the country, over 90 percent of which are managed by local communities.[44]

In India, meanwhile, the world's largest democracy is undertaking a massive experiment in citizen empowerment. India has long suffered under the weight of its famously unwieldy bureaucracy, where hundreds of thousands of public employees spin a vast web of red tape that thwarts even the most basic requests for public service.[45] In 2005, following a decade of lobbying efforts by public interest organizations, India's legislature passed a Right to Information law that now requires government employees to respond to requests from citizens in a timely manner. The law has real teeth, and intransigent bureaucrats who drag their feet are hit with stiff fines. The impact of the new rule has been nothing less than revolutionary. In the first two and a half years, over two million Right to Information requests were filed by Indian citizens. For the first time, poor people can demand accountability when waiting for agencies to deliver public services ranging from food rations to drinking water wells. Over 7,000 of these requests have been directed at the Ministry of Environment and Forests. Turnaround time has vastly improved, ensuring a level of government responsiveness that was previously reserved for those who could afford to pay bribes.[46]

WHEN GOVERNMENTS DO WRONG

"Why should reasonable men adopt public policies that have harmful consequences for the societies they govern?" In 1981 Robert Bates, a political scientist at Harvard, posed this question in an influential book titled *Markets and States in Tropical Africa*. Bates was trying to understand why decision makers in developing countries routinely put in place rules that destroy their agricultural sectors. By mandating low prices for food staples, they undercut incentives for farmers to invest in their own businesses. For countries that depend on the agricultural sector for export income and food self-sufficiency, this policy makes no economic sense. Politically, however, it makes all the sense in the world to appease urban consumers with low food prices at the expense of farmers and rural communities. Bates found that when citizens in a nation's capital are upset, politicians quickly feel the effects as the crowd swells at the palace gates. Policies that hurt the countryside, while not trivial in their political impact, are nonetheless easier for national leaders to handle.[47]

William Ascher of Claremont McKenna College has followed this line of investigation and applied it to questions of sustainability, probing the political logic behind seemingly illogical national environmental policies. In his book *Why Governments Waste Natural Resources*, Ascher asked questions like why Honduras destroys its forests, why Mexico undercuts its own irrigation systems, and why Nigeria squanders the proceeds from its oil exports. Ascher reviewed dozens of cases of government mismanagement, shortsightedness, and reckless behavior in an effort to make some sense of these seemingly senseless decisions. He discovered that governments often act against their own national interests for reasons that are more nuanced than the usual diagnoses of greed or corruption. Sometimes leaders rapidly sell off their country's natural resources to create secret discretionary accounts through elaborate laundering schemes. They use these slush funds to support social programs—sometimes quite laudable ones—that are opposed by their political adversaries and therefore unlikely to receive funding through the official state budget. In other cases, clashes among competing land management agencies leave property rights claims in disarray in forest frontiers. When settlers have no idea whether the land is truly theirs, they have little incentive to care for the land over the long term. In still other cases, officials shore up political support by subsidizing workers' wages in grossly inefficient mining operations.

The political logic of environmental destruction takes many forms. In Southeast Asia, when money rushes into government coffers after a boom in commodity prices, politicians can be found rewriting the rules that govern forests to gain access to the income streams. In the Malaysian state of Sabah, which occupies the northern portion of the island of Borneo, forests were managed reasonably well until the state gained independence from Britain in 1963. The transition stirred the political pot in ways that concentrated power in the hands of the chief minister, giving the holder of that office wide-ranging and largely unchecked control over state resources, including timber harvesting licenses. As Michael Ross documents in his book *Timber Booms and Institutional Breakdown in Southeast Asia*, political parties in the young democracy were eager to garner the allegiance of timber operators, and so they rewrote the rules to allow the extraction of ever-greater quantities of valuable tropical hardwood. Eighty-year rotational cycles for forest concessionaires were ended, replaced with policies that encouraged timber operators to cut and sell wood as quickly as possible. The new rules had tragic consequences: Today less than half of the tropical forests of Southeast Asia remain.[48]

Other political logics can be found in Africa. In his book *Politicians and Poachers*, Clark Gibson of the University of California at San Diego tries to solve the puzzle of why governments in sub-Saharan Africa often pursue wildlife policies seemingly at odds with national interests. President Jomo Kenyatta, the founder of modern Kenya, stood by while illegal hunting decimated wildlife throughout his country in the 1970s. "Politicians at every level needed new sources of patronage during the country's economic downturn," argues Gibson, "and gladly turned to wildlife when its value increased internationally."[49] In Zambia, the rules created by wildlife officials in the 1980s were inspired by their fear of the country's dictator, Kenneth Kaunda. "Their decisions regarding budgets, personnel, and expansion had less to do with promoting conservation than with protecting themselves from the political uncertainty caused by Kaunda's dominant policy making position."[50]

To catalogue the full array of impulses that lead governments to harm the environment would require a diagnostic manual a foot thick. Anthropologist

James Scott argues that governments tend to centralize information and ignore valuable local knowledge concerning things like soil fertility and smart farming practices. In extreme cases, this has produced tragedies like China's Great Leap Forward, when state officials demanded that farmers boost food production so rapidly that the soil was soon depleted, provoking the greatest famine in recorded history.[51] Still other researchers have documented how states use violence and coercion to control natural resources for the benefit of economic elites and against the interests of local communities.[52]

Whether countries do right or wrong, whether their leaders rule in a spirit of public service or plunder, whether they take the form of democracies or dictatorships, are stable or unstable, capable or weak, the plain truth is that the earth is now, and will remain for the foreseeable future, ruled by nations. We like to think of our colorful political maps as artifact and conceptual convenience, and the floating blue globe as the real deal. But the rules created by national governing bodies are as tangible in their physical effects as Earth's gravitational pull. It takes a nation to save a planet because that's where most of the rulemaking power resides. With this observation as a backdrop, let's conclude our inquiry into nationhood by examining the environmental record of the most powerful nation ever to rule the earth—both where it has been and the unresolved question of where it is heading next.

WHICH WAY, USA?

There was a time when America was a global leader—arguably *the* leader—in efforts to promote sustainability. This was true both domestically and in its commitment to tackling global problems. Beginning in the late 1960s, American policymakers approved a succession of ambitious new rules to address the nation's most pressing environmental issues. The Clean Air Act and Clean Water Act vastly improved environmental quality nationwide. The Endangered Species Act took a tough stand to conserve the country's biological diversity, protecting not only charismatic wildlife like the American bald eagle, but hundreds of lesser-known plants and animals—the salamanders, desert flowers, sea snails, bees, field mice, and other members of the nation's biological heritage, the green tips of an evolutionary process stretching back three-and-a-half billion years. One of the most far-reaching rules dating from this era is the National Environmental Policy Act (discussed in greater detail in chapter 10), which requires agencies and developers to undertake environmental impact assessments when their decisions are likely to have a significant effect on the environment. Whether building a new hospital wing in a crowded urban corridor, or converting a decommissioned military base to commercial use, for the first time developers had to think through the environmental consequences of their actions. The approach was subsequently copied by dozens of additional countries.[53]

Additional laws ensured that rulemaking proceeded in a transparent manner, with information made available to ordinary citizens. India's Right to Information law of 1995 was inspired by the US Freedom of Information Act of 1966. Indonesia's PROPER program drew directly on the experience of the US Toxics Release Inventory, created under the federal Emergency Planning and Community Right-to-Know Act of 1986. The inventory shares with the public detailed information on polluting activities taking place in local neighborhoods. In the previous chapter, we saw that the United States also took the lead in banning leaded gasoline in the 1970s and 1980s, an initiative that was followed only a decade later in most of Europe. The difference between the United States and Europe was part of a larger pattern. Political scientist David Vogel has followed the history of policymaking on both sides of the Atlantic for the past three decades. Vogel writes that America was the clearly the trendsetter: "From the 1960s through the mid 1980s American regulatory standards tended to be more stringent, comprehensive and innovative."[54]

American environmental leadership was equally impressive on the international stage. The United States was a driving force behind several treaties including the Convention on International Trade in Endangered Species, and it spearheaded major diplomatic initiatives like the 1972 UN Conference on the Human Environment. The momentum continued throughout the 1980s, surviving the rise of an anti-regulatory agenda in the White House. Early attempts by the Reagan administration to dismantle the institutional achievements of the 1970s failed. As the environmental movement grew in strength, lawmakers from both parties worked together to expand the scope and scale of environmental initiatives, including new rules to protect the ozone layer. The United States had already moved ahead of other countries in regulating ozone-depleting activities in the 1970s, when it banned the use of chlorofluorocarbons (CFCs) in aerosol sprays and held congressional hearings on the potential atmospheric impact of supersonic jets. During Reagan's second term, the White House made a strong push for an international treaty to protect the ozone layer—a move that was initially opposed by several European countries—and eventually succeeding in building support for a treaty that was subsequently strengthened (again under Reagan) with the Montreal Protocol of 1987.[55]

That was then. Over the past 25 years, rulemakers in Washington have made few serious attempts to promote sustainability either at home or abroad. The 1990 amendments to the Clean Air Act were pretty much the last hoorah for bipartisan cooperation on a major environmental initiative. Today, while the European Union races ahead with innovative policies to promote sustainability, America's early accomplishments are starting to look like faded photographs from a time gone by. In an article titled "Trading Places," David Vogel and his colleague R. Daniel Keleman find that "a dramatic and systemic shift from U.S. to EU leadership has occurred."[56] The reversal is most glaring on the international stage, where the United States has shunned several major treaties. The United States rejected the climate change treaty even though American scientists have long been at the forefront of research on global warming. (Within the UN Intergovernmental Panel on Climate Change, the world's leading authority on global warming, the United States has more scientific contributors than all of Europe combined.) The United States has an even worse record of ratifying the treaties it has signed, failing to translate them into domestic law. In 2012, the United States was one of the only countries in the world that had not ratified the treaty on Persistent Organic Pollutants. It would be an exaggeration to claim that American policy is completely in the dumps, but undoubtedly a major shift has occurred. Once a leader in rulemaking for the environment, America is now on the sidelines while Europe sets the pace for environmental policy.

So what happened? To make sense of this role reversal, we need to appreciate the link between domestic and international rulemaking. When a country pursues an ambitious environmental agenda at home, it has a strong incentive to push other countries to do the same. Elizabeth DeSombre of Wellesley College calls this the "Baptists and bootleggers" phenomenon. During the Prohibition Era, two very distinct groups pushed for rules forbidding alcohol consumption. Religious organizations wanted to ban booze because of its perceived corrupting influence on the soul. The bootleggers, who sold alcohol illegally, had a commercial interest in prohibition laws, which enabled them to command a higher price for their goods. In a similar manner, American environmentalists and corporations have a shared interest in ensuring that companies elsewhere are subject to the same environmental regulations that are in place at home. Consider the role of DuPont in the creation of the ozone treaty. During the 1970s, while Europe refused to ban ozone-depleting CFCs, strict domestic regulations in the United States caused DuPont to lose half its market for the chemicals. This prompted the corporate giant to develop a new class of ozone-friendly refrigerants to replace CFCs. DuPont soon emerged as a strong advocate for an international treaty that would provide a market advantage for its new product.[57] International and domestic policy also trend together for a simpler reason: When environmental concerns hold sway over domestic policymakers, this is likely to affect their foreign policy decisions as well. The US has been marginalized on the international stage because environmental issues have been marginalized at home.

If America's stance toward the planet is a fair reflection of what's taking place on the home front, what can account for the shift in priorities within the United States? Researchers looking into this question have arrived at essentially the same conclusion as those working on the front lines of political struggles to protect the planet: The impact of the environmental movement has declined in Washington, while those opposed to environmental regulation have consolidated their influence.[58] This has been accompanied by a shift within the leadership of the Republican Party. Sociologist Riley Dunlap and his colleagues have run the numbers, showing the growing split in congressional votes for environmental laws. In the 1970s, the typical Democratic lawmaker voted for environmental legislation just under 60 percent of the time, and Republicans just under 40 percent, for a 20 percent cross-party difference. By the 1990s, support from Democrats grew by a third, hovering around 75 to 85 percent, while that of Republicans plummeted to under 20 percent—for a gaping 60 percent difference in party support for environmental laws.[59]

To appreciate the significance of this shift, recall the crucial role that Republican leaders have played historically in promoting an environmental agenda. America's renowned national park system and national forests were created by Republican President Teddy Roosevelt. The Arctic National Wildlife Refuge, today a focus of bitter partisan divide, was established by Republican President Dwight Eisenhower. Richard Nixon created the EPA and the White House Council on Environmental Quality and signed into law the Clean Air Act, Clean Water Act, Endangered Species Act, and

National Environmental Policy Act. The Reagan Administration gave full-throated support to the ozone treaty. Reagan's successor, President George H. W. Bush, signed the Clean Air Act amendments of 1990, putting in place the world's first large-scale cap-and-trade system for controlling pollution.

The Republican turn away from environmental issues is neither uniform nor absolute. At one extreme, the administration of George W. Bush filled many environmental regulatory posts with lobbyists from the very industries they were supposed to regulate, and joined the auto industry in a lawsuit attempting to prevent states from taking action on climate change. At the other end of the spectrum, Republican presidential contender John McCain co-sponsored a bill (ultimately defeated by members of his own party) that would have taken steps to address global warming. In California, Republican Governor Arnold Schwarzenegger made a priority of passing state Assembly Bill 32, which takes aggressive action to reduce carbon emissions. And important segments of the conservative American evangelical movement are organizing to increase the salience of environmental issues both within their churches and in Washington, DC. But on balance, whether due to philosophy or political calculation, conservative politicians in the United States appear to have concluded that they have more to gain by railing against the Environmental Protection Agency, disparaging environmentalism as a fringe concern, and downplaying the seriousness of issues like climate change. If the United States is to return to its once venerable position as an innovator in environmental policy, concerned citizens from across the political spectrum will need to band together and increase their influence in Washington. The success of such a movement will depend in no small measure on a movement for change from within the ranks of the Republican Party, led by people with an environmental vision, strategic smarts, and influence within party circles.

SHIFTING SANDS

Whether in the corridors of the White House or the jungles of West Africa, the race to save the earth will be won or lost one country at a time, as a result of political decisions made in almost 200 sovereign nations and their willingness and ability to implement reforms. We truly live on a planet of nations. Yet as with all rulemaking systems, things evolve. While nations are not declining in influence, other sources of rulemaking power are on the rise. The answer to the question *Who rules the earth?* is undergoing a metamorphosis as a result of two political trends that will have planet-wide impacts over the coming decades. The first is the rise of the European Union, a first-of-its-kind supranational governing body that is embracing environmental goals and flexing its muscle on the world stage. The second major trend—political decentralization—is funneling power downward, increasing the rulemaking power of the hundreds of thousands of states, provinces, and local governments scattered across the earth's surface.

PART III

Transformations

7

Scaling Up

THE RULEMAKERS

José Delfín Duarte rises at the crack of dawn in a neighborhood on the outskirts of San Isidro, Costa Rica. He grabs his machete and rain parka, puts on his black galoshes, and heads out in a flatbed truck up a series of muddy roads surrounded by lush forest interspersed with farms. Eventually he arrives at a small water-distribution facility located at the top of a hill overlooking the surrounding watershed. He checks the station's tanks, carefully noting the water levels. Duarte is the elected leader of a group of local citizens who have been given responsibility for managing water resources in their community. They decide how much water is used and how it will be allocated among families and farms in the area. They collect user fees, purchase equipment, and make numerous daily decisions affecting water use. Their role stems from a power-sharing arrangement with the Costa Rican government, which in recent years has crafted similar agreements with hundreds of local water associations throughout the country.[1]

Six thousand miles to the east, Claudia Olazábal begins her day in the outer suburbs of Brussels. She takes the subway to her office in the European Commission, a sleek modern glass and steel building where she heads the Biodiversity Unit of the European

Union's Directorate General for the Environment. On this particular day, her attention is focused on the design of new rules for the control of invasive species, which pose a major threat to ecosystems worldwide. Six years in the making, this rule came about after extensive consultation with stakeholders throughout the twenty-seven member countries of Europe—farmers unions and botanic gardens, prime ministers and pet shop owners. Working with a professional staff of Swiss and Germans, Poles and Portuguese, and many other nationalities, Olazábal is preparing for a lengthy negotiation involving lawmakers throughout the continent in a complex dance that will hopefully produce a new European policy on invasive species.

Claudia Olazábal and José Delfín Duarte operate worlds apart, yet they have much in common. Both are creating rules that will shape our planet for decades and even centuries to come. Each is also constrained by the rules laid down before. Olazábal must work within the operating rules established by European Union treaties and must adhere to the policy priorities identified by European heads of state. Duarte must play by the rules of the Costa Rican Water and Sewer Institute, which required his association to sign a contract detailing the community's role in managing water and the rights and responsibilities attached to that role. Yet within these bounds, each is quietly reshaping the fabric of society, cutting and reattaching the threads of the elaborate rulemaking structures that make it possible to do things like provide potable water to a rural community or protect endangered species across the European continent.

Juxtaposing these two worlds, the international and the local, serves another purpose—for they are not just parallel experiences, reflections of similar things happening in different places. They are directly linked. The rules that Costa Ricans make with respect to their natural resources—like the rules deciding whether people can cut down trees in the vicinity of water sources—influence how much carbon is released into the atmosphere, contributing to global processes that will affect water levels in the canals of Venice over the coming century. Environmental problems are literally un-*ruly*—they meander across political borders and agency jurisdictions, challenging our ability to launch a coordinated response. Tackling these unruly problems requires that we include in our field of vision multiple levels of political organization, from the corridors of the United Nations, with its lawyers, translators, and technical experts, to national governments ruled variously by parliaments, crown princes, and generals, continuing downward to subnational governments (provinces, states, cantons) and countless cities, towns, and villages that populate the planet from Durban to Devonshire.

It's not just the physical characteristics of problems like water pollution and climate change that require a multilevel view of the world. Political strategy calls for it. If we are serious about promoting sustainability—if we wish to move beyond the current infatuation with green consumerism, and tackle the underlying causes of environmental problems—we cannot confine our energies to one level of governance. Rulemaking is like a multitiered chess game, with the outcome at one level often hinging on how the game is played at another. Sometimes caring for the earth requires smaller political units to band together. This logic has led American cities to form regional air pollution districts, where local officials pool resources and coordinate their efforts. On other occasions, the smartest course of action is for larger political units, such as national regulatory agencies, to scale down, empowering cities or provinces to take the lead and tailor responses to local conditions. The old adage to think globally and act locally is just plain wrong. It is far too simplistic in its portrayal of the sources of environmental problems and the solutions at our disposal. We need to think and act at multiple levels if we are to make progress on vexing social and environmental problems. We need to think vertically.[2]

Claudia Olazábal and José Delfín Duarte share something else. Until recently, their rulemaking systems simply did not exist. It is only within the past quarter century that the European Union has developed the capacity to respond to environmental problems in a coordinated fashion. Costa Rica created new rules to strengthen water associations only in 2000, as part of a broader push to increase the power of local governments in this traditionally centralized country. One reason why we must ask ourselves, *Who rules the earth?* is that the answer changes over time—sometimes for the better and at other times for the worse, in some cases gradually and in others through an unexpected burst of innovation. The question is not merely who occupies the role of decision maker—the farmer, the judge, the mayor—but who defined the scope of their decision-making authority in the first place. When the rules change, suddenly we find that our labor, our voices, and our resources (wild rivers and tax revenues alike) are channeled in new ways and toward new ends. These are questions that environmentally-minded citizens cannot afford to ignore.

In this chapter and the next we will focus on two of the most important changes now underway in who rules the earth. The first is the rise of the European Union, an experiment in international cooperation that, in its sheer scale and ambition, is without historical precedent. The second is a global trend toward political decentralization, in which central governments are giving subnational regions and local communities a greater role in managing their affairs. Both developments are surrounded by a whorl of power struggles and contentious debates, as everyone—from the seasoned political operative in Paris to the family eking out a living in rural Zimbabwe—tries to figure out what these changes portend for their future.

If all of this scaling up and scaling down business sounds familiar to American readers, it should. The world is now struggling with many of the same questions that have confronted the US federal system over the course of its history. How much power should be granted to the central government versus the regions? How can jealously independent political units be brought together for the purpose of union? The changes underway in the distribution of decision-making power around the globe also carry consequences for American citizens. If a woman living in New York applies makeup imported from Europe, she can be sure that it was not tested on animals because the EU has banned the practice; if it was produced in the United States, she can't even be sure that it's safe for use on humans because it is essentially unregulated.[3] American businesses hoping to compete in areas like alternative energy technology must contend with the fact that Europe's regulatory environment is in many ways more conducive to green innovation. The implications of decentralization are equally profound. Anyone who hopes to engage with the wider world beyond their borders—investors, conservation groups, researchers, retirees, diplomats, human rights activists—will encounter growing assertiveness by local governments around the globe as they test the boundaries of their new authority.

POWER UP

International treaties occupy a prominent place in political discourse about the future of our planet. Admittedly, there is something thrilling about the idea of nations overcoming their differences, even for a moment, to think collectively about what the world needs and how this might be advanced through closer international cooperation. The historical record suggests that environmental cooperation is possible not only in spirit, but in deed, as the number of environmental treaties has grown rapidly over the past half century (Figure 7.1).

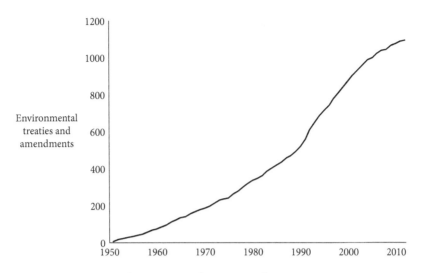

FIGURE 7.1 International environmental cooperation has grown

Data are from Ronald B. Mitchell, 2002–2013, International Environmental Agreements Database Project (Version 2013.1). http://iea.uoregon.edu/.

But do all of these treaties really mean anything? The short answer is sometimes yes, sometimes no. The longer answer is that global cooperation, and the system of law that underpins it, take place in circumstances that bear little resemblance to lawmaking as you and I understand it. For one thing, there is no international government to enforce the law. Quite apart from whether you consider the absence of an international government to be a good or a bad thing (and reasonable arguments can be marshaled on both sides of that debate), it is a reality that colors every aspect of international rulemaking for the environment.[4]

Take the United Nations, that august organization that embodies the idea of internationalism. The UN might appear to have some of the features of that species we call government. There is the UN flag, with its famous olive tree wreath. The UN has a president and a general assembly, and dozens of professionally staffed agencies. But the UN has none of the essential characteristics of a sovereign government, such as the power to tax or raise an army. The budgets of its humanitarian agencies are supplied purely at the discretion of member nations. The same is true of the peacekeeping forces dispatched with their light blue helmets to trouble spots around the globe. The United Nations website address ends in dot org.

For all the talk of a global community, there are hardly any functional courts at the international level. The few that exist, such as the International Criminal Court or the UN's International Court of Justice, are rarely able to stop a country or a corporation from breaking the law. A Nigerian family harmed by the polluting activities of a multinational oil company has no legal recourse in a world court. The will of any court ultimately relies on police power, yet there is no global police force relevant to environmental matters, with the exception of limited efforts by INTERPOL to halt the illegal trade of endangered species and banned chemicals.

In short, the international system of governance is characterized by anarchy, in the literal sense of the term meaning "without government." This is not a controversial diagnosis among political scientists, nor is it a well-kept secret among international diplomats. But what does all of this mean for efforts to protect the earth? For one thing, if a country wishes to not be bound by a treaty, it has every right to back out of the agreement. Imagine what domestic law would look like if people could ignore any law not to their liking, and you can begin to appreciate the challenge of rulemaking at the planetary level. International environmental cooperation is an attempt at governance without a government.[5]

Against this backdrop, what is happening in Europe today is truly remarkable. Overcoming centuries of hostility, Europeans have banded together to coordinate their approaches to economic growth, foreign affairs, social welfare, and sustainability. From its modest beginnings in the 1950s as a handful of countries coordinating coal and steel production, the EU now includes twenty-seven nations with over 500 million people. Presidents and prime ministers confer together in a European Council, where they issue joint decisions that affect daily life across the continent. Their deliberations are informed by the European Commission in Brussels, where a staff of 33,000 issue regulations to ensure that national environmental agencies across the region, from Spain to Switzerland to Slovenia, coordinate efforts and work from the same set of overarching rules. Gone are the national currencies of francs and deutsche marks, in are the new euros. European citizens travel freely throughout the region without national passports or cumbersome border checks. They vote for elected representatives in an all-Europe Parliament. Europe stands as a unified block when negotiating with the United States and China over everything from wheat prices to climate change. The combined economy of the European Union is larger than that of the United States.

The unification of Europe is the sort of historical phenomenon that, when viewed up close and over brief time spans, appears to move like molasses. The news headlines leave the impression that narrow economic agendas and short-term political calculations will forever trump grand ambitions for international cooperation. Yet when viewed from a wider perspective, spanning even just a couple of decades—a mere blink of an eye, by the metric of European history—the pattern is unmistakable: Europeans have made steady progress toward the most ambitious multi-nation rulemaking system ever to govern a piece of the earth.

How did such a momentous change come about? To make sense of the new rules governing the old continent, we need to step outside of a purely "environmental" history and take a peek at the broader sweep of events that got us to where we are today. Untangling the cause-and-effect relations that precipitate major shifts in the direction of human affairs is never an easy feat. But all observers agree that the European Union would not exist in its current form, and quite possibly not at all, were it not for the efforts of an enterprising cognac merchant by the name of Jean Monnet, who saw in the ruins of the continent's battlefields the possibility for changing the rules of the game in Europe.

A FRACTURED CONTINENT

In many ways, Europe is the last place you would expect to find a rulemaking system designed to promote harmony among national governments. This is, after all, the continent that invented words like "provincialism" and "fiefdom," and where conflicts carry names like The Hundred Years' War. The turreted castles scattered across the European landscape today stand testament to centuries of territorial animosity. Recall that national sovereignty—the bedrock concept underlying the modern nation-state—is a quintessentially European invention. At the Peace of Westphalia in 1648, warring factions came together—Hapsburgs and French and Dutch, Swedes and Spanish, independent cities and the Holy Roman Empire, all of them armed to the teeth—and agreed to the principle that each country should be left alone to conduct its affairs as it pleases, without foreign interference.

In retrospect, the birth of national sovereignty as a central organizing principle for political life was obviously insufficient to prevent further bloodshed. Over the next three centuries, the mobilization of people and equipment for repeated wars—more than 100 distinct European wars in all—required tax collection systems and capable government administrations, which further strengthened individual nations.[6] By the time the 20th century rolled around, a tinderbox of age-old rivalries, modern states, and deadly new technologies twice ignited into major conflagrations, the first sparked by the assassination of Archduke Franz Ferdinand in Sarajevo, precipitating World War I, and the second by the rise of Adolf Hitler.[7] By 1945, the continent lay in ruins. In the words of historian Louis Snyder, "Throughout Europe the land was laid waste as if by a gigantic scythe."[8] In World War II alone, more than 53 million people died, two thirds of them civilians.[9] (I must mention, even if in passing, that like other Jews of European descent, this included essentially all of my relatives in Russia and Poland.) Half a million homes in England were flattened. In Germany 2,000 bridges and a third of all buildings were damaged or destroyed.

In the aftermath of the wars, the European Union was created not all at once, but through a gradual process of widening cooperation. Like a cautious bather entering a chilly Alpine lake, each small step proved the value of the undertaking and emboldened Europeans to take the next. At the center of this process was Jean Monnet.[10] Monnet was born in the town of Cognac on the Atlantic coast of France in 1888, the same year that the other Monet—the fellow with the beard and paintbrushes—was completing his famous paintings of grainstacks at Giverney. With a short and sturdy physique and no-nonsense attitude, Jean Monnet was often described by his contemporaries as farmer-like in his demeanor. Remarking on his tenacious pursuit of European unity, Hubert Beuve-Méry, founder of the French newspaper *Le Monde*, noted "his obstinacy, like a peasant determined to sell his cow."[11] While his roots were firmly established in the French countryside, Monnet was decidedly more merchant than *paysan*. His father ran a successful cognac company and Jean, as the elder of two sons, was expected to eventually take over the family business. The cognac trade required extensive international travel, and the Monnet family home was frequented by visitors from the far corners of the globe. The cosmopolitan world of fine brandy would play an important role in Monnet's meteoric rise as the architect of modern Europe, where his success stemmed from an unusual ability to bridge disparate cultures—between international commerce and domestic politics as much as between France and the English-speaking world.

At the tender age of sixteen, Monnet left school and traveled to England, where he took up residence in the financial district of London with the hope of gaining fluency in the language of the international cognac trade. In the ensuing years he made frequent trips abroad on behalf of the family business, visiting North America, Egypt, China, and Russia. (It was during this time, according to Monnet's biographer François Duchêne, that the young cognac salesman tried unsuccessfully to secure a spot on the maiden voyage of the *Titanic*.[12] It is a testament to the challenge of prediction in the social sciences that, had the young Monnet boarded the ship, it is entirely possible that the European Union would not have come into being, and thus Europe would have no common policies on issues like climate change today.) With the outbreak of World War I in 1914, Monnet's global outlook and precocious skills as an international negotiator would be put to the test.

If you trace your finger down one of the concrete and brass war memorials found in every French town today, you will encounter a long list of the deceased from World War II, but a much longer roster of names from its predecessor, the tragically titled "war to end all wars." During World War I, France would lose 1.5 million people. As the war progressed, the country was growing desperately short of wheat for bread, the staple of the French diet. Immersed in the minutiae of international commerce, the young Monnet was surprised at how seldom France and Britain actually cooperated on economic matters, despite their formal military alliance to defeat the Axis Powers. Using his father's connections in the French government, Monnet turned for help to the Hudson Bay Company of Canada. He forged an agreement in which the giant trading company would ship wheat from North America to Europe, and drew up detailed plans for rationing supplies among the Allies. Building on this early success, Monnet enlisted help from high-ranking officials in the French and British governments who shared his commitment to economic cooperation. They expanded wartime trade through a new Allied Maritime Transport Council, an early testing ground for the idea of European integration.

By the conclusion of the war, Monnet's dogged efforts to build a rule-making structure to promote international cooperation had attracted the attention of French Prime Minister Georges Clemenceau. In 1919, with support from France and Britain, Jean Monnet was appointed at the age of 31 to be Deputy Secretary-General of the League of Nations, which was created at the end of the First World War to promote the elusive goal of world peace. Before the League's demise at the outbreak of World War II, Monnet proved adept at brokering agreements and designing innovative rules to manage postwar challenges such as the reconstruction of Austria. It would be several decades before the idea of "the environment," as we understand it today, would come to the fore, huddling together disparate issues like industrial emissions, nature appreciation, and energy efficiency under the same conceptual banner. But in these early days, the League of Nations demonstrated the potential benefits of international environmental cooperation. Responding to concerns raised by health officials, the League sponsored a treaty in which members agreed to ban the use of lead from paints applied to the interior of homes where people could be exposed to its neurotoxic effects.[13] The measure was soon adopted by a number of countries throughout Europe. The American government, meanwhile, chose to ignore the new rule, permitting the use of leaded paint in homes for another half century; as a consequence, to this day lead exposure in older American homes poses a major health hazard for children, who are most susceptible to its effects. Parents who nervously wipe dust from the windowsills, and forestall home improvements for fear of mobilizing lead dust in the air, can trace their dilemma to the rulemaking path chosen by American officials a century ago.

After stepping down from his position at the League of Nations in 1923, Monnet's trajectory zigzagged between commerce and government, but never strayed from the goal of strengthening cooperation among nations. This included a position as vice president of European Blair, an American investment firm that specialized in making large loans to European countries. It was during this period that Monnet solidified his status as a political insider at the highest levels within Europe and the United States. Monnet learned to craft agreements that delicately balanced economic and political goals, and in the course of this highwire act, Monnet built up a network of personal contacts that would prove critical to the creation of the new union. His assistant at the investment firm was René Pleven, the future prime minister of France. Monnet also became close friends with John Foster Dulles, who would later become secretary of state under Eisenhower. In November 1932, Monnet was even approached by the government of China, which sought his help with a complex negotiation. Chiang Kai-shek's finance minister hired Monnet as his representative to Europe and North America in an attempt to secure funds for the modernization of Chinese railroads (an initiative that was ultimately scrapped at the insistence of the Japanese). Having witnessed Monnet's creativity and grit throughout the ordeal, Chiang Kai-shek observed that he would make a good Chinese general.[14]

TAKING ON THE "EUROPEAN PROBLEM"

With the outbreak of World War II, Monnet quickly found himself at the center of the maelstrom, as European leaders enlisted his help to broker international deals with the hope of defeating Hitler. In 1938 the Nazis were on the move. European powers capitulated to Germany's annexation of Czechoslovakia at the same time that they sought to slow Hitler's advance. Of special concern were warplanes. France had dominated the skies as recently as the late 1920s, but Germany bypassed the French with the invention of aluminum-bodied aircraft that could travel faster and carry heavier loads. In October 1938, French Prime Minister Édouard Daladier arranged for Monnet to travel in secret to America to request help from President Roosevelt. This was a time of intense isolationist sentiment in the United States. Reflecting the public mood, officials in Washington (including military leaders) were loath to get involved in foreign entanglements. Roosevelt, however, had become convinced that America could not sit on the sidelines. He agreed to meet with Monnet at the president's private residence in Hyde Park.

Using his trademark number-filled charts, Monnet explained to the president the need for a massive increase in warplanes. Roosevelt, deeply wary of the political reaction at home, proposed a covert plan. They agreed to establish manufacturing plants in Canada, out of the reach of German bombers and out of sight from American news reporters. The planes would be built with American parts and then shipped to Europe.[15] The plan was later abandoned due to objections from US Treasury Secretary Hans Morgenthau. Then in the spring of 1940, Nazi forces invaded and occupied the northern half of France as well as a strip along the Atlantic that included Monnet's home town of Cognac. (In a bizarre twist of history, the cognac industry was saved only through the efforts of Lieutenant Gustav Klaebisch, the Nazi administrator for the region. Klaebisch was a German brandy merchant who had been born in Cognac and took a special interest in protecting the region from looting and destruction.[16])

In Paris, the French government under Marshal Philippe Pétain made the infamous decision to accept the Nazi takeover, wagering that collaboration with the occupying force was their only salvation. With Paris no longer willing to oppose the Germans, Monnet turned to his contacts in England. He obtained help from Winston Churchill, who knew of Monnet's efforts and helped him obtain a British passport so that he could relocate with his family to the United States. There the Frenchman would serve as an employee of the British government, brokering a deal with the Americans to send planes to Europe for the defense of France. A century and a half after Benjamin Franklin traveled to France to secure crucial support for the Revolutionary War (a collaboration commemorated by the French gift of the Statue of Liberty), Monnet crossed the Atlantic in the other direction, placing orders for hundreds of thousands of bombers. In so doing, Monnet helped convince the United States to ramp up its war production capacity well in advance of Pearl Harbor, at a time when few Americans supported the war.

In 1943, Monnet briefly joined General Charles de Gaulle and the French Resistance in their government in exile in Algiers, across the Mediterranean. At 6 foot 5, de Gaulle was an outsized character, fiercely nationalistic and steadfast in his ambitions for a free France. Three years earlier, just after Pétain's government surrendered to the Nazis, de Gaulle broadcast his famous speech to the French people from a secret BBC studio in London, exhorting them in his mesmerizing voice to hold fast against the Nazis.

FIGURE 7.2 Jean Monnet, father of the European Union
Historical Archives of the European Union, European Union Institute.

Making an end run around Pétain's government, de Gaulle and his Free French forces wrested control of the French colonies, including Algeria, using these as the base for their planned return.

In Algiers, the young general ruled a resistance movement spanning several continents and focused on restoring the glory of France. Jean Monnet, meanwhile, was engaged in quiet contemplation on the long term. Having witnessed how economic cooperation could help in times of war, Monnet became convinced that it might also hold the key to a lasting peace. In a note written on August 5, 1943, he stated his conclusions with characteristic bluntness. "There will be no peace in Europe," he wrote, "if States are reconstituted on a basis of national sovereignty with all that implies in terms of prestige politics and economic protectionism." National interests, however, figured very much in his calculations. "The British, the Americans, the Russians have worlds of their own into which they can temporarily retreat. But France is bound to Europe. France cannot opt out, for her very existence hinges on a solution to the European problem"[17] (see Figure 7.2).

A UNITED STATES OF EUROPE?

After the conclusion of the war, with the Nazis defeated and the Soviet empire looming large in the east, many leaders were reflecting on the "European problem" to which Monnet referred. The idea of Western European integration, though still ill-defined, was gaining traction at the highest levels. At a speech at the University of Zurich in September 1946, Churchill proclaimed "the need to build a kind of United States of Europe." Churchill's reference to American federalist institutions was not an anomaly. Today European leaders and analysts canonize Monnet as one of the "founding fathers" and invoke the phrase "an ever closer union," mirroring the language of "a more perfect union" in the preamble to the US Constitution. There is no small irony in this, given that the drive for European unification is motivated to a significant degree by the desire to counterbalance American power. But after the war, America's role in European integration was considerable, and not just in providing a potential model for political union. The Truman administration insisted on economic cooperation among European countries as a precondition for receiving the billions of dollars in reconstruction funds promised under the Marshall Plan, announced by Secretary of State George Marshall in a solemn commencement speech at

Harvard in June 1947.[18] Weary of battle, wary of the Soviets, and nudged by the Americans, European leaders were soon debating not whether to have closer integration, but how much and what sort.

Jean Monnet provided the answer. To be sure, his was not the only voice for change. In May 1948, nearly 1,000 political leaders and intellectuals from across Europe descended on The Hague in the Netherlands to attend the Congress of Europe, which was convened for the purpose of promoting European unity. The delegates dreamed large, advancing lofty proposals for integration to end centuries of hostilities. But it was Monnet's narrower and more down-to-earth approach—emphasizing a gradual expansion of cooperation in specific areas of the economy—that provided the winning formula of national self-interest and international unity that is the foundation of Europe today.

Most observers at the time expected that England would be at the center of any new arrangement to unify Europe. England already had a track record of cooperation with France under the earlier accords designed by Monnet and his British cohorts. England also enjoyed a certain prestige during this period, as the one Western European country that successfully resisted the Nazi onslaught. But despite Churchill's pan-European rhetoric, England had—and still carries today—deep reservations about a supranational authority governing the region. For France and Germany, on the other hand, a European alliance offered a potential solution to some very thorny problems. The French, fearing future hostilities from their neighbor, embraced the wisdom of keeping your friends close and your enemies closer; they saw Franco-German cooperation as a way to prevent future German aggression. German leaders, still under Allied occupation, watched as the Soviets settled into the countries liberated from Nazi control and then tried to grab Berlin. Following his victory over the Nazis, Joseph Stalin left no doubt about his intentions when, at 6:00 A.M. on June 24, 1948, he canceled all train and car travel to West Berlin in an attempt to starve the city's population of 2.5 million into capitulation. Stalin's strategy was rendered ineffective when US General Curtis LeMay led an effort by the Western Allies to airlift 5,000 tons of food per day to Berlin's citizens. Fresh from these experiences, and under the able stewardship of Konrad Adenauer, the Germans concluded that the key to postwar prosperity lay in establishing stronger ties to the rest of the West.

As a first step, Monnet drew up plans for a European Coal and Steel Community, announced on May 9, 1950, by French Foreign Minister Robert

Schuman (himself a unique blend of French and Germanic backgrounds). The Schuman Plan would coordinate production of these vital inputs to the global economy and establish shared control over the strategically vital Ruhr region of Germany. The plan was a more politically palatable alternative to earlier French moves to annex the region entirely. It is hard to imagine, in retrospect, that such a mundane sounding agreement would eventually shift the course of European history. But its success speaks to the genius of Monnet's sober approach to the question of international cooperation. The limited scope of the agreement lent a pragmatic flavor to integration, allowing the idea to advance in a piecemeal fashion while keeping alive the hopes of those with a grander vision for European unification. On April 18, 1951, delegates gathered to sign the Treaty of Paris and the new community was formed. It boggles the mind to consider that a mere six years after the end of World War II, France and Germany—joined by Italy, Belgium, the Netherlands, and Luxembourg—decided to form a union. The new Europe was born.

GREENING EUROPE

How did all of this lead to Europe's emergence today as the global pacesetter for environmental policy? What is the thread of history that connects those grainy black-and-white images of war-torn Europe to today's efforts to promote alternative energy, sustainable agriculture, and greener manufacturing processes?

Jean Monnet was no environmentalist.[19] But for Monnet and others, the need for a regulatory arm of the European body politic was obvious from the start, and flowed logically from the goal of economic integration. To facilitate the free exchange of goods, Europe needed rules that were reasonably consistent from one country to the next. The idea that coordinated social policy should advance in lockstep with economic integration found clear expression in 1957, when the six founding countries came back together to propose a common market that would expand beyond coal and steel. The rules undergirding this new European Economic Community were spelled out in the Treaty of Rome, which sought "to approximate national economic policies, and to develop common policies, most specifically in agriculture." Like the Steel and Coal Community it replaced, the new structure featured a European Commission that would perform the daily functions we commonly associate with a national bureaucracy. It also created the other three structural pillars of modern Europe, including a Council of Ministers, where heads of state agree on policy directions for the union, as well as a European Parliament and a European Court of Justice.

Europe experienced unprecedented economic growth from 1950 to 1973, due in part to the coordination of economic policy achieved under the Treaty of Rome.[20] But during this period there was little forward movement toward the broader goal of unification. The original group of six were joined by Denmark, Ireland, and the United Kingdom in 1973; but without a clear mandate for cooperation, this mattered little. European politicians, and in particular the ultranationalist Charles de Gaulle, who served as president of France from 1959 to 1969, had no appetite for Monnet's larger European vision. (Indeed Monnet and de Gaulle developed a famous dislike for one another. Monnet thought the towering general arrogant and narrow minded; de Gaulle had little patience for pint-sized Monnet and his number-filled charts.) The pace and direction of events changed in the 1980s, however, when two historical developments—greening and unification—proceeded along parallel courses like two branches of a river picking up momentum before rejoining.

In the environment stream, the 1980s saw the rise of green parties throughout Europe. The greens would have a definitive influence on the priorities of the new union. Green parties had popped up sporadically throughout Europe since the first green parliamentarians won seats in Switzerland in 1979.[21] But in the mid-1980s, the green party concept took off. West Germany was again on center stage. Young and well educated, the German Greens were disenchanted with national politics, where traditional political parties ignored a host of public concerns, from nuclear disarmament to acid rain, that did not fit neatly into conventional distinctions of left versus right. The political scientist Herbert Kitschelt, whose early study of green parties remains a classic in the field, describes the greens as "left libertarian," in the sense that they promote social justice but are suspicious of centralized power in any form.[22] With Soviet missiles to the east, American missiles to the west, and the haunting legacy of the Nazi regime and the Holocaust behind them, it is little wonder that German activists mistrusted the grand designs of governments.[23] Under the slogan "Neither Left nor Right, but Forward," the Greens sought nothing less than to redefine politics.

To do so required finding a way to blend grassroots participatory democracy with effective electoral strategies that would win the Greens a seat at the table of government. This was no easy task. The environmental movements that arose throughout the industrialized world in the 1970s were received in different ways by different governments. As John Dryzek and his colleagues document in their book *Green States and Social Movements*, in the relatively open and porous political system of the United States, leaders of the environmental movement were quickly welcomed into government positions in the new agencies created as a result of their efforts. In Germany, in contrast, environmentalists were excluded from power. As a result of their isolation, Germany's grassroots movements, increasingly coalescing around the Green Party, developed strong critiques of the status quo that were untempered by the necessities of compromise facing those who participate shoulder to shoulder within the political establishment.[24] Many of these concerns converged on nuclear energy, which the Greens saw as a risky technology that concentrated power (both physically and politically) in the hands of large corporations and government bureaucrats.

The Greens brought their anti-establishment brand of environmental thinking with them into parliament for the first time in 1983, when German voters awarded the Greens 5.6 percent of the vote. On their first day in parliament, while politicians from the traditional parties attended church in their three-piece suits, the Greens marched to the Bundestag carrying on their shoulders pine trees that had been damaged by acid rain. Green parties were beginning to score similar electoral successes throughout Europe. Then on April 26, 1986, fifty miles north of the city of Kiev in Ukraine, a power surge started a fire at a little-known nuclear power plant called Chernobyl. The resulting explosion produced a massive cloud of radioactive material that rained down across the continent. I was a student in Paris at the time of the Chernobyl disaster, and I recall the prohibition against consuming dairy products, as health officials were concerned that radioactive material that settled on pastures had passed through dairy cows and into the milk supply. Chernobyl generated a public outcry across Europe. The disaster seemed to confirm the warnings of the green parties, who pointed to the perils of drawing energy from technological systems that require infallible government and corporate behavior. Green parties soon scored successes in France, Spain, England, and Belgium. In Sweden the Environmental Party was the first new political party to enter parliament in seventy years.

While voters throughout Europe were putting green parties into positions of rulemaking authority for the first time, a separate stream of events in the 1980s quickened the pace of European unification. The election of pro-Europe leaders like François Mitterrand in France and Helmut Kohl in Germany opened a window of opportunity for new initiatives. Jacques Delors, the dynamic leader of the European Commission, wasted no time in taking advantage of the opening. His staff of experts churned out analyses envisioning what a larger and more powerful European alliance might look like and laid the legal groundwork for expanded cooperation. Despite opposition from Britain's Margaret Thatcher, the deliberately cautious and limited forms of cooperation spearheaded by Monnet in the 1940s and 1950s served as the foundation for a larger and more elaborate structure in the 1980s, one capable of housing more members and a wider range of policy concerns. Greece joined the European group in 1981, and Portugal and Spain followed suit in 1986. That same year, leaders signed the Single European Act, which widened the scope of cooperation into new areas including environmental protection. This was due in no small measure to the influence of the new green parties, who were quick to form cross-border alliances and won seats in the European Parliament in 1984. Europe-wide environmental initiatives were also pushed by countries like Germany, Denmark, Sweden, and the Netherlands that had strong environmental constituencies.

The Single European Act tackled the greatest remaining impediment to cooperation. The most vexing problem thwarting joint decision making in Europe was not inward-looking leaders or provincial publics, economic crises, or cold war conflicts. It was a rule. This particular rule belongs to a category I call "super rules" (discussed in chapter 10) that govern how other rules are made. Under the old rules of the European Economic Community, most voting in the Council of Ministers (or European Council, as it is known today) took place by consensus. This gave any country the power to sink a proposal that otherwise enjoyed broad support. Under the Single European Act, this was replaced with a new rule—the qualified majority vote. The specific requirements of the rule have evolved as membership has grown, but as of 2014, it consisted of a double majority vote: major new policies had to garner support from at least 55 percent of the member countries (15 out of 27) representing at least 65 percent of the European population.[25]

With this procedural logjam cleared, unification moved on a faster course. When these two streams combined—the one for a grander Europe, the other for a greener Europe—the result was a force that would shift environmental practices throughout the continent and the world.

By the 1990s, European officials were working at a feverish pitch to raise environmental standards across the continent. These efforts were driven by three priorities. The first was to identify win-win solutions that would benefit the economy and ecology alike. The European Ecolabel program, launched in 1992, exemplifies this approach. The program gives an official European seal of approval for green business practices, certifying their authenticity for consumers and thereby rewarding companies that exercise environmental stewardship. The second priority was to adopt a more holistic approach to sustainability, thinking not just in terms of end-of-pipe pollution controls, but the broader systems of production that give rise to pollution. Under the "polluter pays" principle, for example, new rules governing packaging standards place the responsibility not on consumers and taxpayers to recycle more, but on producers to create less packaging in the first place. The third priority was to use environmental issues as an opportunity for enhancing European influence on the world stage, demonstrating to its members the benefits of the whole unification idea. Foreign policy coordination was especially timely with the end of the Cold War, which produced new uncertainties and saw the emergence of the United States as the sole superpower. (It was during this time that French Foreign Minister Hubert Védrine characterized America as an out-of-control "hyper-power.") At the 1992 Earth Summit in Rio de Janeiro, European leadership on climate change stood in stark contrast to the tepid response from the United States. With the vacuum in American leadership, and the whole world watching, Europe deftly cast itself as a force for positive social change.

A RELUCTANT EMBRACE

This newer, stronger Europe took most Europeans by surprise. In 1992, when the Maastricht Treaty formally launched the European Union, solidifying the common market and creating a common euro currency, it was as if the continent suddenly awoke from a slumber. Suddenly politicians and the public were up in arms about this new European Union that ruled in their name. If the advantage of Monnet's go-slow approach was that it made unification politically acceptable, it also merely postponed the day of reckoning. Paul Magnette, a Belgian political scientist and politician, summarizes the public anxieties, then and now, surrounding the new rulemaking structure.[26]

> Its strictures seem all the heavier and less legitimate as they come from remote and anonymous authorities: the Commission in Brussels,

whose members' faces are familiar to few people; conclaves of ministers locked away behind heavy Council doors; the judgment of the Luxemburg-based court, cut off from the rest of the world on the Kirchberg plateau; or the decisions of the central bankers scrutinizing monetary Europe from atop their Frankfurt towers, not to mention an array of agencies and committees whose exact functions and numbers are a mystery even to EU insiders.

These are hardly characteristics that appeal to green parties and others who feel that power should be decentralized; unsurprisingly, EU institutions soon became a favorite target of populist leaders from across the political spectrum. Reflecting the unease that many countries have about tying their fortunes to the fate of the new European experiment, a number of EU members (including the United Kingdom and Denmark, among others) refused to adopt the euro as their common currency. But the benefits of unity were equally clear. Few could argue with the logic of banding together to advance common ends, and it was obvious that the European Union was capable of delivering results—economic, diplomatic, and environmental—that would be impossible for countries to achieve on their own. The union continued to grow as Austria, Finland, and Sweden joined in 1995. The logic of unification had special appeal for the newly independent states of Central and Eastern Europe, which were just emerging from decades of communist rule. Between 2004 and 2007, another twelve countries joined the union, expanding the total to twenty-seven. At this writing, another eight countries are waiting in the queue for European accession, reviewing the requirements like hopeful actors practicing their lines before an audition.

According to David Benson and Andrew Jordan of the University of East Anglia, "Virtually all environmental policy in Europe is now made in, or in close association with, the EU."[27] This includes more than 500 directives, regulations, treaties, decisions, court rulings, and recommendations covering a wide range of environmental topics. But the inventiveness of the European Union is not a matter of creating more rules; notwithstanding the fears of its detractors, the EU is not some bureaucratic thunderstorm raining red tape down on the people of Europe. As often as not, greater coordination has the effect of streamlining dozens of national policies into a smaller and more coherent set of guidelines. A crucial part of this process—the part that has aspiring EU members rehearsing their lines—is the environmental *acquis communautaire*, a legal principle at the heart of the whole European project. The acquis states that any country wishing to join the union must change their domestic laws to adopt the entire set of EU policies. New members don't have the luxury of cherry picking only the policies they like. For the environment, the acquis includes about 100 distinct rules.[28]

One of these rules, the Directive on End-of-Life Vehicles, requires automakers to take responsibility for recycling or reusing 95 percent of the weight of the cars they produce at the end of their life cycles. Another rule holds manufacturers of electronic equipment responsible for recycling or safely disposing of the hazardous waste in their products. In 2006, the EU adopted a major new policy called REACH that changes the way synthetic chemicals are evaluated and managed.[29] These rules are not only changing how business is conducted within Europe. Katja Biedenkopf of the University of Amsterdam has been documenting how these rules diffuse around the globe. She reports that Europe's new rules on electronic hazardous waste, for example, have been adopted by the Chinese government, which is eager to keep up with the latest design standards so that China can remain competitive in the electronics industry.[30]

Many of the EU's environmental rules were first incubated at the national level. After a country proves the worth of a particular strategy for promoting sustainability, the union serves as a conduit for speeding the transfer of ideas and expertise throughout the region. Tanja Börzel of the Free University of Berlin has traced the movement of new environmental policies as they ricochet from one European capital to the next. She finds that the EU directive on urban wastewater was inspired by a model developed by the Danes. Rules governing pollution from small cars and trucks were based on ambitious standards in the Netherlands. Europe's integrated pollution prevention measures were first developed in Britain. In each case, these first movers have discovered that they can leverage large-scale change by innovating at home and then pushing for the adoption of their innovations at the European level.[31] They think vertically.

Environmental rulemaking in the European Union is by no means an unqualified success. Harmonizing national standards with European rules is an arduous process, hampered by wide variation in the structure of national industries and the diverging political preferences of member states.[32] Europe's ambitious cap-and-trade scheme for carbon has floundered, as an excess of permits has caused the carbon market to crash.[33] And EU rules designed to "green" agricultural practices have done little to benefit biodiversity.[34] But there is no doubt that amid its many success, and bountiful failures, the new Europe has changed who rules the earth.

SOMETHING NEW ON THE CHARENTE RIVER

Back in Monnet's hometown of Cognac, today a stern bronze plaque marks Jean Monnet Plaza, a tribute to the favorite son who became the father of Europe. With a population of 19,000 today, Cognac has not changed much in size since Monnet first left for London as a young man over a century ago. But as the people of Cognac go about their daily routines, with euros jangling in pockets and purses, their lives are shaped in many ways by the new rules of the game emanating from the European Union. Europe-wide food standards now govern cognac production. International trade rules, secured with help from the EU, ensure that the Cognac name cannot be appropriated by generic brands. EU water quality regulations protect the quality of the Charente River that runs through town, and which first allowed Cognac to emerge as a major center for international trade in the 17th century.

But today life for the people of Cognac is shaped by a second global trend, unfurling right alongside European unification, that carries equal significance for the people and their environment. While Europeanization is scaling power up, a worldwide movement toward decentralization is pushing decision-making authority in the opposite direction, conferring on local communities new powers that were unthinkable a generation ago. We will see that these trends are not contradictory, nor are they diminishing the power of national governments. Like some sort of fun house mirror, these new developments are reapportioning global power in ways that carry important implications for who rules the earth. Unlike the rules that gave birth to modern Europe, decentralization is by its very nature more dispersed and even difficult to track, as new laws and policies under the decentralization banner reconfigure politics in cities, provinces, and villages around the world. To demystify the process of decentralization, let's begin by returning to the Costa Rican countryside, where local communities are just beginning to come to terms with their new roles in ruling the earth.

8

Scaling Down

PIECES OF PARADISE

Dominical is a small town nestled on the Pacific coast of Costa Rica, where tropical forests spill onto the sandy shores of its world-renowned beaches. Dominical has a laid-back atmosphere of surf shops, open-air restaurants, and children in school uniforms weaving between the puddles and rocks on their way to class. But behind the scenes, something else is going on in Dominical. A clue can be found alongside the dirt road that runs through the center of town, where a billboard for Century 21 Real Estate depicts a happy couple overlooking their oceanfront property, accompanied by the English-language caption "Your Piece of Paradise!" The sign provides a glimpse of the larger forces at play in this remote corner of Central America. A frenzy of speculative real estate development is underway, led by foreigners vying for their own piece of paradise before the remaining lots are all sold by the local farmers whose families have inhabited the land for generations.

One such farmer is Juan Carlos Madrigal. I visited Juan Carlos with a group of students in 2008 during one of my annual trips to Costa Rica, to learn more about how local landowners are coping with these pressures.[1] This land has been in his family for a long time, its towering tropical forest encompassing tree plantations, bean, and cocoa crops, and sweeping views of the ocean. After a hike across the property, we cooled off in a swimming hole below a large waterfall, one of many in the area, which thundered down from the lush jungle above, the water volume swollen by seasonal rains. After toweling off we sat down and began the interview, discussing his vision for the future of this land. A humble yet dignified man with wrinkles deepened from decades of farming, Madrigal reported that a group of Americans had recently approached him with an offer to buy his property for a million dollars. Shaking his head, he said that of course he refused. I know from my many years of research in Costa Rica that *Ticos* (as Costa Ricans call themselves) are proud of their agricultural heritage and have a strong commitment to conservation. Still I was impressed, and nodded knowingly at my students. I wondered whether the commitment of people like Juan Carlos Madrigal might stem the tide of land speculation in the region and the social disruption that comes with it.

"I'm holding out for two million," he added.

It is too early to discern the long-term consequences of land transfers from Costa Rican hands to North Americans and Europeans. Some foreign buyers are making exemplary efforts to conserve the forests and are even restoring lands degraded by years of local cattle ranching. Others have leveled hundreds of acres of forest for luxury resorts catering to foreign elites. Meanwhile landowners like Madrigal are torn between the allure of easy money and the prospect of seeing their community unravel. These decisions are often shrouded in far-out fantasies of imminent riches. A forest official described this to me as the *Gringo Imaginario*, who will arrive with pockets full of cash to fulfill the dreams of every farmer.

The outcome will depend on whether locals muster the will to enforce the existing rules governing land-use planning, as well as their ability to enlist help from the national government in confronting moneyed interests. But something else is changing in Dominical. As part of a global trend, local governments throughout Costa Rica are gaining more power and resources than ever before. Historically, Costa Rica has been one of the most centralized political systems in Latin America, with tax dollars and rule-making authority flowing to the capital.[2] But today new rules are allowing local governments to collect property taxes, increasing their percentage of the national tax base to 10 percent[3]—hardly a revolution, but more than enough to make a real difference in people's lives. This allows the local municipality to hire more planners and property inspectors and—perhaps—to get a handle on growth.

The fate of Dominical echoes a larger point about the politics of particular places. William Clark, a pioneer in research on global environmental governance, points out that no one really lives on planet Earth. We live in Dominical. Or the suburbs of Chicago. Or the Cognac region of France. These are the places we identify with and can most readily affect. The rules that we build or dismantle locally determine whether we protect green spaces or cover every inch with pavement, whether we have attractive urban centers where people actually want to hang out, or sterile strip malls designed with little consideration beyond the short-term priorities of real estate developers. Local outcomes are also shaped by the rules in place at higher scales of organization, such as national policies; we will see that local champions for change are most successful when they think vertically, tapping resources and expertise at multiple levels of governance. But the undeniable *concreteness* of local concerns means that towns, cities, states, and provinces are a logical focus of political action for those who are ready to move beyond feel-good Earth Day celebrations and make a sustained push for sustainability.

DOWNSHIFT

Today local politics are becoming more relevant than ever as a result of a global trend toward decentralization. This historic shift is taking place in dozens of countries, in different ways and at different paces, but everywhere with profound consequences for who rules the earth. So what exactly does this mouthful of a word "decentralization" mean?

Decentralization can be thought of as a giant political centrifuge whirling political power outward from the center—the national government with its seat of power in the capital—to far-flung regions and cities. (To extend the centrifuge metaphor with a bit of political realism, we might imagine politicians hanging on for dear life as they spin about, trying to resist the dispersion in their decision-making prerogatives—but more on that later.) In its more timid version, decentralization keeps decisions in the hands of national officials but locates more of their offices outside the capital with the promise of bringing government closer to the people. In countries that are ruled by authoritarian governments, this version of decentralization also ensures that locals are more easily monitored and controlled.[4]

Under a more robust version of decentralization, tax revenues are reapportioned so that local governments take a larger slice of the pie. That means more staff and more capacity to deliver social services. In its strongest version, decentralization gives local communities unprecedented new responsibilities for managing forests and water and wildlife. Increasingly, it is up to local decision makers—chiefs and mayors, town councils and village elders, governors and provincial officials—to decide the purposes to which the earth's resources are put. Even more important than the shift in *where* decisions are made are the new rules governing *how* they are made. Moving in tandem with the spread of democratic reforms, decentralization is allowing ordinary citizens to participate in the business of governance as never before. Jonathan Rodden, a political scientist at Stanford University who has been tracking these developments around the globe, concludes, "The basic structure of governance is being transformed."[5] The outcome is anything but certain, but we know one thing for sure: A change is underway in who rules the earth.

POWER PLAY

To grasp the origins of the decentralization craze, you might reasonably ask how countries became so darned centralized in the first place.[6] One answer is that nation-building has always involved a pull of power from the hinterland to the capital. The architects of new countries—from Napoleon to Nelson Mandela—use combinations of persuasion, economic incentives, and coercion in their attempts to integrate diverse regions and peoples into a cohesive whole. The United States, which is far less centralized than most, is in many ways an outlier in the history of country-craft. The founders of America's democratic experiment concluded early on that the surest protection against tyranny of the sort experienced under the British crown was to break up power and scatter the pieces far and wide. The federalist system they created reserved vast rulemaking authority for the states. The prerogatives of presidents were further diluted through the system of checks and balances that spreads power across the different branches of government. Today there are a couple dozen of these federalist systems in the world, including those of Canada, Brazil, and Germany, where provinces, estados, and Länder wield real influence.[7] Everywhere else, nation-building has historically gone hand in hand with concentrating power and resources in the capital.

The centralizing trend really took off after World War II. Wartime mobilization made clear the benefits of strong central governments capable of commanding vast economic and military resources. Today we ridicule the economic failings of communism, but the Soviet Union showed impressive economic growth throughout the 1950s and 1960s, averaging almost 4 percent per year. Many national leaders, particularly in poor countries, took notice. State socialism and its centralized approach to economic planning seemed to offer a viable strategy for achieving prosperity; it also provided a rationale, for those in search of one, for concentrating power in the hands of the few.

But top-down planning wasn't only popular in regimes of the Left. No one better represented the verve for centralized governance than Jean Monnet, the founding father of the European Union, whose story we followed in the previous chapter. Monnet crafted trade agreements and other rulemaking structures to promote international cooperation and to mobilize economic resources. In the United States, even before the war, antipoverty initiatives under Franklin Delano Roosevelt's New Deal lent credence to the idea that in countries embracing capitalism, strong central governments are in the best position to deliver vital social services and to shore up the more self-destructive tendencies of unfettered markets. In the post-war period, Eisenhower built a national highway system that integrated a sprawling spaghetti mess of local roads into a cohesive transportation network spanning a continent—showing yet again that central governments can accomplish feats that smaller political units cannot.

The enthusiasm for central planning was transferred to Europe's colonies—those places that today we gather under the unwieldy title of "developing countries." Recall that by the conclusion of World War II, almost all of Asia, Africa, the Middle East, the Pacific, and the Caribbean was still occupied by colonial powers, most notably Britain and France. The colonies, it turns out, were even more centralized than the colonizers. Colonial administrators set up hierarchical economic systems that enabled them to extract resources like timber and rubber for export. This logic of economic control was reflected in the top-down political structures that colonial powers put in place to ensure compliance from local populations.

By the mid-20th century, as the moral and economic logic of colonialism came under attack, France and England made a last-ditch effort to promote the well-being of the colonies by investing in areas like education and agriculture. It was during this period that "development" emerged as a star concept on the world stage, giving rise to cadres of professional development experts (agronomists, engineers, economists, health experts, and others) and mobilizing billions of dollars in overseas aid from governments and private donors. Development aid further reinforced centralization because donors preferred to channel resources to national officials in the capital rather than haggle with countless communities.

Beginning in the 1940s, revolutionary movements for independence swept across the colonized world. Nationalist leaders ousted the centralized regimes of their colonial occupiers—and went straight to work replacing these with highly centralized independent governments. In India, Mahatma Gandhi's mass civil disobedience campaigns captured world attention and helped precipitate the end of British rule in 1947. But Gandhi's preferred model of small-scale development and self-sufficiency was quickly discarded in favor of Prime Minister Jawaharlal Nehru's vision of top-down, large-scale modernization.[8] Following India's lead, dozens of countries fought for and won independence from the 1940s through the 1960s. As colonial powers retreated, these new nations—places like Jamaica and Botswana, Bangladesh and Kenya, Morocco and Indonesia—faced enormous challenges, not the least of which was how to integrate remote regions and kick-start economic growth. Early priorities emphasized dams, ports, roads, and other large-scale infrastructure projects that were beyond the capability of local actors, and had the effect of further concentrating resources and decision-making power in national capitals. India had an important influence on the thinking of the new leaders, who emulated Nehru's state-heavy model of national socialist development.

Central planning was still wildly popular in the West in the 1960s, and development agencies from industrialized countries encouraged the centralizing trend. And let's not forget that most of these newly minted countries were not democracies. Charismatic leaders like Nasser (Egypt), Kenyatta (Kenya), and Suharto (Indonesia) instilled national pride in their countries, but did not extend their embrace of independence to include voting rights for their own people. Central governments expanded their staff and filled the world's capitals with imposing structures housing new ministries of education, planning, health, and transportation. Local institutions, meanwhile, withered on the vine. During the colonial period, local governments and tribal chiefs were often appointed or co-opted by colonial administrations; after independence, national leaders viewed them with suspicion and saw little reason to share power. The new nation-building project was centralized and large-scale, promising economic modernization and political autonomy.

In Latin America, the centralization story followed a different path leading to the same destination. Most Latin American countries achieved political independence over a century before the rest of the developing world, having given the boot to Spain and Portugal in the early 1800s. Economically, however, the region still depended heavily on American and European markets. Latin American leaders desperately needed foreign investment, and yet resented the influence of powerful multinational corporations, which controlled enormous swaths of land throughout the region and did not hesitate to meddle in domestic politics.[9] In an effort to break the bonds of economic dependence, from the 1930s through the 1960s, Latin American leaders embraced a new strategy, called Import Substitution Industrialization, that called for replacing foreign goods with domestic products. The idea was to sharply reduce imports while managing economic growth for the benefit of domestic populations. This required—you guessed it—strong government involvement in economic decision making, and it further concentrated power in the hands of populist leaders throughout the region. As is often the case, economic and political structures mirrored each other. Stirring up nationalist sentiments, leaders built elaborate political systems to mobilize poor and underrepresented groups of peasants and laborers, but did so in ways that were designed to tightly control them.[10] By the late 1960s, the world was a highly centralized place.

RESHUFFLING THE POLITICAL DECK

Today dozens of countries are starting to change course, redressing the imbalance created by centuries of centralization and giving a measure of rulemaking power to local and regional governments. What can account for this radical reversal? The first moves toward decentralization began, interestingly enough, in Europe. There the power shift was largely a response to separatist movements, led by people living in culturally and linguistically distinct regions like the Basque (in Spain and France), Sardinians (in Italy), and Flemish (in Belgium). These groups never bought into the whole national integration idea. Beginning in the 1970s, separatist movements became increasingly vocal, expressing their grievances through tactics ranging from peaceful marches to terrorist bombings. These events placed enormous pressure on national leaders, who sometimes responded with escalating violence, and at other times with political concessions. In the

United Kingdom, the new rules that resulted included the creation of the Scottish Parliament (*Pàrlamaid na h-Alba* in Scottish Gaelic) and the Welsh Assembly (*Cynulliad Cenedlaethol Cymru*). First proposed in 1979, the two legislative bodies were officially launched in 1998, with new powers to legislate on a variety of matters independent of the British Parliament, from fisheries management to the preservation of historic buildings.[11]

In Spain, Francisco Franco, victor of the Spanish Civil War and Europe's longest ruling dictator, left behind a legacy of highly centralized power. When democracy was restored to Spain in 1977, regional demands for autonomy exploded. Catalonia, Andalusia, Alisha, and the Basque Country called for new rules that would decentralize political and economic power. In the Basque Country and Navarra, all income and corporate taxes now go directly to regional governments, which then reimburse the central government for services rendered.[12] Today Spain is a collection of seventeen Autonomous Communities with wide-ranging environmental responsibilities, from the regulation of air pollution emissions to urban development. Local governments have wasted no time in testing their new powers. Andalusia is promoting forest conservation through its *Regional Strategy for Climate Change*; Barcelona requires solar water heating systems in all new buildings.[13] Similar events are unfolding in Italy and Belgium. Even France, traditionally a bastion of top-down governance, launched a major decentralization initiative in 2004, handing a greater share of power to the 54,000 local governments that dot the French countryside.[14]

Notice that the push for decentralization across Western Europe is occurring just as European countries are centralizing power within the new European Union discussed in the previous chapter. Europe is scaling up and scaling down at the same time. Notice too that national governments have not been weakened by either trend. The changes underway in who rules the earth are not a simple zero-sum game in which beefing up one level of governance impoverishes another. Like the corporation that is strengthened by sharpening its focus and dropping unprofitable investments, governments can increase their capacity to act by letting go of some responsibilities and allowing power to flow downward and upward.[15] Meeting local demands for autonomy can enhance the legitimacy of national governments, shoring up public support while letting local officials take the heat when citizens are dissatisfied with the provision of social services. Handing power upward to an international organization like the European Union, which is tasked with promoting trade among its members, can bolster economic growth, which in turn increases the tax revenues available to national policymakers. Scaling up also enhances the international bargaining power of national governments through the new alliances supporting their efforts.

TILTING THE EARTH'S POLITICAL AXIS

For a while the model of centralized control seemed capable of promoting prosperity in developing countries. Economic growth in wealthy countries during the 1950s and 1960s buoyed the export economies of the Third World. This allowed leaders of these countries to expand government programs and to maintain power by rewarding political supporters with jobs, new roads and schools, and control over lucrative industries. But the heyday of fast-growing economies steered by heavy-handed governments came to an abrupt end with the oil shocks of the 1970s and the global recessions that followed.[16] By the time another recession hit in the early 1980s, citizens ruled by cash-strapped and unaccountable governments quickly grew impatient with the top-down status quo. Dilapidated government agencies proved incapable of delivering social services, while portraits of exalted national leaders in ceremonial sashes made a mockery of the public trust. Adding to the political pressure cooker, many of these governments experienced a string of corruption scandals and blatant displays of incompetence, such as the Mexican government's botched response to the 1982 Mexico City earthquake. The crisis of legitimacy ran deepest in authoritarian regimes, where unthinkable human rights abuses were committed by national leaders in a desperate bid to hold onto power.[17]

From the mid-1980s onward, the push for local power spread rapidly across the planet. Ironically, the initiative came not from mass uprisings, but from top-down initiatives orchestrated by small groups of political elites. Why, you might ask, would political leaders willingly give up power? At first glance, this would seem to go against everything we've come to believe about politicians. But what makes research so thrilling is we get to move beyond first impressions and take a closer look at the inner workings of things. In her book *Audacious Reforms*, political scientist Merilee Grindle does just that. Based on interviews with reformers in Latin America, Grindle finds that decentralization was spearheaded by national leaders who could see that their systems were in crisis and figured out ways to benefit politically from strengthening local governments. In Argentina, President Carlos Menem agreed to share power with the regions in exchange for a constitutional reform allowing him to run for another term. Other researchers have discovered that in Colombia and Bolivia, presidents with a precarious hold on power calculated that their political parties could win more elections at regional levels, and fought for new rules that would enhance the power of the offices they were likely to occupy.[18]

Once local governments and grassroots organizations had a taste of their new powers, they demanded more.[19] Within a decade, people everywhere—in dictatorships and democracies, in regimes of the left and right, in rich and poor countries—were calling for a change in who rules the earth. In the industrialized world, decentralization laws were passed in Japan, South Korea, Australia, and New Zealand. Today at least sixty developing countries have decentralized some aspect of natural resources management.[20] In the Philippines, following the overthrow of the Marcos dictatorship in 1986, responsibility for managing national parks was handed over to local authorities.[21] Local governments in Mexico, Brazil, and Costa Rica have a greater say in the management of water resources than ever before.[22] Decisions governing the forests of Bolivia, Honduras, Guatemala, and Nicaragua are now made in large part by locals.[23]

IS LOCAL CONTROL A GOOD THING?

It is all too easy to be swooned by the populist rhetoric surrounding decentralization and to imagine that greater local control over environmental decisions is necessarily a positive development. What could be better than bringing power closer to the people? As E. F. Schumacher argued in his famous book title, "Small is beautiful"—isn't it?

To begin, let's remember that the Civil Rights era in the United States was fundamentally about promoting social justice by *centralizing* decision-making power—refusing to allow cities and states to decide whether and when to end racial discrimination, and instead letting the federal government take charge.[24] The same is true of federal requirements that women are allowed to vote, that children are required to attend school, and that city buses accommodate people with disabilities. So even before we introduce the environment into the equation, let's be clear that empowering local governments does not necessarily mean more local democracy.

As with these earlier episodes, there is a lot at stake in the debate over decentralization and the environment. Can local communities be expected to do right by the planet? Should village leaders decide how many trees to harvest from the surrounding forests? Should mayors and town councils set acceptable air pollution levels? Do communities really have the wherewithal to effectively manage local drinking water?

In the debate over local environmental governance, no idea has exerted more influence than the "tragedy of the commons" metaphor, popularized by the American ecologist Garrett Hardin in a paper published in *Science* in 1968.[25] The tragedy of the commons holds that natural resources shared by a community are doomed to be overexploited. In casual conversation, I find that even folks who claim to know little about the environment have often heard of the tragedy of the commons. If a natural scientist decides to plunk one social concept into her environmental studies course, this is the one. Given its popularity, the tragedy of the commons provides a logical starting point for us to consider whether local control is a good thing and therefore what decentralization portends for our future.

The real tragedy is that most people don't realize that Hardin was wrong. Or to put a finer point on the matter, the tragedy of the commons applies to a far narrower set of circumstances than Hardin claimed. This has been documented in hundreds of books and research articles published over the past quarter century. But once again, those of us working on the front lines of environmental social science have failed to spread the word beyond our inward-looking circles of research specialists. No wonder that an idea with such clunky wheels has managed to roll so far. To begin, let's take a closer look at what, exactly, Hardin was arguing.

"Picture a pasture open to all," wrote Hardin. "It is to be expected that each herdsman will try to keep as many cattle as possible on the commons." With growing human populations, this pressure will increase. When it does, each herdsman faces the decision of whether to add an additional animal to graze on the pasture. As he ponders this question, it occurs to the herdsman that he will reap all the benefits of doing so, in harvesting dairy and meat products; he needn't share these with any other herdsmen. On the other hand, the cost of adding another animal—increasing pressure on the land and eventually reducing its ability to produce grass—is spread over the whole group. After all, the herdsman can just move his cattle to greener pastures within the shared resource as needed. So he adds another animal. And yet another. Other herdsmen make the same decision until the grassland ecosystem, under the stress of unsustainable grazing pressures, collapses. "Each man is locked into a system that compels him to increase his herd without limit—in a world that is limited." Hardin claimed that this occurs anytime people own a resource in common, from the deep sea to the atmosphere, and this carries dire consequences. "Ruin is the destination toward which all men rush," he warned, "each pursuing his own best interest in a society that believes in the freedom of the commons. Freedom in a commons brings ruin to all."[26]

At one level, Hardin's idea is perfectly intuitive. If you have ever shared a refrigerator with a group of people, you know—by the smell alone—that when everyone is responsible for caring for a shared resource, too often no one takes responsibility, leading to its moldy demise. This idea has a long pedigree. Writing in the 4th century BC, Aristotle wrestled with the question of common ownership in his masterwork *The Politics*. "Should the citizens of the perfect state have their possessions in common or not?" he asked. Earlier in the century, Socrates had made the rather shocking suggestion (as Socrates was wont to do) that wives and children should be owned in common by male citizens to foster a stronger sense of community. Aristotle condemned the proposal for a number of reasons (among them, love would be diluted like a "small amount of sweetening dissolved in a large amount of water"). But he saved his sharpest rebuke for the idea of common property itself. "The greater the number of owners, the less the respect for the property. People are much more careful of their own possessions than of those communally owned; they exercise care over public property only in so far as they are personally affected."[27] Writing 2,304 years after Aristotle and fourteen years before Hardin, the economist H. Scott Gordon published a paper in the *Journal of Political Economy* that made the same point. "There appears," he wrote, "to be some truth in the conservative dictum that everybody's property is nobody's property. Wealth that is free for all is valued by none because he who is foolhardy enough to wait for its proper time of use will only find that it has been taken by another." Overfishing, he argued, results because the oceans are owned in common, in contrast to the private property rights that dominate modern farming.[28]

THE TRAGEDY OF THE OVERWORKED METAPHOR

So where does Hardin's tragedy of the commons fall short? The basic error in Hardin's argument is he confuses two distinct arrangements: open access and common property. With open access, there are no social rules governing human behavior—it is truly a free-for-all. Whoever deploys the biggest fishing boat or the most logging trucks reaps the greatest harvest, while the costs of unsustainable resource depletion are borne by the whole group. (In the videogame *Law of the Jungle*, which my students created to accompany this book, the player begins in just such a situation.[29]) Open access situations typically confirm Hardin's pessimistic prognosis.

But there is a second possibility that Hardin overlooked: Shared ownership can work quite well when community members put in place rules to govern their resources—rules like who can harvest, how much they're allowed to take, and what penalties users face if they break the rules. H. Scott Gordon, the fisheries economist, noted that there were examples in "primitive communities" where locals live by rules that ensure a more measured approach to mining nature's riches. In the discussion of property rights in chapter 4, we saw examples of local communities sustainably managing shared forest resources in Mexico and Colombia. These communities have carefully crafted systems of rules that preserve habitat for the benefit of rural families, migratory birds, and global climate stabilization—the sort of sustainable arrangements that make industrial resource extraction sound rather primitive by comparison. Sustainable harvests are ensured through rules that regulate access, sanction offenders, and specify clear procedures for modifying the rules when needed. But Garrett Hardin, who seemed to relish his role as the outspoken contrarian defending politically unpopular positions (including his subsequent, highly controversial work on "lifeboat ethics"), made no mention of community-level institutions for managing shared resources. He suggested that the only way to avoid the tragedy of the commons is to pursue one of two options: Carve up the commons into private property or let a central government take control.[30]

As the great evolutionary biologist Stephen Jay Gould pointed out, it is not enough to be contrarian—you also have to be right. Soon other researchers discovered the flaw in Hardin's argument. In 1975 two agricultural economists, Sigfried von Ciriacy-Wantrup of the University of California at Berkeley and Richard Bishop of the University of Wisconsin, published a stinging rebuttal. "Institutions based on the concept 'common property,'" they observed, "have played socially beneficial roles in natural resources management from economic pre-history up to the present."[31] Ciriacy-Wantrup and Bishop noted that even England's grazing commons, the subject of Hardin's famous metaphor, were managed sustainably for centuries as a result of community rules that established quotas and limited overgrazing. "The historical reduction of the commons in Great Britain is well-documented in the voluminous literature on enclosure," they wrote. "Overgrazing was not a cause." At the time of their publication, an impressive 1.5 million acres of grazing commons persisted in England and Wales.

The definitive critique of Hardin's tragedy of the commons came from Elinor Ostrom, a political economist at Indiana University who was awarded the Nobel Prize in Economics in 2009—the first woman to receive this distinction—for her work on the commons. A towering figure in the social sciences until her untimely death in 2012, no one did more than Elinor Ostrom to advance our understanding of how social rules affect the planet.

Ostrom's attack proceeded along two fronts. First, she exposed the flaw in Hardin's logic, which relied on an oversimplified portrayal of the decision facing a herdsman or any other resource user. Hardin assumed that it was impossible for resource users to cooperate, ignoring the possibility of entering into mutually binding contracts, exchanging information, and monitoring behavior. Ostrom used game theory—a type of mathematical model, popularized in the film *A Beautiful Mind*, that analyzes how groups of people make decisions—to show that noncooperation is but one of many possibilities facing the user of a natural resource. The choices that people make about whether to harvest a resource in a sustainable or reckless manner change when the surrounding community puts in place rules that reward cooperation and punish harmful behavior.

Ostrom's second, and most convincing argument on behalf of local governance for sustainability came from research documenting real-world examples.[32] For Ostrom, the first clue appeared during her Ph.D. research in the 1960s, when she was studying water management in her home town of Los Angeles. In the 1940s and 1950s, groundwater in the region was being drawn down at alarming levels. Cities, counties, regional water wholesalers, and private companies (most notably the oil industry) were engaged in a free-for-all, pulling water from a shared basin that seemed to carry all the characteristics of Hardin's doomsday scenario. "Individuals in one agency viewed individuals in other agencies as competitors," she observed, and they deliberately withheld information from one another while extracting what they could. It soon became apparent that unless water use was brought under control, saltwater from the adjacent Pacific would be pulled into the aquifer. Stakeholders in the area formed the West Basin Water Association, which set a regional cap on water use, allocated quotas to local water users, and pooled their resources to build a barrier against sea water intrusion. In other words, the association crafted rules to promote cooperation in the use of a common-pool resource.[33]

While presenting her results in a seminar at the University of Bielefeld in Germany in 1981, Ostrom was challenged by Paul Sabatier, a prominent policy scientist, to see if the rulemaking system that she had documented as a graduate student was still in place. Ostrom discovered that indeed it was. With funding from the US Geological Survey, she and her colleagues expanded their focus to include the management of shared groundwater basins throughout California. They wanted to know whether and how local communities had protected scarce water resources over the prior half century. Again they found that effective rules made all the difference in a community's ability to govern a resource sustainably.

As her work attracted attention and greater funding, Ostrom and a growing number of collaborators around the globe launched an initiative to collect all known case studies of local communities managing commonly owned forests, lakes, irrigation systems, coastlines, and other resources. By 1989, these researchers had amassed 5,000 case studies. By 2013, the number reached almost 9,000—with an average of five new studies arriving every week. Dipping a ladle in their online Digital Library of the Commons, you will find a breathtaking array of local institutions to promote sustainability. In Iceland, local farming communities called *hreppar* have managed common mountain pastures through a 1,000-year-old rulemaking system that carefully controls its use, principally for grazing sheep.[34] Outsiders wishing to use the pastures must first receive the consent of the entire community. A community member can call for an assessment of the grazing capacity of the pasture at any time (an important matter, given the highly variable environmental conditions) and sheep quotas are established on this basis; those who break the rules are subject to hefty fines. Similar examples of local environmental governance can be found throughout the world. In the Philippines, communal irrigation systems called *zanjeras* have managed water for at least 500 years. According to Ostrom, several hundred of these communities of irrigators "determine their own rules, choose their own officials, guard their own systems, and maintain their canals."[35] In the Indian Himalayas, researchers have found that local communities often do a better job of managing forests than do national agencies.[36]

In all of these cases, the rules created by and for local communities are essential. Some rules help to resolve the inevitable conflicts that arise among users of the resource, by creating local forums for dispute resolution. Other rules spell out strategies for monitoring and sanctions, to prevent the temptation to cheat and to foster overall confidence in the system. Still other rules specify who gets to participate in voting assemblies when it comes time to make or modify rules.

These researchers have discovered many other cases where the outcome is less sanguine. Many localities lack functioning rulemaking systems and have experienced resource degradation due to the tragedy of open access. So why doesn't every community create rules to overcome the problem and manage its resources for the long term? A big part of the answer is that effective governance carries a cost. It takes a lot of time and resources to establish boundaries, convene meetings, monitor behaviors, resolve conflicts, and sanction those who break the rules. Locals may decide that it's just not worth the effort, instead harvesting what they can while there's still something left. Often rulemaking systems that worked well for centuries have crumbled when confronted with market pressures from expanding international trade in commodities like timber and mining. In other cases, central governments have deliberately undermined these age-old rulemaking systems, which they see as a threat—often with disastrous consequences for local resources and the people who rely on them.

Whether resources are sustained or squandered, these local outcomes carry global consequences. Most of the world's marine fish catch, estimated at 100 million metric tons annually, is undertaken by small-scale, local fishing communities.[37] In a survey of eighteen countries, Jonathan Mabry reports that on average, 62 percent of any given country's irrigation is accomplished through local water management systems.[38] Local communities control about 800 million acres of forests worldwide.[39] With decentralization, this trend will only accelerate.

THE BUS RIDE TO NOWHERE

Research by Ostrom and others has cleared the air of a lot of misconceptions regarding the ability of local communities to care for the earth. But research on common property paints only a partial picture, leaving important questions unanswered. First, how much purchase does the study of these commonly owned resources really give us on the larger question of local environmental management? In practice, natural resources are governed by complex mixtures of common property and government and private property, each with specific rules governing who has access, how property changes hands, and the rights and responsibilities of participants. To understand what decentralization will bring—what it means in practice to turn over control of the earth's resources to locals—let's first consider where people actually live. Most of us are not Icelandic shepherds, but rather live in cities and their surrounding suburbs. In 2011, just over half of the world's population lived in urban areas. By 2050, the number is projected to reach 67 percent. People in the industrialized world tend to carry rather romantic notions of rural life in developing countries. When I was a Peace Corps volunteer in tropical Africa, I often received letters from American friends asking about life in my "village," though I lived in a remote city of 20,000 nestled in the jungle. The level of urbanization in South America (83 percent) already surpasses that of Europe and North America. Even in southern Asia, the most rural of the world's major population centers, half of all people will live in cities by 2025.[40]

So what are these places we call cities, and what will the future bring as they gain greater power and responsibility for ruling the earth? Megacities of 10 million or more have received the lion's share of media attention. Today there are twenty-nine of these behemoths across the planet—places like Tokyo, Cairo, Mexico City, and Mumbai, where a crush of humanity gathers with the hope of taking advantage of everything that urban life brings. But only one in ten urbanites lives in a megacity. Fully half of the earth's urban population live in urban settlements smaller than half a million. It's one thing for Shanghai or São Paulo or Chicago to pursue sustainability, with their throngs of urban planners and large tax base, and quite another for a small- or medium-sized city to do so.

Many urban governments are ill-prepared to take on the new environmental responsibilities expected of them under decentralization reforms. In poorer countries, cities are often chaotic agglomerations of human life, where property rights are insecure and lawlessness is only kept at bay, if at all, by community self-help organizations trying to cope with inefficient government services and unplanned urban settlements. The size of these cities can be deceiving. I came to appreciate this during an unexpected bus ride while living in Santa Cruz de la Sierra, Bolivia, a pleasant city where I rented an apartment near the city center for six months as part of my graduate research. I knew that the official population estimate of Santa Cruz de la Sierra was over a million, but there was little in the main part of town to suggest an expansive metropolis.

I came to understand the true meaning of sprawl, however, when one day my wife and I hopped on the wrong bus, and soon learned that the only way to return home was to endure a journey to the outer reaches of the city and back. Hundreds of these small buses zip around the city, their passengers bouncing along in creaky seats while drivers make the sign of the cross before flying headlong through blind intersections. Opening my eyes every so often from my slumber, I witnessed block after block, neighborhood after neighborhood whiz by as the bus coursed through concentric circles moving farther out from the city center. Cracked pavement and stray dogs, narrow storefronts and huddles of children, street vendors and women carrying bags of vegetables, and young men leaning against walls. An hour into the journey, when I was sure we must be at the city's outer limits, it kept going—a panorama of never-ending neighborhoods like some surreal looping dream. It was a Santa Cruz I never even knew existed. The experience provided a glimpse of what life must be like for those who take this bus to and from work every day.

Virtually all population growth over the coming decades will occur in cities in developing countries, often in expansive urban settings like the outer reaches of Santa Cruz de la Sierra. As decentralization pushes rulemaking authority downward, it will be up to local leaders to decide whether public transit is safe and convenient and whether investments are made in water and sewers. So if we want to understand who rules the earth, we need to understand something about how cities are governed. Yet we know relatively little about how urban governments function around the globe and whether they are willing and able to take on the new environmental responsibilities expected of them under decentralization. In practice, research is a lot like spraying pheromones in patches on a blackboard and releasing a batch of bees. Researchers cluster around questions that have managed to attract some buzz, while completely ignoring others. I would not be surprised if for every hundred publications on the local commons, there is at best one rigorous study of environmental rulemaking in cities and states.[41]

We do know that rulemaking in cities bears little resemblance to that of the small rural communities featured in studies of the commons.[42] Shared resources like forests or coastal fisheries may or may not be governed by local rules, and this goes a long way toward explaining whether they are managed sustainably or subject to the tragedy of open access that Garrett Hardin describes. In contrast, virtually every urban area has a government. They may be visionary or shortsighted, their leaders honorable or corrupt. But even struggling towns with a weak tax base and low levels of citizen participation still have some sort of decision-making structure in place.

The actual power of local governments varies quite a bit depending on where in the world you're standing. In the United States, where local governments have significant power over land-use planning, many communities have raised money to permanently protect farmland and open space for recreation and wildlife habitat. In the Northeast, between 2000 and 2004, 395 local voter referenda were approved for the purpose of open space protection (Figure 8.1).[43] In Sweden, one out of every three tax dollars goes to

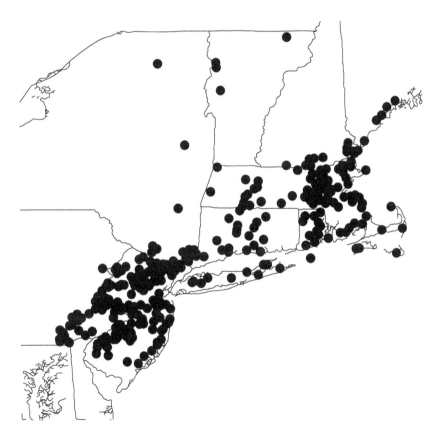

FIGURE 8.1 Voter-approved referenda to protect local open space in the northeastern United States, 2000–2004

From Erik Nelson, Michinori Uwasu, and Stephen Polasky (2006) Voting on Open Space: What Explains the Appearance and Support of Municipal-Level Open Space Conservation Referenda in the United States?, Ecological Economics 62:580–93.

local governments; in Greece, local governments must make due with a paltry 1 percent of the tax revenues collected each year. In Australia, about 8 percent of all public employees work for local governments; in Japan, the figure is 60 percent.[44] In Belgium, local leaders are appointed by the national government, an arrangement that would be unthinkable in the decentralized United States. In France, if you are so politically ambitious that you just can't decide whether to run for mayor or the national parliament, you have the option of occupying both offices at the same time.[45]

Despite their diversity, local governments share something in common: Almost everywhere, their powers are in flux as a result of decentralization. Now more than ever, we need to pay attention to the powers that local governing bodies hold in both urban and rural areas and how those powers are used.

A RACE TO THE BOTTOM?

Is your local government up to the task of caring for the earth and its people? Why do mayors and city councils behave as they do with respect to the environment? Why are some states and provinces ahead of the curve and others far behind, and what will happen as their rulemaking powers increase?

We can get a glimpse of the future by looking at the past on a stretch of the earth where local governments have long played a pivotal role in managing the environment: the United States. There basic pollution standards are set in Washington, DC, while the states decide how to implement those standards and whether to exceed them. When American pollution laws were first put into place in the 1970s, observers worried that the states would engage in a regulatory race to the bottom, lowering their pollution standards to the bare minimum required under federal law in an effort to lure industrial jobs away from states with more stringent standards. In 2001, Matthew Potoski of Iowa State University published a paper that tested the race-to-the-bottom hypothesis. Potoski found that almost a third of the American states have adopted air quality standards that are more stringent than those of the federal government; fully two-thirds of the states have stronger requirements for monitoring pollution than is required by Washington. "There is no evidence of a race to the bottom," he concludes.[46] What might explain this outcome, and why are some states greener than others? You might guess that some places simply depend more on polluting industries, but Potoski controlled for this in his statistical analysis. After running the numbers up and down, he ultimately concluded that people power makes all the difference. "Overall, the results indicate that citizen demands—particularly the strength of green groups in each state—play the most important role in states' decisions."[47] He also found that states with a large professional legislative staff are more likely to act to protect the environment. Other studies seem to confirm the importance of local citizen activism. Kent Portney and Jeffrey Berry of Tufts University find that American cities with the greatest overall commitments to sustainability are those with the highest levels of citizen participation in politics.[48]

It is fine and well to observe pro-environment behavior on the part of local governments in a wealthy industrialized country. But what about in poorer countries, which rule the greater portion of the earth and house most of its people? It turns out that people in poor countries are just as concerned about environmental quality as are people in rich countries. The notion that environmental concern is a "full stomach" phenomenon, which people embrace only after satisfying other needs, is a mere stereotype that has never been supported by the facts. Public opinion polls consistently show equal levels of environmental concern across rich and poor countries and across income levels within those countries. Thousands of citizens have mobilized for environmental protection in poor countries, where legal reformers have been trying, against great odds, to put in place rules to protect the environment.[49] There are active environmental movements and countless citizens' environmental groups throughout the developing world. But does this public concern translate into local policies to promote sustainability?

A team of researchers including Krister Andersson, Clark Gibson, and Fabrice Lehoucq approached this question through lengthy interviews and follow-up surveys with the mayors of 200 randomly selected communities throughout Guatemala and Bolivia. The team wanted to figure out which local governments are doing a better job of protecting nearby forests and why.[50] Their findings do not inspire confidence in local stewardship of tropical forests. On the whole, the mayors rank forestry low on their list of priorities, and this attitude is reflected in their staffing and budgets. Still, some mayors have bucked the trend and have made forest management a priority. So why do some mayors devote resources to local forestry while others ignore it?

The research team analyzed a range of possible explanations, from the mayor's education level to population density, political party affiliation, central government oversight, and the community's reliance on income from timber sales. They found only one explanation that can account for variation from one town to the next on all three measures of local commitment to forest management (staffing, budgets, and the mayor's stated preferences). That explanation shares a lot in common with studies of local environmental outcomes in the United States. The mayors of these tropical locales are responding to pressure from their constituents. Among mayors who perceive little pressure from their citizens to regulate forest use, only 11 percent actually make it a budgetary priority. Among those who perceive a lot of pressure from their citizens, 68 percent make it a priority. Stepping back from the numbers and putting it plainly: The research suggests that when local citizens mobilize, they can and do have a positive impact on environmental quality.

INCREASING THE SPREAD

Citizen pressure matters, but the forces that decide whether local governments help or hurt the planet are numerous and complex. The further you dig into the research, the more explanations you discover, and it soon becomes apparent that the winning recipe for greening local politics varies from one place to the next and changes over time.[51] What we can say with certainty is this: The current decentralization trend will increase the range and variability of local environmental outcomes. What was once a national decision will become dozens, and sometimes thousands, of local decisions. As this happens, the rules that local communities live by will produce results ranging from a dismal disregard for the environment to awe-inspiring efforts to promote local sustainability.

Two centuries ago, James Madison, the activist intellectual who was in many respects the brains behind the American federal system, warned of the wild swings and extremist tendencies of small governments. Madison laid out his argument in the *Federalist Papers*, a remarkable series of open letters in which he and his colleagues, writing under the pseudonym Publius, argued that democracy is not only possible in a large country but also carries distinct advantages over small-scale democracy. Madison observed that local leaders can assemble a majority to advance radical agendas that would never survive the moderating tendencies of a larger and more diverse political community. Madison was concerned about preventing the spread of an "improper or wicked project."[52] But his argument also implies something less nefarious: When power is decentralized, we can expect some local governments to adopt beneficial policies that are more ambitious than those of their national governments. This carries some clear implications for the strategies we use to advance a sustainability agenda. As rulemaking power spreads downward and outward—as the fate of the Amazon is placed in the hands of thousands of local governments rather than one president, and water management in Egypt is decided by countless local towns rather than a handful of central planners—we must search out those local governments at the forefront of sustainability, support their efforts, and showcase their accomplishments by facilitating peer-to-peer meetings with leaders of other localities who might be inspired by their deeds.[53]

CHINESE BOXES

Increasingly, the rules of the game are being decided at local levels. But local politics is never truly local. The activities of mayors and farmers and suburban commuters are embedded in larger forces that shape their behaviors. The renowned geographer Piers Blaikie described this phenomenon in his book *The Political Economy of Soil Erosion in Developing Countries*. After many years of observing local conservation practices around the globe, Blaikie concluded that "soil erosion problems can be analysed in a framework of Chinese boxes, each fitting inside the other. The individual within a household, the household itself, the village or local community, the local bureaucracy, the bureaucracy, government and [the] nature of the state, and finally international relations all represent contexts within which actions affecting soil erosion and conservation take place."[54] The same is true of a bicycle rider in Portland, who can cross the city safely thanks to widening circles of social rules, from local zoning ordinances creating bike lanes, to statewide policy that devotes funds for this purpose, and a national constitution that empowers states to make such decisions.

As the world moves toward a more localized model of decision making, the ability of communities to shape their destiny will continue to be influenced by these larger forces. This includes the decisions of national power brokers who may or may not be fans of local control. These are political struggles, and these contests are taking place alongside the fight for democracy in countries where authoritarian rule has long been the norm. In many parts of the world, democracy and decentralization have proceeded hand in hand. In Latin America, the rule-making power of local governments has grown in the fertile soil of democratic reforms. In a sharp break from the past, local officials are now directly elected by the people (Figure 8.2).[55]

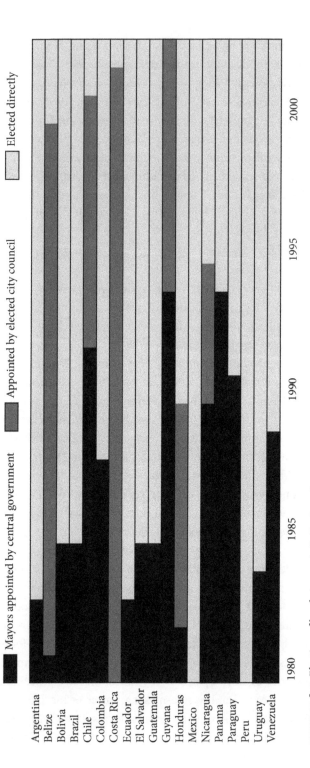

FIGURE 8.2 Election of local mayors in Latin America

See note 55 for data sources.

Without democracy, decentralization can backfire. Jesse Ribot of the University of Illinois at Urbana-Champaign has spent many years studying the democracy-decentralization link. Ribot finds that in West Africa, where there are few opportunities for citizen participation in decision making, decentralized forest management has merely reinforced the power of local elites.[56] National support for local democratic institutions is essential.

So how much power should be given to local communities? Clearly, national officials must set and enforce a regulatory floor of environmental standards rather than allow a local free-for-all. But most researchers find that in developing countries, the scales are still tilted too far in favor of centralized control, and national leaders often fail to live up to their promises of empowering local communities.[57] This hurts the cause of democracy, because elections are not enough; leaders of local communities must also have the power to allocate tax dollars and make real decisions affecting their communities. Ribot summarizes the challenge: "Transferring power without accountable representation is dangerous. Establishing accountable representation without powers is empty."[58]

THE POLITICAL ELEVATOR

Back in Dominical, Costa Rica, with its patchwork of forests, farms, and foreign-owned luxury homes, we can see why local environmental problems are never really local in any strict sense. It is not just that land speculation in Dominical is powered by economic forces originating outside the community. New rules at the national and international levels are opening up new opportunities for the farming families of Dominical. As we saw in chapter 4, today Costa Rican farmers can sell the "environmental services" provided by their land to the national government and foreign investors, receiving annual payments in exchange for growing and preserving forests that protect water and wildlife and help avert global warming.[59] Costa Rica's Payment for Ecosystem Services program cannot compete head-to-head with an overheated real estate market; if a local landowner is fixated on selling the farm for windfall profits, this won't stop him. But the income stream from the new program makes it a lot easier for those who would like to keep the land in their family and manage it sustainably for the benefit of future generations. Thousands of local landowners throughout Costa Rica are taking part.

This new program was created by rule changers working at multiple levels of governance—local, national, and international—like passengers on a political elevator. Locally, nonprofit organizations have spread the word to encourage landowners to participate. Nationally, Costa Rica's Payment for Ecosystem Services program is the capstone accomplishment of decades of efforts by conservation advocates who built up a national park system, passed environmental laws, and created an environmental agency with a penchant for innovation.[60]

The international story behind the program is perhaps the most interesting one. How is it that local landowners can now receive payments from foreign companies and governments to reduce carbon dioxide by protecting trees? As is true of markets generally, the international market for forest-based carbon storage did not arise spontaneously out of thin air. It was the result of a lot of hard work by enterprising Latin American policy reformers who understood that to improve environmental outcomes, we must change the rules shaping those outcomes. It is a story that few people know, perhaps because it departs from the tired old script in which poor countries are thought to care less about the planet than their enlightened counterparts in the industrialized world.

As part of the climate change treaty signed at the Earth Summit in 1992, new rules were created to allow industrialized countries, which bear most of the responsibility for reducing greenhouse gases, to meet some of their commitments under the treaty by investing in cleaner industries in developing countries and the former Soviet republics. But the treaty never allowed industrialized countries to invest in forest conservation abroad. This was a serious shortcoming. Forests store vast amounts of carbon through photosynthesis—pulling carbon dioxide out of the atmosphere, chopping off and releasing its two oxygen atoms (lucky for us breathing folk), and incorporating the carbon atoms into the loops and chains we know as carbohydrates. Forests absorb an astounding one-third of all fossil fuel emissions each year; the destruction of forests today, primarily in the tropics, releases more carbon dioxide into the atmosphere than is produced by the entire transportation sector.[61]

In the late 1990s, a group of Latin American policymakers decided it was time to change the rules underpinning the international climate change treaty so that it promotes forest conservation. They accomplished this through an obscure and short-lived diplomatic alliance called the Latin American Initiative Group, or GRILA. The story of GRILA was shared with me by Christiana Figueres, who served on Costa Rica's climate change negotiating team for fifteen years. In 2010, Figueres was appointed executive secretary of the climate change treaty, making her the world's top climate policy official. But Figueres's role in changing the rules governing our climate began much earlier. Throughout the 1990s, she worked with Latin American countries to lay the legal groundwork for international trade in carbon credits. Using her political connections throughout the hemisphere (Figueres hails from Costa Rica's most famous political family, which includes two former presidents), she traveled from one country to the next, sharing experiences, convening workshops, and spreading the word that there was money to be made in forest conservation—but only for countries that would build the institutional architecture to take advantage of the new market.

By the late 1990s, a number of Latin American countries were eager to position themselves as leaders in the global carbon market, which offered a way to save the region's fast-disappearing forests while providing income for rural landowners. Creating a market for forest conservation required riding the political elevator to the international level, where these reformers hoped to modify the rules of the climate change treaty to include forests. The Latin Americans were blocked, however, by leaders of the Group of 77, the official diplomatic coalition that negotiates on behalf of 131 developing countries during the intense wrangling that surrounds international treaty-making. The G-77 speaks with one (and only one) voice in international forums and is dominated by a handful of powerful countries including Brazil, India, and China. These countries feared that forest conservation projects would divert foreign investment away from their own projects to reduce industrial carbon emissions.

Poorer countries with small economies are often marginalized in international negotiations, and deliberations within the G-77 are no exception. Diplomats tell me that there is little tolerance for dissent from smaller countries when the G-77 is formulating its official negotiating positions. To counter this, every Latin American country except for Brazil and Peru joined together to form GRILA. Figueres compared GRILA to "a bunch of little people standing on each others' shoulders" to take on the heavyweights of international diplomacy. Pooling their resources, they hired a Washington, DC, law firm to help them write the legal briefs needed to influence international treaties. Delegates from Colombia and Guatemala tell me that the Latin Americans would split up to represent the region's interests at the numerous meetings taking place during complex climate change negotiations. They were joined by the United States, Canada, Australia, and Japan in pushing for forest conservation.

When the foreign minister of Argentina caught wind of this renegade environmental alliance and its daring departure from conventional foreign relations protocol, he sent faxes to his counterparts throughout the hemisphere encouraging them to shut it down. GRILA was soon disbanded without ceremony. But it set in motion a process that put new rules in place to enable an unprecedented global effort to address climate change.[62] Although Costa Rica was the first to experiment with payments for forest carbon storage, the world is now embracing the approach as a centerpiece of international efforts to combat global warming through a United Nations program called Reducing Emissions from Deforestation and Forest Degradation (REDD+).[63]

Whether scaling up or scaling down, changes are underway in the architecture of rules shaping our planet and our lives. Navigating these changes requires thinking vertically—moving beyond folksy prescriptions to "think globally and act locally," and instead thinking strategically about the reforms needed to promote sustainability at multiple levels of governance. In chapter 3, we considered the question of whether significant social change is even possible. Hopefully at this point in the book, the answer will strike you as obvious. Change is not only possible, it is pervasive. The only question is whose interests are served by these changes and whether you and I take part in the process. With this in mind, in the next chapter we will take a closer look at how social change works.

9

Keep the Change

On a fall morning in 1980, Pitzer College freshman George Somogyi walked out of his dormitory, looked up, and froze in his tracks. In front of him was something incredible. An enormous mountain, over 10,000 feet tall, stretched up to the sky in the near distance. What made this sight so bizarre is that the mountain wasn't there before. Somogyi had been at college for three months and had never laid his eyes on Mount Baldy, a five-million-year-old formation that stands just a few miles from this campus on the eastern edge of Los Angeles County, because it was shrouded in smog so thick that it obscured the view for months at a time.

Air pollution is a problem well known to the people of Los Angeles. In the 1970s their city became an icon of urban air pollution, as photos of brown haze choking downtown LA circulated worldwide. The air was so hazardous that people were hospitalized by the thousands. Yet today the air around Los Angeles, while far from perfect, is markedly improved. The amount of smog has been sliced in half since the 1970s, even as the population has doubled in size. More impressive still, the amount of particulate pollution—the small dust particles that lodge deep in the lungs and are especially harmful to human health—has been reduced to one-fifth the levels experienced in 1955.

How did a change of this magnitude come about? This physical transformation was precipitated by a political transformation, as the people of Los Angeles joined together and fought for new rules to clean up the air (Figure 9.1). Beginning in the 1940s, citizens demanded that city officials look into the causes of the problem, which were not obvious at the outset. Their efforts led to the creation of the Los Angeles Bureau of Smoke Control in 1945. Soon the movement spread throughout California, where in 1947 state legislators passed the Air Pollution Control Act—a full quarter century before national policymakers adopted similar legislation. By the time the 1990s rolled around, decades of research and increasingly stringent regulations had produced technological changes in automobile designs and industrial practices that helped to clear the air breathed by Angelinos.

As the nation awoke to the broader dimensions of the environmental crisis in the 1960s, the environmental movement grew in size, emboldened by protest movements of the period and expanding the range of concerns to include the ecological foundations of human well-being.[1] Following the lead of students on college campuses across the country, on April 22, 1970, an estimated 20 million Americans turned out for the first Earth Day, demanding nothing less than an alternative approach to civilization based on respect for the planet. This movement catalyzed substantial changes in the rules we live by (Figure 9.2).[2] The flurry of new legislation was not the work of some narrow constituency, but resulted from efforts by leaders in both the Democratic and Republican parties who put practical problem solving ahead of ideology.

The environmental movement that began in the United States and a handful of European countries eventually spread across the planet. In the early 1970s, only a few countries had national agencies devoted to controlling pollution; today, virtually every country does. In 1900, you could count the world's national parks on one hand. Today there are about 200,000 protected areas, covering approximately 15 percent of the earth's land surface.[3] "Environmental management has now become an essential component of state activity," observes James Meadowcroft, a political scientist at Carleton University. "It is publicly recognized as a fundamental part of what a civilized state should do."[4]

Social change is not about creating more rules than existed before. For every new rule put in place to protect the environment, others were dismantled. America's new pesticide laws granted new regulatory powers to the Environmental Protection Agency, but also took that power away from the US Department of Agriculture, whose rules ignored health concerns. Every new national park is accompanied by rules governing the way people treat the land, but these dislodge older rules, such as homesteading laws that required

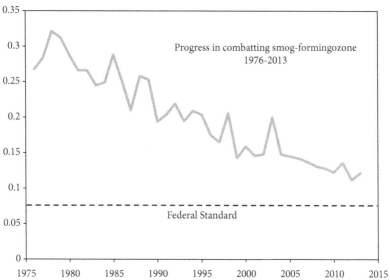

Units show the maximum concentration detected over 8 hours (parts per million)

FIGURE 9.1 Los Angeles citizens fight for—and win—cleaner air
Top: LA Times *Collection, UCLA Library.*
Bottom: Data courtesy of the South Coast Air Quality Management District.

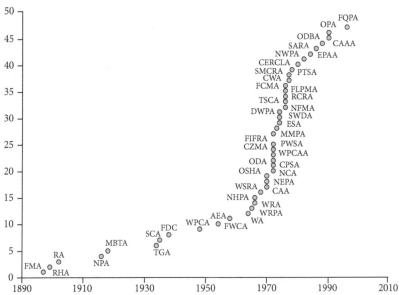

FIGURE 9.2 People power and policy change

Top: New Yorkers fill Broadway, closed to traffic by Mayor John Lindsay in honor of the first Earth Day in 1970. Bettmann/Corbis/AP Images.
Bottom: Growth in US environmental laws. Figure modified from Choucri, 1993. Acronyms refer to specific laws. See note 2 for details.

farmers to cut down forests to prove they were making use of the property. The same is true of the rules imposed on industries by government regulators. It takes a lot of rules to get thousands of people to work together in a coordinated fashion to design, assemble, test, and market an automobile. Technical specifications and standard practices, patents and labor contracts, production quotas and supplier agreements are all part of the mix. When clean air legislation came along in the 1970s, requiring the use of catalytic convertors, automobile manufacturers replaced one set of rules with another.

It is essential that we recognize and celebrate these past achievements, even as we remain cognizant of how very much remains to be done. I would argue that anyone who cares about the environment but does not believe that large-scale change is possible is wearing blinders as crippling in their effects as those of people who deny the existence of environmental problems in the first place. But what does it actually take to bring about change? Indeed, what does change consist of? Is it a matter of changing minds, changing laws, switching personal habits, or something else? Fortunately, the inner workings of social change have been studied intensively by generations of researchers from across the social sciences. Although the answers are not simple—when you have billions of intelligent people interacting strategically with one another and their physical surroundings, the answer is never simple—nevertheless some clear lessons emerge. Before skimming the cream from that research tradition, let's begin by dispensing with some popular misconceptions about the engines that propel improvements in social and environmental conditions.

FOUR IDEAS THAT WILL NOT CHANGE THE WORLD

A couple of years ago, I was hurrying through the Denver Airport on my way to a research convention, when I spotted an advertisement by the University of Denver that read "Politicians won't break our addiction to oil. Engineers will." Though I had to admire the can-do spirit behind the message, it reflects a widespread myth about the nature of scientific and technological progress. Major breakthroughs in areas like alternative energy, wastewater treatment, crop production, and disease prevention do not just happen by themselves. The inventor does not toil in isolation in a laboratory, shout "Eureka!," and emerge with a new creation for society's benefit. Rather, inventors respond to social signals. Those signals come from investors, customers, professional peers, funding agencies, social movements—and yes, politicians.

Langdon Winner, a science historian at Rensselaer Polytechnic Institute and pioneer in the field of science, technology, and society, coined the term "autonomous technology" to describe the misperception that scientific progress results from a mysterious and self-directed process of discovery, separated from politics and social priorities. In practice, science does not proceed along a straight path from the unknown to the known.[5] Scientific research is more of an infinitely branching tree, with some branches well developed and strong, and others mere stubs, their growth truncated for lack of nourishment in the form of dedicated researchers and the time and equipment they require. Science is costly, and so there are far more questions than there are resources to pursue them. Choices must be made. Molecular biologists investigating diabetes are not working on malaria vaccines. Tax dollars spent on developing the latest generation of military drones are not devoted to wind power. We like to talk about how technology changes our society, and this is obviously true. But political change is a major force behind technological change. We saw this in chapter 5 with the development of scrubbing technology to reduce air pollution from factory smokestacks. The technology languished for half a century until new air quality rules prompted a burst of innovation. Without the kind of political mobilization that points creative minds in new directions, many of the inventions needed to help the environment will simply never come to be. Political innovation is often a prerequisite for technological innovation.

A second idea that will not save the world, closely related to the first, is the assumption that environmental conditions automatically improve as societies grow richer. A leading proponent of this brand of thinking is John Tierney, a columnist at *The New York Times*. In his article "Use Energy, Get Rich and Save the Planet," Tierney argues "The richer everyone gets, the greener the planet will be in the long run."[6] His argument rests on a misunderstanding of a phenomenon known in the research community as the Environmental Kuznets Curve.[7] In 1991 two economists from Princeton University, Gene Grossman and Alan Krueger, wrote a paper in which they compared environmental outcomes across countries of different income levels. They found that for some pollutants, such as sulfur dioxide, economic growth seemed to initially harm and then later help the environment. To gain an intuition for this dynamic, think of a rural society (Zimbabwe), a rapidly growing country with dangerous industrial practices

(China), and a wealthy country deploying state-of-the-art technologies (Switzerland). Later research revealed that the Environmental Kuznets Curve only applies to a few pollutants—and even there, the relationship is murky at best. Some things, like urban wastewater treatment, get better as economies grow, with no particular dip; others, such as carbon dioxide emissions, grow steadily worse; and still others, like deforestation and biodiversity loss, have no straightforward connection to the trajectory of a country's economy. Some low-income countries are environmental leaders (think Costa Rica and biodiversity conservation), while some wealthy countries (the United States and climate change) are doing poorly despite the considerable technological resources at their disposal. Even in those rare cases where the data assume the shape of an Environmental Kuznets Curve, the relationship is merely a statistical curiosity. It tells us nothing about what actually causes environmental outcomes to improve or worsen. So many things change as countries grow richer—increases in education, scientific investment, and political participation, shifts in agricultural production, accelerating urbanization and automobile use—that we are more likely to help the world by directly targeting those factors that appear to make things better or worse, rather than waiting for economic growth to run its course and hoping for a brighter tomorrow.

A third idea that will not change the world is the notion that solving environmental problems merely requires that we unleash human inventiveness through the operation of markets unfettered by environmental regulation. This model of change has attracted many devotees over the years, including leading critics of the environmental movement and its legislative achievements, such as Bjørn Lomborg. It is curious, given their faith in the workings of the economy, that proponents of this model of change pay so little attention to the published research in economics. As we saw in chapter 5, a generation of research by leading economists has established beyond any doubt that markets, while outstanding at delivering some types of goods and services, are woefully inadequate at providing others—including, unfortunately, environmental quality and sustainability. The economics literature also shows quite clearly that the private sector chronically underinvests in research and development of all sorts.[8] Undaunted, writers like Lomborg and others at places like the Hoover Institution and the Cato Institute, who don impressive-sounding titles like "Senior Fellow" but never show the independence of mind to veer from the ideological line of their host institutions, continue to churn out papers claiming that markets will come to the rescue of the planet, if only we would abolish environmental regulations.

Adherents of this school of thought frequently point to the impressive growth in global agricultural output over the past half century, which defied the more alarmist predictions of some leading environmental spokespeople in the 1960s, as evidence that environmental scarcity is merely a short-term phenomenon that can be alleviated by private enterprise. In his best-selling book *The Population Bomb*, biologist Paul Erlich warned that the combination of unchecked population growth and finite natural resources constituted a ticking time bomb, one that would soon produce a hellish world of widespread famine and population collapse. Instead, what occurred was a series of scientific breakthroughs known as the green revolution, which facilitated unprecedented gains in crop production through plant varieties that produce more food on a given plot of land.

Erlich was clearly mistaken in his predictions. But did market incentives avert a global food crisis? Let us leave aside the question of whether world hunger is the result of insufficient food production, rather than factors like unequal land distribution, inadequate access to credit, war and conflict, and the political marginalization of rural communities. Let us also leave aside the environmental impacts of the green revolution, which significantly expanded the use of fertilizers and pesticides. That is admittedly quite a lot to put to the side. But it turns out that even if we accept the free marketeers' version of events—that the near-disappearance of famine from the latter part of the 20th century onward resulted from increases in agricultural productivity spawned by the green revolution—it turns out that free enterprise cannot take the credit. A closer look at the history of the green revolution reveals a very different picture.

I learned a great deal about that history during the summer of 1993, when I was part of a research team at Harvard tasked by the US Congress to help ensure that climate change science remains relevant to the needs of decision makers. My job was to write a case study on agricultural innovation as a model for making science relevant to developing countries. As part of this research, I spent long hours poring through archival materials in the library of the World Bank in Washington, DC, reviewing the publications and meeting notes from the organizations that launched the green revolution—most notably through something called the Consultative Group on International Agricultural Research, which was created in 1971 with the goal of eradicating world hunger. The movers and shakers behind that effort included national government agencies, international organizations like the United Nations and the World Bank, and private nonprofit foundations such as the Ford and Rockefeller Foundations. Individual ingenuity also played a role, most visibly in the work of Norman Borlaug, the renowned agricultural scientist who traveled the world collecting and cross-breeding plant varieties that had been developed by generations of local farmers, an effort that earned him the Nobel Peace Prize in 1970. But those who point to the green revolution as proof that free enterprise alone will forestall environmental disaster are missing the point and misrepresenting the history. These innovators were not responding to market signals. They did not consult price data and quarterly profit sheets when deciding how to proceed. Motivated by a desire to end world hunger, they used combinations of governmental, nonprofit, and (to a much less extent) private sector initiatives to create a global research infrastructure that persists to this day, producing cutting-edge research on topics like tropical forest conservation and agricultural policy.

RECYCLING ALONE

Before turning to the question of how social change actually works, let's consider one more idea that will not change the world. This is the notion that we can save the planet if we all do our part through lifestyle changes and greener consumer choices. As I have argued throughout this book, these isolated individual actions are fine and well, but are simply inadequate given the size of the challenge. If today we breathe more easily, or can jump into a cool river without breaking out in a skin rash, or have more natural spaces in which our children can run around and explore, it is because previous generations took seriously the political dimensions of sustainability. They were not content to quietly switch laundry soaps or change the composition of their garden. They joined together and changed the rules.

In his book *Bowling Alone*, Harvard political scientist Robert Putnam reports that the tendency to isolate ourselves socially, avoiding participation in group activities, is part of a larger trend in the United States. Putnam analyzed a quarter-century of data from the General Social Survey, which includes half a million interviews with American citizens, on questions ranging from personal preferences to political behaviors. Putnam found a disturbing trend of weakening social bonds and reduced participation in public life.⁹ Between 1973 and 1994, attendance at public meetings in the United States declined by nearly half. In the early 1960s, over half of all American citizens said they trusted other people; by 2010, the figure was less than a third.¹⁰ There are some encouraging signs, too. Young people are more interested in politics than ever, discussing political topics with even greater frequency than did their counterparts in the 1960s. Putnam and other authors conclude that it is too early to tell whether Internet-based social networking can help to reverse the decline in civic participation or whether it will simply reinforce social isolation.¹¹

The retreat from public life is nowhere more apparent than in the way most people express their environmental concern today, like green hermits working in isolation—in essence, recycling alone. This is perplexing to social scientists who study what it takes to bring about meaningful change. Michael Maniates of Allegheny College expresses this frustration in his marvelous essay "Ride a Bike, Plant a Tree, Save the World?," which has become a clarion call for researchers in this field. "Although public support for things environmental has never been greater, it is so because the public increasingly understands environmentalism as an individual, rational, cleanly apolitical process that can deliver a future that works without raising voices or mobilizing constituencies."¹²

Sociologist Andrew Szasz of the University of California at Santa Cruz explores the consequences of environmentalism-without-politics in his book *Shopping Our Way to Safety*. Szasz observes a pattern of behavior he calls the "inverse quarantine." People purchase organic produce with the hope of isolating themselves and their families from the consequences of larger environmental threats. "Their sense of being at risk diminishes. The feeling, correct or not, that they have done something effective to protect themselves reduces the urgency to do something more." The result is a form of political amnesia, forgetting that major improvements in environmental quality have come about through active engagement in politics and civic life. Today a concerned citizen reaches for the eco-friendly shampoo and is done with it. "What is the likely consequence when a significant fraction of the public imagines that they have successfully bought their way out of a collective problem?"¹³ They—we—ignore the root cause of the problem, make few demands on politicians and others who make the rules, and environmental conditions deteriorate as a result.

These researchers are not suggesting that consumer choices are unimportant or that we should stop recycling our bottles. It's that recycling is not enough. So, you might ask, what is enough? What will it really take? To answer these questions, let's begin by taking a closer look at the surprisingly close relationship between social change and social stability.

THE RUTS WE CHOOSE

We are creatures of habit, not just as individuals but as civilizations. The way we conduct business, our choice of transportation, the energy sources we rely on to light our streets—these are not the result of continually updated calculations about how best to achieve our goals. We follow the rules, our wagon wheels bouncing along in well-worn ruts carved deep by those who passed before. There is a certain sameness over time in social behavior, and this continuity is achieved in large part through institutions—the rules that pattern how we interact with one other and with the natural systems that sustain us. "History matters," writes economic historian Douglass North, "not just because we can learn from the past, but because the present and the future are connected to the past by the continuity of a society's institutions."[14]

Often we get stuck in ruts. Consider, for example, the astronomical amount of money spent on the military in the United States. Americans spend as much money on defense—about $700 billion per year—as the next fifteen top-spending countries combined. More than one out of every three dollars spent on defense worldwide comes from the American taxpayer.[15] This seems like a pretty irrational course of action for a people with a centuries-old revulsion against taxes. We collect a third less tax revenue than our European allies, as a percentage of our economy, and yet we spend double the percentage on the military. As a result there is little left for things like high-speed trains and higher education. Here's a sobering calculation: If Americans were to reduce defense spending by 20 percent, that would free up enough tax dollars to provide free college tuition for every full-time college student in the United States.[16]

To be sure, national security is a serious matter. But why do other wealthy countries spend less? Surely Italians do not care less about their families' safety, nor the British about freedom. But European leaders—and those of Australia, Japan, South Korea, Mexico, and dozens of other countries—know that the United States will come to their defense if need be. In taking on the role of global security guard, America inevitably becomes the target of resentment by those living in the shadow of its power. So why not share the globo-cop role with our allies among the world's wealthy democracies, asking them to bear their share of the burden—especially given that many of them express deep reservations about US power and domination in world affairs? The current course of action does not seem to be in America's own best interests.

Any attempt to change this situation will require that we dig out of a rut several decades in the making. The rut itself was shoveled deep by a number of forces. The first of these is a vast economic constituency that has come to depend on a steady stream of tax dollars devoted to war. During the military expansion of the past half century, the Department of Defense deliberately dispersed factories, shipyards, military bases, and weapons research laboratories throughout the fifty states to shore up political support. Rebecca Thorpe, a political scientist at the University of Washington, conducted a detailed analysis of the geographic spread of federal defense contracts to figure out who gets what and why. She concludes that "defense industries spread out their operations across multiple [congressional] districts to stimulate greater political demand for weapons systems."[17] This particular rut has also been dug deep with help from an idea—the idea that the United States must always be Number 1, without peers in our ability to project power abroad. Americans have come to believe that if we were to revamp NATO and really share responsibility for global security with the likes of France, Canada, and Australia—curtailing our monopoly over decision making in the process—this would represent an existential threat to our way of life.

There is yet another reason why we are stuck in this rut, and it has more to do with entanglements outside the United States. For better or worse, the entire world order is constructed on the assumption that American soldiers and taxpayers will continue to pay the price. If the United States were to change course, then laws and budgets and economies around the globe would need to follow suit. This does not make such a change impossible. But it makes any change very slow, and fraught with uncertainty—even difficult to imagine.

Yet ruts are not all bad. Consider America's national parks. If someone had the idea to sell off places like Yosemite and Yellowstone and turn them into shopping malls, there would be a huge public outcry. As is the case with defense spending, economic and political constituencies would play a central role in preventing any such reversal; hundreds of communities and businesses have come to rely on billions of dollars in tourist revenue for their livelihoods, and would rise to defend the parks. Just like the military establishment, the national parks also owe their existence to an idea—"America's best idea," in the words of documentary filmmaker Ken Burns. The parks symbolize a deeply ingrained cultural norm regarding the importance of nature conservation. This idea was forged through 150 years of writing and advocacy by the likes of John Muir, Henry David Thoreau, Rachel Carson, and David Brower. As the historian Roderick Nash documents in his book *Wilderness and the American Mind*, early in the country's history, the vast North American wilderness was a symbol of freedom and beauty that the new country could hold up as equal in grandeur to the cathedrals and cultural accomplishments of Europe. By the mid-19th century, wilderness had become "a cultural and moral resource and a basis for national self-esteem."[18] The protection of wild places became part of what it means to be American.

Social change, it turns out, entails two very different things. It requires both moving away from the old set of rules, and putting in place new rules that endure. The process is more nuanced than "out with the old, in with the new." It requires longevity, so that the new can grow to a ripe old age. Advocates of gay rights are not arguing that marriage should be available to all for merely a few months or years. Similarly the push for sustainability needs to be sustained. Change involves both switching and sticking—picking up one set of tracks, but also laying down new ones. The challenge is to put into motion self-reinforcing trends and to create new assumptions of normality. Permanence is part of the machinery of change.

The need for longevity has important implications for any effort to change the world for the better. If you shake a group of environmental studies students awake in the middle of the night, they'll shout "Awareness! We have to raise awareness!" But education and consciousness-raising efforts, while essential, are only part of the equation. The relationship between ideas (the usual focus of awareness campaigns) and rules is an important one. A social rule is an idea with an anchor attached. You don't want the anchor to be so light that it can be uprooted with the slightest shift in the tide of public opinion and social affairs. Nor do you want the anchor to be so massive that it can never be hoisted up again, mooring you to whatever spot your predecessors thought fortuitous. As in life, we must choose our commitments carefully. With this in mind, let's take a closer look at the relationship between changing ideas and changes in the policies and practices that project these new ideas into the future.

THE COLLECTIVE "SHOULD"

While conducting research in China in the late 1990s, Lily Tsai, a political scientist at MIT, noticed a striking pattern: Some villages had access to running water, roads, and decent primary school facilities, while others did not. Officials from the governing communist regime are responsible for providing these services. But she found that the disparity could not be explained by which villages had high-level political connections. Nor was it the case that villages with better public services were especially rebellious, which might inspire official efforts to placate them. This led Tsai to ask a very interesting question: Why would an authoritarian regime, which lacks mechanisms for democratic accountability, provide more than the minimal amount of public services required to maintain social stability?[19]

Tsai spent the next four years doggedly pursuing this question. She traveled throughout seven Chinese provinces, and then undertook a systematic study of 316 villages in the southern province of Fujian. Tsai discovered that the difference among villages can be explained by a form of peer pressure. In those locations where government officials were deeply embedded in local social networks, such as temple organizations and family lineage groups, they made sure that the national government delivered the promised goods and services. These local officials were immersed in repeated face-to-face interactions with villagers who were their social peers. With their social reputations at stake, these government officials valued public approval and were more susceptible to the sting of social isolation. They faced more pressure to do the right thing.

Shared ideas about right and wrong can influence our actions with as much force as a police baton. These include ideas about what is valuable and appropriate and true, and whether a physical condition like a trash-filled stream poses a minor inconvenience, or a major problem that calls for swift government action. When these shared ideas change, they can cause profound shifts in human behavior. Most days we take these social norms for granted. But we need only stray beyond the comfortable boundaries of our home cultures to see our stock of assumptions on stark display. I was reminded of this some years ago, when I spent a pleasant day at the beach with a French friend and his family near the city of Bayonne, in the southwest of France. As we settled in with our blankets and picnic basket, and struck up a conversation, my friend's seventeen-year-old sister paused midsentence and nonchalantly whipped off her bikini top. Her mother followed suit, so to speak. I knew, of course, that this is the norm at many European beaches and did my best, after a meditative blink, to wear the cool expression of someone noticing absolutely nothing out of the ordinary. A few minutes later, however, a nudist couple sauntered carefree along the shoreline. The reaction of mother and daughter was one of total shock. "Why do they do it?" mused the daughter. The mother averted her gaze. "It's disgusting," she muttered, as she leaned half-naked over the basket and unwrapped the cheese.

WATCHING SOCIETY CHANGE ITS MIND

Perhaps more remarkable than cross-cultural differences in norms is how these ideas shift over time. Not so very long ago, agricultural chemicals were viewed in the industrialized world as a symbol of progress, the triumph of scientific ingenuity over untamed nature. This was helped along by aggressive public relations campaigns sponsored by chemical companies, one of many combatants in a contest to influence our ideas about what is right, true, and normal. (One of my personal favorites is a 1947 advertisement by Penn Salt Company, published in *Time* magazine, that features a housewife surrounded by joyous farm animals singing "DDT is good for me-e-e!") Today pesticides carry a very different connotation in much of the world. This change was precipitated by cultural combatants like Rachel Carson, whose 1962 bestseller *Silent Spring* raised public awareness of the dangers of pesticides, and by the United Farm Workers, who led a boycott of California grapes to protest worker exposure to agricultural poisons.

Changes in social values occur at different speeds, as the result of two very different processes. The first source of change is generational turnover. This phenomenon has been explored by Ronald Inglehart of the University of Michigan, who runs the world's largest ongoing survey of public opinion, the World Values Survey. This ambitious effort enlists teams of researchers in over fifty countries who painstakingly document how social priorities shift across cultures and over time. With the benefit of this massive dataset, Inglehart has uncovered a consistent pattern that helps explain how changes in cultural norms work in industrialized societies.

226　WHO RULES THE EARTH?

Every generation adopts certain worldviews during their young adult years. These mental frameworks are based partly on what we learn from our elders, but are also shaped by events like wars or the experience of relative peace and prosperity. Once formed, these mental attitudes are surprisingly resistant to change throughout our lifetimes. For example, generations born from the 1950s onward show a greater concern for quality of life issues, including environmental protection, than do those born in the 1920s to 1940s.[20] When public opinion polls show growing support for environmental protection over time, this is not necessarily because anyone has changed their mind. Earlier generations and their stubborn views are dying off, replaced by a newer, greener set of stubborn attitudes.

HIGH-SPEED CHANGE

Generational turnover produces a very gradual drift in social priorities. But change often happens much more quickly than that, as a result of a second process in which individuals actually change their outlook. We see this in public attitudes toward pesticides, which took a 180-degree turn in the United States. In the summer of 1965, rural sociologists Robert Bealer and Fern Willits of Pennsylvania State University surveyed 1,075 residents of their state to learn about public attitudes toward pesticides. In 1984, a team of researchers led by Carolyn Sachs decided to repeat the same survey (again in Pennsylvania, but using different respondents) to see whether attitudes had changed.[21] When asked "How much have you personally been concerned or worried about the possible dangers of farmers using pesticides?," only 32 percent of those surveyed in 1965 indicated they had either "some" or "a great deal" of concern. In 1984, fully 76 percent of respondents expressed concern over pesticides. The change in public sentiment was even more significant with respect to farmworker safety—the focus of the United Farm Workers' grape boycott. When asked "How much danger from pesticides do you feel there is for the farmer who handles and applies them?," only 15 percent of respondents in 1965 perceived some or a great deal of danger; in 1984, the figure was 79 percent.

In his book *Culture Moves*, political scientist Thomas Rochon explores these rapid reorderings of right and wrong to understand what produces shifts so sudden that they cannot be explained by generational turnover. As an example, Rochon documents how the concept of sexual harassment became familiar to Americans practically overnight, precipitating a shift in shared assumptions about appropriate workplace behavior. He discovered that rapid changes in social norms can occur when small groups of intellectuals operating outside the mainstream (in this case feminist legal scholars like Catherine McKinnon) produce new ideas that are adopted by broader movements for change. The women's movement was up in arms over the stunning congressional testimony of Anita Hill, a former employee of US Supreme Court nominee Clarence Thomas, who accused him of sexually offensive workplace behavior. In their efforts to raise public awareness, leading organizations within the women's movement framed the issue using McKinnon's new concept of sexual harassment, which recast this type of behavior as a fundamental violation of equality rights. Soon the idea became part and parcel of American public discourse.

I had the good fortune of reading an early draft of Rochon's book when I was in graduate school. His work inspired me to see if I could trace a similar sort of change with respect to public support for the environment in Latin America. I knew from my field research that you can quickly sense the presence of environmental concern in Central and South America, particularly among young people, if you just spend a little time on a college campus there. But where did these pro-environment attitudes come from, and when? Can we even measure such a thing?

There are no surveys of environmental attitudes in Latin America that go back far enough to detect when the shift occurred. But changes in social norms leave behind traces in the words that we use to communicate through popular media, advertisements, letters, and other places where these words are captured in print. Just as paleontologists follow the evolution of life by examining the clues deposited in bedrock, we can use the word-fossils of newspapers and historical archives to measure changes in social priorities. I assembled a team of a dozen Latin American university students, who scoured the archives of the major daily newspapers of Costa Rica and Bolivia, covering the period 1960 to 1995. Remarkably, the team found over 3,000 environmental news events during this period—citizen protests over deforestation, high school environmental art contests, environmental science conferences, letters to the editor complaining about industrial pollution in local waterways, and government initiatives to protect endangered species. But the level of attention given environmental issues was not constant. We found a rapid increase in environmental concern in both countries—far too rapid to be explained by generational turnover (Figure 9.3).[22]

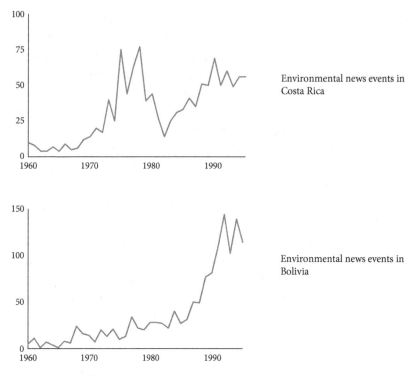

FIGURE 9.3 Growing environmental concern in Latin America

We also discovered a shift in the *content* of environmental ideas in these societies—in what it means to care about the environment. In the early 1960s, you can find the occasional news event reflecting concern over an issue like deforestation or pollution. But these different topics almost never appeared side-by-side in the same news story. An article reporting on concern over radiation fallout from above-ground nuclear testing in the 1960s would not mention water shortages or pesticides. No one thought to discuss so-called brown issues (relating to pollution) and green issues (species conservation, forests, national parks) side-by-side at the same social event. The idea had not yet arrived that these issues are different parts of an interrelated, larger phenomenon described today with terms like "the environment," "ecology," or "sustainability." Over time, however, this changed and these issues were increasingly bundled together. In the 1960s, about 5 percent of all articles featured both green and brown topics together. By the mid-1990s, a quarter of all articles did so.

This change in thinking, emphasizing connectedness among environmental problems, led to changes in the rules governing how people interact with the earth. Here is where the relationship between ideas and social institutions comes into play. Every public policy has embedded within it specific ideas—ideas about what we value, how we define a given problem, and how best to address it. When the idea emerged that there are links among issues like energy, water scarcity, and pollution control, government officials in Latin America and throughout the world started to gather together agencies that had previously worked in isolation, placing them under the umbrella of new environmental agencies with mandates for environmental protection. A similar evolution occurred within civil society, as the agendas of activists, educators, philanthropic foundations, and research communities expanded to incorporate new concerns. A switch in perspective brought new actors to the table, excluded others, and redirected funding and staff—it changed who rules the earth.

MIND AND MUSCLE

Social change happens when innovative ideas are encoded in the rules we use to coordinate our activities—be they laws, contracts, urban plans, constitutions, or industrial design standards. It takes both mind and muscle. Without winning over minds, new rules are unlikely to be followed (and when they are, it may be the mark of an effective tyrant rather than enlightened governance). Some rules, like driving on the proper side of the road, are self-enforcing; people don't need to be convinced of its importance. More commonly, new rules must be actively enforced, and a rule that clashes with predominant social norms is likely to be ignored. Even in societies ruled by thugs and autocrats, where citizens dare not confront the authorities, people often devise clever ways to subvert the will of government officials trying to implement unpopular proposals, using what anthropologist James Scott calls weapons of the weak.[23] These subtle acts of sabotage can take the form of disabling a pollution control device in a taxicab, or "accidentally" shooting an endangered animal during a hunting trip.

Just as official rules need the support of cultural norms, the reverse is also true: If you want to bring about a lasting change in your community, it is not enough to merely inspire people to adopt a new outlook. Despite the shift in American perceptions of pesticides, for example, the United States has made almost no progress on this issue, applying over a billion pounds of pesticides (one-fifth of the world total) every year.[24] We saw in the opening pages of this book that a major change is underway in Canada, with hundreds of communities and entire provinces banning the use of nonessential pesticides in homes and public spaces. In the United States, meanwhile, pesticide industry lobbyists fought for and won new rules that make it illegal for local communities to attempt to regulate these poisons. New ideas about how we should treat our planet must be encoded in rules that give these ideas force and staying power. It requires digging new "ruts" that protect the public interest.

CREATURES OF HABIT

An important part of the digging-in process involves the creation of new routines, habits, and standard operating procedures that guide the daily activities of consumers, commuters, companies, and governments. Our routines make it easier to accomplish tasks by establishing a default mode of operation, so that we needn't constantly rethink and renegotiate how we do things, like how to tie our shoes, when to arrive at work, and which flower species to plant along the highway. It's a little like breathing. You can take conscious control any time you choose, deviating from your normal breathing pattern to blow out a birthday candle or touch the bottom of the pool. But when you need to devote your attention to some other task, you don't suffocate because your body has a default routine that keeps the system running. To put it crudely, routines allow us to be dumb and effective at the same time. Our frustration comes when we find people pursuing an unreasonable course of action in this semiconscious, zombie-like state because they are a "slave to routine," like the customer service representative who refuses to depart from the company script and set things right. But there is an intelligent method to the mind-numbing madness of routine. Neuroscientists have discovered that human intelligence relies on the brain's ability to be selective in its attention. We block out some sensory data (the leaves on that tree are shimmering) so that we can focus on others (stoplight ahead). Rules and routines allow us to move through a complex world without subdividing our attention to the point of mental paralysis.

Consider the behavior of a multinational corporation that has decided to set up a factory in a developing country. We often hear that these giant companies are a source of great harm, exploiting local labor and destroying the environment. The reality is much more complex and reveals a lot about the sheer force of habit, and how it affects our planet in good ways and bad. Over the past two decades, researchers have tracked the activities of multinational companies to see if they relax their environmental standards in developing countries, where the enforcement of environmental laws is often sporadic at best.[25] As is frequently the case with good research, the results contradict some widely held assumptions. It is true that there are many documented cases of multinational firms conducting their business in deeply irresponsible ways, colluding in human rights violations while destroying the forests of Malaysia, polluting the Niger Delta, or creating toxic dumps in the jungles of Ecuador.[26] But many multinationals maintain the environmental standards of their country of origin, even though they are not required to do so by the host country. Glen Dowell of Columbia University and his colleagues measured the environmental standards of eighty-nine US-based multinational manufacturing and mining companies to see if they lowered their standards when operating in the poorest developing countries. They found that "defaulting to local environmental standards is by no means the most common practice.... Rather, the most common strategy in this sample is to adopt a stringent internal standard that is applied globally."[27]

Why would a company go beyond the minimum level of compliance required by the host country? The research suggests that some do so to maintain positive relations with local communities, similar to the peer pressure that Lily Tsai found to be a motivating force for Chinese officials providing villagers with public services.[28] Other firms appear to be responding to the hot spotlight of international activist organizations that are eager to publicize their deeds.[29] But there is yet another reason for environmentally responsible behavior on the part of multinational corporations: They've simply grown accustomed to it. Like other organizations, they are creatures of habit.

In an influential article published in 1985, sociologists Paul DiMaggio and Walter Powell observed that as industries mature, they develop predominant ways of doing business that have little to do with rational calculation or up-to-the-minute analyses of benefits and costs, and more to do with "This is just how things work in our industry." These routines exert a pull on new entrants into the industry; new firms must adopt the standard practices if they want to be taken seriously.[30] In other words, organizations mimic those they interact with, and this further reinforces the shared routines. For multinational corporations, new environmental practices in their countries of origin, as well as new industry-wide environmental standards, have become so deeply embedded in their operating procedures that it would be costly to modify their practices to take advantage of the lenient regulatory environment found in poor countries. These companies have their preferred suppliers and well-established design standards. Their mechanics and engineers are trained to work with certain types of machines. In the chemical industry, environmental standards have become so deeply entrenched in their business practices that multinational firms are precipitating the spread of improved environmental practices within domestic industries in developing countries.[31]

Copycat behavior among competing organizations, and the industry-wide routines that result, can be found in a wide range of settings. Consider the environmental behavior of city governments. In the early stages, a city may adopt an innovation such as a bike-sharing program because visionary leaders take advantage of fortuitous political circumstances; at the later stages, other cities adopt the practice because "everyone's doing it"—it's part of what it means to be a forward-looking metropolitan center.[32]

The development of new routines is an important part of what makes a social change stick. Rules and routines are different things. When a concert violinist methodically stretches her fingers before a performance, following essentially the same pattern each time (touch pinky to thumb, now ring finger to thumb…), that is a routine. When she removes her chewing gum before taking the stage, that is a rule. The difference is easy to tell: If you break a social rule, it provokes a negative response from those around you, with the sanction ranging anywhere from mild scorn to life imprisonment, depending on the nature of the transgression. In practice, as rules and routines intertwine and take root, new patterns of human behavior emerge. Consider the people on the front lines of any professional practice—the carpenters, gardeners, real estate developers, auto body mechanics, and others whose daily routines shape the planet in so many ways. If you listen to an experienced carpenter advising a young apprentice, you will hear a mixture of official and off-the-record advice about the right way to do the job. The conversation includes references to formal rules like the building code—we have to use a 16-inch laminated support beam here—but it also includes a deep well of wisdom about best practices and rules-of-thumb. You might want to use this saw blade rather than that one. Wrap some tape around that pipe before you attach it. And don't dump that toxic solvent down the drain—we have a disposal bucket on the truck.

EMBEDDED RELATIONSHIPS

To create lasting change, we need to embed sustainable practices within broader sets of relationships involving people whose principal goals may have little to do with sustainability. Consider LEED certification, the private rule-making system that has transformed the construction industry by creating a new market for green buildings. While I was conducting research for the discussion of the green building movement in chapter 3, Malcolm Lewis, the former chair of LEED's scientific advisory committee and a dear friend, discovered that he had advanced-stage cancer, which took his life several months later. Soon after learning of his illness, we had a long phone conversation in which he conveyed anxiety about the future of green buildings. He was particularly adamant about the need to raise the bar for environmental performance, arguing that much more progress is possible than most people realize. But equally noteworthy was what Malcolm Lewis did not mention. He expressed no concerns about whether the green building movement would last. In interviews, I find that baby boomers—the generation that launched the modern environmental movement—are thinking a lot about legacies these days, and whether their innovations will have staying power. But this didn't come up. What did Malcolm Lewis know that might teach us something about how change really works, resulting in new patterns that will outlast their inventors?

If you want to test the staying power of a new social invention, imagine future scenarios in which it comes under threat. What if tomorrow the US Green Building Council, which created LEED, were to announce that, due to a financial scandal or other mishap, it was closing up shop and would no longer be available to assess the environmental impacts of buildings? What would happen next? A large social constituency has come to depend on LEED. In recent years, American college presidents and other nonprofit leaders have been tripping over one another to brag about the environmental credentials of their new buildings. LEED standards have been incorporated into the building codes of dozens of American cities, and are part and parcel of federal building standards. Thousands of firms have come to specialize in green design to take advantage of the new market. These constituencies would surely come to the rescue of the Green Building

Council, revamping or quickly replacing it with a new organization capable of picking up the torch. The founders of the Green Building Council took measures that will help to ensure the resilience of their creation. They created a professional credentialing program that has trained tens of thousands of engineers, architects, and others throughout the building industry. These people have come to see LEED certification as part of what it means to work in their industry. Every time these professionals highlight their green credentials to potential clients, and share glossy project portfolios featuring low-impact designs, they dig a little deeper the channel that leads toward a more sustainable built environment.

THE PERILS OF PERMANENCE

Although a social change must stick if it is to amount to anything, durability has its pitfalls. Take, for example, the US Forest Service. Created by Teddy Roosevelt and the US Congress in 1905, the Forest Service carries the daunting responsibility of managing over 190 million acres of land in trust for the American people in perpetuity. If ever there was a social structure designed to last, the Forest Service is it. The agency's success in protecting its original mission from external pressures provides a cautionary tale about the downside of designing rules to resist change.

At its birth, the Forest Service marked a sharp departure from past practices. By the end of the 19th century, America's forests were in serious trouble. Logging companies were clearcutting the land with abandon. Huge swaths of timber were felled to meet the country's growing demand for wood, which was needed for everything from homes and bridges to fuel for the steam trains traveling over thousands of miles of wooden railroad ties. Rather than plant new trees and preserve the forests around watersheds, as was the practice in Europe, the timber companies would remove all tree cover from the land, sell the barren landscapes to settlers, and quickly move on to the next plot to avoid paying taxes. Government policy only made the situation worse. The General Land Office, the agency in charge of forests, was known to be incompetent and corrupt. Between 1850 and 1910, almost 200 million acres of American forests were destroyed.[33]

A confluence of new ideas and new rules provided the opportunity to turn the situation around. In 1864, George Perkins Marsh's groundbreaking book *Man and Nature* appeared and was read widely in intellectual circles, raising awareness about the destruction taking place in America's forests. On the political front, Teddy Roosevelt's Progressive Movement was trying to root out cronyism and instill greater professionalism within government agencies. The Progressives were on a mission to ensure that government bureaucracies were guided by ideals of science, efficiency, and commitment to the public good.[34] This and Roosevelt's well-known love of the outdoors provided an opportunity for change. At Roosevelt's prodding, Congress transferred responsibility for federally owned forests to the new US Forest Service under the leadership of Gifford Pinchot. A charismatic and tireless reformer, Pinchot laid down new rules that, for better or worse, guide American forestry practices to this day.

Pinchot brought with him a new idea. The mission of the old General Land Office was to quickly dispose of federal lands as part of a broader effort to help people settle the frontier. The new Forest Service had the bigger picture in view, borrowing from British utilitarian philosophers, who argued that the proper aim of government is to ensure "the greatest good for the greatest number." Pinchot adopted this as his guiding principle and gave it a sustainable twist, imbuing the Forest Service with a new mission to provide "the greatest good for the greatest number in the long run." It is no accident that the first US government body to embrace sustainability was the one concerned with trees. As any gardener knows, growing trees requires a long-term perspective. "That's what foresters do," explains historian Char Miller. "They think out across time."[35] In the forests of North America, growth and harvesting cycles often span 100 years or more. But Pinchot's guiding idea is equally noteworthy for what it did not include: wilderness. Pinchot's contemporary, the white-bearded outdoorsman and essayist John Muir, advocated the preservation of wild places, where natural processes could run their course and the human spirit could take respite from the stress of urban life. Pinchot, on the other hand, embraced values of sustained economic production. With an emphasis on timber and the protection of vital water resources, Pinchot set out to hardwire his new idea within rules and organizational structures that would have the resilience of a sturdy old pine capable of surviving the ever-changing political seasons.

"Pinchot possessed the zeal and purpose of a prophet," writes historian James Lewis. "He chose his missionaries carefully and assembled the men and machinery to deliver the message."[36] Pinchot faced two distinct problems in his effort to make the change stick. First, he had to ensure that the Forest Service would resist the forces of politics and short-term profits, which he knew would exert relentless pressure on the agency to shift course. His predictions proved accurate; the Forest Service was the target of repeated political attacks over the coming years, from ranching and mining interests to politicians seeking to privatize the public lands. Second, Pinchot needed to make sure that Forest Service employees working in remote rural locations across the country would be faithful in implementing the new rules, translating general policies and principles into new management practices on the ground. This included everything from proper land restoration techniques to the exercise of due diligence when reviewing applications for harvesting permits.

In his classic book *The Forest Ranger*, published in 1960, political scientist Herbert Kaufmann described the methods that the Forest Service used to promote conformity among the field staff, keeping them true to the mission. First among these was the revered Forest Service Manual, "the forester's bible" as it came to be known, which specifies in great detail the philosophy and practices to be followed by every ranger. Conformity with the mission was further reinforced through an intense socialization process to ensure that foresters identified with the agency and acted in accordance with its wishes. The first stage of this socialization involved "recruiting the right men"—incorruptible characters who could handle the solitude and rigors of remote fieldwork, yet comport themselves as gentlemen. New recruits were then immersed in extensive training sessions with supervisors and peers in order to build a strong esprit de corps. Promotion occurred strictly from within the ranks. Agency leaders graduated from forestry schools created by Pinchot himself, ensuring fidelity to the agency's founding idea of sustainable timber production.

The process worked. When Kauffman undertook his research a half-century after the agency's creation, he reported, "Rarely does one hear it said that Rangers behave in a fashion inconsistent with Service policy; on the contrary, they are sometimes described as too zealous in their conformance."[37] But the durability of the new rulemaking regime for forests stemmed not just from strategies internal to the organization. The Forest Service plunged its roots deep in the political soil through the Twenty Five Percent Fund Act of 1908, which guarantees that states receive a quarter of all proceeds from nearby forests when timber and other natural resources are sold. Rural communities came to rely on the proceeds from timber sales as a lifeline for financing schools and roads. In effect, this rule locked in a constituency for the continued exploitation of timber from the national forests.

Pinchot's success in designing for permanence would turn out to be a major barrier to change when a new set of concerns arose in the final decades of the 20th century: ecosystem health and the conservation of biological diversity.[38] From the 1940s to the 1980s, the Forest Service approved ever-increasing harvests in response to state demands to "get out the cut" to generate jobs and timber revenues. The harvesting method preferred by the Forest Service was to clearcut the land, completely destroying all aboveground vegetation in a quilt-like pattern alternating with patches of timber stands. Unlike the practices of the prior century, timber companies were required to plant saplings after the harvest, so these clearcuts were arguably consistent with the agency's original mission of "preserving a perpetual supply of timber."[39]

By the 1960s, Forest Service practices came under withering public scrutiny, as the concerns of research ecologists combined with an increasingly assertive environmental movement. These critics pointed out that there's a lot more to a forest than just marketable timber; forests house a complex ecosystem of plants, animals, and fungi. Forested landscapes also provide opportunities for recreation and for people to reconnect with the sublime beauty of wild places. As the public rancor grew, it was as if the spirit of John Muir had returned from the grave to do battle with Pinchot's legacy.

These concerns accelerated throughout the 1980s and 1990s as the extent of the global extinction crisis came to light. Forests cover only 6 percent of the earth's land surface but are home to about 90 percent of all terrestrial species. When an old and diverse forest ecosystem is mowed down and replanted with a few commercially valuable tree species, the new forest sustains only a portion of the biodiversity of its predecessor. In the national forests of the Pacific Northwest, this chop-and-plant routine threatened valuable salmon fisheries, which are disrupted by the soil sediments that pour into streams following clearcuts. The pressure to extract ever larger quantities of wood also caused an alarming drop in numbers of the northern spotted owl, which is considered an "indicator species" that gives early warning signs about an ecosystem's overall health. Widely distributed photos of clearcuts came to symbolize all that is wrong with American land management agencies that prioritize industrial profit over ecological vitality.

About 94 percent of the original forest cover was cut down during the first two centuries of American development. Logging companies wanted access to the rest. In what came to be known as the timber wars, thousands of Americans engaged in waves of protest and civil disobedience with the hope of protecting the old-growth forests that remained (Figure 9.4). State lawmakers pushed for continued harvests, while timber-dependent communities experienced bitter divides between loggers and environmentalists. The Forest Service was sued by all sides. In an effort to resolve the conflict, on April 2, 1993, President Clinton and Vice President Gore presided over a Timber Summit held in Portland, Oregon, bringing together fifty representatives from the logging industry, environmental groups, and local governments. In a clear signal of his dissatisfaction, Clinton did not invite Forest Service officials to speak at the event. But even with this political storm swirling around it, the Forest Service refused to bend. So in an effort to move the agency off its path of destruction, Clinton appointed wildlife biologist Jack Ward Thomas as the new director. Thomas was the first non-forester ever to lead the agency, and many hoped that he would imbue the agency with a new sense of ecological responsibility. But Pinchot's creation was still dominated by foresters, who were ill equipped to undertake the species surveys required under the compromise agreement that emerged from the Timber Summit. This led to long delays that produced less timber than originally

FIGURE 9.4 The US Forest Service was designed to resist change

Left: The forest ranger, a symbol of stewardship and fidelity to a timeless mission. Photo courtesy of the Forest History Society, Durham, NC.
Right: It took three decades of social mobilization to convince the Forest Service to incorporate ecological health into its mission. Photo courtesy of Evan Johnson.

envisioned in the agreement, much to the dismay of timber-dependent communities.

The Forest Service is still a work in progress. Today Pinchot's creation places a greater emphasis on ecological health, but this has produced exasperation for many within and outside the agency who argue that the pendulum has swung too far in the direction of preservation, at the expense of sustained use. Regardless of where the right balance lies, the Forest Service story illustrates the perils of permanence. After soaking for a century in a philosophical brine that predated the rise of scientific fields like ecology and conservation biology, the Forest Service immunized itself against new ideas at a great cost to our natural environment.

TEFLON AND TAR PITS

So just how sticky should our social rules be? The story of the US Forest Service suggests that if rules are too difficult to dislodge, there is no room for learning and adaptability, which are sorely needed when dealing with complex long-term problems in a democratic society. Conditions change, preferences evolve, and new information comes to light. The original rules may not even have their intended effects. But if our rules are too adaptable, tossed in the trash bin with every transient shift in public opinion, market conditions, or governing coalitions, then progress becomes impossible. Investors will lack confidence, polluters will just wait it out, and agencies will lack the long accumulation of experience needed to diagnose and tackle challenging social problems.

The question of durability versus change was fiercely debated by the founders of the United States, who created one of the world's longest lasting political systems. In 1788, Alexander Hamilton argued that constitutions should include only very general provisions because they "must necessarily be permanent and they cannot calculate for the possible change of things."[40] Thomas Paine made the case against permanence in his famous book and call-to-arms, *Rights of Man*. "Every age and generation must be as free to act for itself, *in all cases*, as the ages and generations which preceded it. The vanity and presumption of governing beyond the grave is the most ridiculous and insolent of all tyrannies."[41] Thomas Jefferson made a similar point in a letter written to James Madison in 1789, on the question of "Whether one generation of men has a right to bind another." Jefferson concluded that "no society can make a perpetual constitution, or even a perpetual law. The earth belongs always to the living generation."[42] Ultimately the founders designed a system that had a little of both: an adaptable constitution subject to the constraint of not violating the "inalienable rights" of its citizens.

Adaptability within boundaries—this is the core challenge when building resilient social institutions capable of steering society in a new direction. And resilience is a prerequisite for lasting social change. A resilient system of rules must establish a hierarchy of permanence: Some rules promote ideals so deeply held that no one should be allowed to revise them. (Those who wish to do so must overthrow the entire system.) Other rules can be overturned only with great difficulty; and still others deal with a level of detail so minor that they should be changeable at the pen stroke of a mid-level bureaucrat. The question of which ideas constitute timeless principles, and which are disposable conveniences, is clearly a matter of great importance. This is true whether we're talking about a national constitution, a city ordinance, or the by-laws for a nonprofit organization.

One of the challenges of designing for permanence is that we can never truly predict what changes the future may bring. Without the benefit of a crystal ball, committing to any particular path carries risks. To mitigate these risks, rulemakers can include in their original designs the requirement that the new rules are reviewed on a regular basis, with opportunities for modification (such as performance reviews and contract-renewal periods). Rules may also include sunset clauses, specifying a date when the rule will expire unless reauthorized. The relative permanence of the rules we live by should also be tailored to match the severity and irreversibility of the problems they address. Given the irreversibility of extinction, for example, it makes sense to create rules that ratchet conservation upward, making it easier to protect habitat (through an act of a president or legislature, for example) than it is to rescind this designation (requiring the approval of both).[43] While not precluding future changes, rules that are difficult to overturn make a modest request of future generations: that they stop and think before rushing headlong down a different path.

DON'T JUST BREAK THE RULES—MAKE THEM

Self-styled social mavericks are fond of portraying rules as the antithesis of human creativity. They exhort us to break free of our hidebound habits with mottos like "throw away the rulebook" or, to quote the title of a business best seller, "First, break all the rules." So when I make a plea that we pay attention to the rules governing our planet, I realize that I must sound quite retrograde in comparison to these popular gurus of change, and more like a schoolmaster, glasses balanced on the end of his nose, waving a wooden pointer at the rules scribbled on the chalkboard. But rules and creativity are not at odds—they are, in fact, close allies. Our greatest sports contests and most sublime musical performances rely on rules that guide the creative spirit down the field and up the musical scale. Likewise our greatest environmental successes (an 80 percent reduction in LA's air pollution) and most abysmal failures (the loss of 94 percent of America's old-growth forests) can be traced to the rules that prodded human initiative along one or the other path.

We don't often celebrate the rules themselves, and for good reason. You won't hear a sports announcer give an excited play-by-play describing how an athlete did a stunning job of following the rules throughout the match. We instead celebrate the creative use of those boundaries for achieving great feats. But when our focus on individual initiative extends to efforts to protect the earth, we are missing the forest for the trees. We laud individual acts of environmental benevolence, like turning down the thermostat in the winter, while giving little thought to the rules (such as national energy policies) that decide whether these individual actions translate into large-scale improvements. Today the inventors and entrepreneurs at the leading edge of green energy development are playing a game in which the rules are stacked against them, and in favor of the fossil fuel industry. The new thinkers may have twice the brains and grit. They may be class valedictorians and brilliant business strategists. But they are in a contest they can't win—until and unless we as a society create rules that level the playing field.

PART IV

Leverage

10

Super Rules

If you watch a group of children at play in an unstructured situation, soon you will be treated to a microcosm of how societies make rules, boiled down to the essentials. After some random running about, the children will eventually seek to build a social structure in the form of a game. The process unfolds with remarkable swiftness and predictability.[1] By definition, every game requires rules, and these are the subject of considerable haggling at the outset. *You have to touch the tree to be safe; no one can go past the rocks.* The participation rules are negotiated with special care, because every child knows intuitively that these will affect the outcome. *You have more people, so we get the big kid.*

It is equally fascinating to observe who makes the rules of the game. Over a chorus of competing ideas, the rulemaker is often the oldest or most assertive child, but not always. Someone may make a credible threat based on her resources and the power that accompanies them: *It's my ball and I don't want to play that game.* Alternatively, she may appeal to a source of moral authority recognized by the other players—*it's my house and my birthday party*. Once settled, all participants in this miniature society must understand and abide by the rules. Those who break them are subject to a collective outcry from the group and even efforts at third-party enforcement: *Mom, Richard keeps cheating!*

The situation is not so very different from the inner workings of our entire civilization, which is built upon a vast infrastructure of rules. Every business and every community, every religion and nonprofit organization, every terrorist network, taco vendor, and art museum relies on social rules to achieve its ends. Throughout this book we have seen how our lives and our landscapes are shaped by these rules, be they policies or property rights, safety codes or shared cultural norms. We are now ready to take a closer look at a special and very powerful category of rules—I call these super rules—that decide how other rules are made.

So what exactly are super rules? In any given rulemaking setting, whether a corporate boardroom, community garden association, or the United States Congress, we must distinguish between two types of rules.[2] The first are those rules designed to affect substantive outcomes in the world. The US Endangered Species Act is an example. Passed by Congress in 1973, this legislation is designed to protect the country's biodiversity by forbidding people to "harass, harm, pursue, hunt, shoot, wound, kill, trap, capture, or collect" any species at risk of extinction. The second type of rules— super rules—govern the rulemaking process itself. This includes the rule in Article I, Section 7 of the US Constitution that says if you want to create a law like the Endangered Species Act, you need approval from the president and a majority of each chamber of Congress. If rules are the levers that move the world, super rules are the lever-making factory. Super rules determine what methods we have at our disposal for bringing about change in the world.

Super rules govern every arena in which people come together to make mutually binding decisions. In the example of children at play, there's the *It's my birthday party* rule, which empowers one imperious child to choose the game and how it's played. In settings where the stakes are much higher—in the management of global fisheries, for example—the design of super rules is of paramount importance. Should decisions be made by a small group of scientific experts or by the general public? What role should commercial fisheries and coastal fishing communities play in the rulemaking process?

If you want to understand how our environment reached its current condition, and what it will take to set things right, these are the sorts of questions that must be addressed. When a town council wants to set aside some public land as protected open space, should this require consensus among its members, or a simple majority vote? Who bears the burden of proof for regulating new chemical compounds introduced into your food and clothing—should the industry have to demonstrate that they are safe, or must regulatory officials prove that they are harmful? What criteria do they use when reaching their conclusions? Do the results of these decisions have to be shared with the public? Can they be contested, and if so by whom, and through what mechanisms?

In the previous chapter we saw that social rules are purveyors of ideas, like the notion that species should not go extinct as a result of human activities. But rules do something more: they lock in place power relations among groups of people and reproduce these relationships down through the generations. This is why it is so important to pay attention to the rules that govern rulemaking. When these super rules undergo change—for example, revising a town charter to enable public scrutiny of government decisions—the effects reverberate throughout the entire system, affecting numerous policies and regulatory standards by revising the very process through which rules are created.

WHO DECIDES?

Participation is fundamental to the exercise of power. As political insiders are fond of saying, "You're either at the table or you're on the menu." If you try to change the rules shaping your world, one of the first things you will encounter are super rules that decide whether you are even allowed to join the conversation. Let's say you wanted to express some concerns to decision makers at the United Nations. Perhaps you feel that climate change discussions are ignoring an important set of considerations affecting your community. Under the super rules governing UN procedures, there is no opportunity for you as a citizen to stand before UN officials and make your case. Instead, you will have to convince a national government to convey your viewpoint. That or you must ally yourself with one of the officially "accredited" nongovernmental organizations that are permitted to attend UN functions, where they have the right to monitor negotiations and distribute information to diplomats. The super rules governing UN procedures forbid direct participation by citizens speaking on their own behalf.

Super rules also determine which voices are heard in national policy debates. Consider the fate of green parties in Germany and the United States. In both countries, a large majority of citizens express high levels of concern for the environment. Both are wealthy, industrialized democracies and both countries have active environmental groups in virtually every major city. But while the German Green Party, *Die Grünen,* is a fixture in that country's national legislature, in the United States, green parties have never managed to gain even a single seat in Congress.[3] Since 1983, when *Die Grünen* were first elected to Germany's Bundestag, there have been about 7,000 elections for seats in the US Senate and House of Representatives. Green party candidates have won precisely zero of those electoral contests. Why would two democracies with similar levels of environmental concern have such radically different outcomes with respect to who rules the earth?

The fate of green parties makes more sense when we consider the super rules at play in the two countries, and in particular the election laws that determine who gets to participate in the rulemaking process. In the United States, it is practically impossible for an alternative political party to have a significant presence in Washington. Using a winner-takes-all voting method, America's rules all but guarantee a two-party system.

To see how this works, let's consider a deliberately oversimplified example that illustrates the general point. Let's assume that 10 percent of the American electorate would be willing to vote for green party candidates in the race for the House of Representatives. Under the current winner-takes-all system, whichever candidate wins the most votes in a given voting district is elected. Reading across the top row in Figure 10.1, in what we'll call District A, the Democratic candidate wins 50 percent of the vote, the Republican 40 percent, and the Green candidate 10 percent, resulting in a win for the Democrats. Under American voting rules, once the outcome is decided within a voting district, any votes cast for the losing candidates are essentially thrown away; they cannot be added to Green or Republican votes in other districts. In District B, the Republican wins and the Green and Democratic votes are thrown away. This logic is repeated in voting districts C to J, with a strange result: The Greens receive zero percent of the seats in Congress even though they received 10 percent of the vote.[4]

Germany, like most European countries, uses a very different voting system called proportional representation. German voters are presented with lists of candidates associated with various political parties. The more votes that a party receives, the more of its candidates win seats in the legislature. But here's the twist: Under this voting system, votes are added up across large regions of the country, so that a party's share of the legislature reflects its share of the nationwide vote. (Germany's system is slightly more complex than this, combining elements of the winner-takes-all method with

WINNER-TAKES-ALL SYSTEM

Percent of Votes Received

District	Democrat	Republican	Green	Winner
A	**50**	40	10	D
B	40	**50**	10	R
C	40	**50**	10	R
D	**50**	40	10	D
E	40	**50**	10	R
F	40	**50**	10	R
G	**50**	40	10	D
H	40	**50**	10	R
I	**50**	40	10	D
J	40	**50**	10	R

End result in legislature: 4 Democrat, 6 Republican, 0 Green

PROPORTIONAL REPRESENTATION

Percent of Votes Received

District	Left	Right	Green
A	40	50	10
Lists	**Ruddy**	**Kidwell**	**Muir**
	Nicholson	**Perez**	Pinchot
	Smith	**Spears**	Brower
	Obonki	**Yee**	Erwin
	Barney	**Rose**	Chavez
	Oliver	Mullen	Carter
	Donghi	Martinez	Newman
	Morrisey	Fein	Moore
	Goldman	Maina	Jenks
	Sanchez	Gerrigold	Kelly

End result in legislature: 4 Left, 5 Right, 1 Green

Coalition politics ensue, enhancing the influence of the much sought after Green legislator.

FIGURE 10.1 Voting rules shape political outcomes

proportional representation in two rounds of voting; but overall, German election rules ensure that the composition of parliament is roughly proportional to the vote.[5]) In our hypothetical example in Figure 10.1, the German Greens received exactly the same percentage of the vote as the US Greens, yet earned 10 percent of the seats in the legislature while their American counterparts received none.

At this writing, the Greens in fact occupy just over 10 percent of the German legislature, while in the United States there is not a single Green Party representative in the capital. A casual observer might conclude that German citizens must place a higher priority on environmental issues than do Americans, given the Green Party's greater success there. But it is a beginner's mistake to assume that social outcomes are a fair reflection of social interests, akin to our physical traits mirroring their underlying genetic code. Rules—in this case the rules governing elections—are the causeways along which personal desires are transformed into collective outcomes.

When the rules ensure dominance by two large parties, it is easy for party leaders to settle in comfortably and ignore innovative proposals emerging from outside their ranks. Recognizing this, a growing number of American cities are revising their electoral laws to make it easier for minor parties to establish a presence. In 2002, the city of San Francisco adopted a new super rule called instant-runoff voting (also known as alternative voting), in which voters rank their top three candidates instead of merely choosing one. Under this system, if your favorite candidate does not garner sufficient votes to win, that candidate is eliminated from the contest and your vote is then reassigned to your second favorite candidate. This allows you to choose the person and the party you would truly like to see in office, without fear of wasting your vote by supporting the underdog.

Every voting system carries a downside, and proportional representation and instant-runoff voting systems are no exception.[6] For example, rules that enable smaller constituencies to participate in government are likely to produce not only fresh insights, but also mean-spirited ideas from the margins. Fascist parties espousing neo-Nazi views have scored electoral successes in a number of European parliaments in recent years—winning 7 percent of the legislature in Greece and a remarkable 17 percent in Hungary—where they espouse racist and anti-semitic views, and influence immigration policies among other rules.[7] The point is that the rules governing participation have an enormous influence on who rules the earth.

TILTED BY DESIGN

The impact of participation rules becomes glaringly obvious when we consider the challenges facing environmental activists in countries ruled by authoritarian governments. Political scientists Timothy Doyle of Keele

University and Adam Simpson of the University of South Australia offer a chilling account of how environmental politics play out in Burma and Iran. Throughout the 1990s, Burmese activists from the Karen ethnic group tried to halt construction of the Yadana gas pipeline, which the government decided to build on their traditional lands. The Burmese military, known as the *tatmadaw*, completed the pipeline through "forced labour, forced relocation, rape and summary execution of Karen villagers" as well as "the burning of villages and surrounding environments, including rice and crop fields and forest areas."[8] As Burma now transitions haltingly toward a less repressive form of government, the first order of business for proponents of democracy and sustainability is to change the super rules that decide who decides.

In Iran, the architecture of repression has assumed a different form. According to Doyle and Simpson, the Iranian government allows and even encourages the formation of environmental groups, but they have to be officially registered with the Interior Ministry, and must take great care to avoid criticizing the government. This poses a major barrier to change in Iran, given the government's extensive role in economic planning and its control over countless policy decisions affecting the people and their environment. Iranian organizations like the Women's Society Against Environmental Pollution have managed to raise public awareness of environmental threats facing the country. But they must portray these as technical problems to be solved rather than political problems to be confronted. As is the case in established democracies, environmentalism stripped of its political content merely plays into the hands of those in power. Those organizations that dare to operate without government approval in Iran must do so clandestinely, forgoing physical offices and expressing their views through unofficial channels like Radio Zamaneh, an opposition station based in the Netherlands. The few environmental protesters who openly defy the government face arrest and the possibility of torture.[9]

You don't need to live under an authoritarian regime, however, to see how participation rules are tilted to favor some interests over others. A town planning process in the United Kingdom may operate in ways that elevate the interests of developers over those of the public. A college board of trustees in the United States may be accountable only to themselves when deciding whether to invest the campus endowment in firms with a commitment to social responsibility and environmental stewardship. Whatever the political setting, when outside perspectives are excluded from the decision-making process, there doesn't even appear to be a conflict, because entire categories of questions never make it onto the agenda. In a famous article published in the *American Political Science Review* in 1962, Peter Bachrach and Morton Baratz argued that if we want to understand political power, we can't just look at who prevails in highly visible political struggles. We also have to consider the non-struggles—the arguments that never take place, because those wielding power have defined the terms of debate in such a way that certain questions simply never come up.[10] To understand who rules the earth, these authors argue, you have to consider how those who benefit from the status quo "tend to limit the scope of actual decision making to 'safe' issues."[11]

When you are denied access to a rulemaking process, you may lack the very information needed to challenge the status quo. As an outsider, you are not privy to the conversations and internal memos, and may not have professional analysts of your own to provide authoritative critiques of the official reports issued by city planning departments, developers, or government health experts. Your attendance at meetings—held at a time and place of the insiders' choosing—may be sporadic. As a result, it is easy for those in power to dismiss your challenge to their authority by claiming that you lack a genuine understanding of the issues. Taking this argument to its logical conclusion, those holding the reins of power then argue that you are simply not qualified to join the decision-making group, furthering a vicious cycle of exclusion.

SHAPING AND SHOPPING

How can you gain access to an exclusive silo of rulemaking power affecting the environment? There are actually a number of options at your disposal, including venue shaping and venue shopping. With venue shaping, you make a push to change the super rules that decide who gets to participate. At the most basic level, you and your allies can attempt to revise a governing body's charter so that seats are reserved for additional stakeholders—adding students to a campus physical planning committee, for example. A more ambitious example of venue shaping can be found in participatory budgeting, which has been introduced in a number of cities like St. Paul, Minnesota. Rather than delegate every budgeting decision to city officials, in St. Paul a committee with eighteen neighborhood association representatives ensures that budget priorities match local needs. These representatives also have a say in zoning decisions affecting their neighborhoods. "The neighborhood associations that are the foundation of these systems have real powers," write Kent Portney and Jeffrey Berry of Tufts University, who study innovations in urban democracy. "They are not advisory bodies; they have substantive powers that make them significant players in city politics."[12] Neighborhood associations and participatory budgeting have been introduced in a growing number of cities in the United States and throughout the world.[13]

If venue shaping proves too difficult a task, an alternative is to go venue shopping. This term was coined by political scientists Frank Baumgartner of the University of North Carolina at Chapel Hill and Bryan Jones of the University of Texas at Austin to explain how advocates for change circumvent entrenched groups of decision makers. With venue shopping, you seek out a different decision-making venue that is more welcoming of your participation, and use this to influence the outcomes currently monopolized by the more exclusive group. Baumgartner and Jones show that changes in super rules are not only possible, they're commonplace. Long periods of stability in a decision-making arena are upended with surprising swiftness and replaced by new rules of the game. The process is more dynamic than you might imagine. "There are no immutable rules that determine which institutions in society will be granted jurisdiction over particular issues," they write. "Depending on the issue and on how it is understood by those potentially involved, it may be assigned to an agency of the federal government, to private market mechanisms, to state or local authorities, to the family, or to any of a number of institutions."[14] You can take advantage of this fluidity, shaping the public agenda by shifting the fulcrum of decision-making power to a new venue.

Venue shopping was an integral part of the strategy used by leaders of the movement for sustainable buildings. In chapter 3, we saw that David Gottfried's efforts to create green building standards were getting nowhere, so long as he had to work through the traditional rulemaking venue of the American Society for Testing and Materials. ASTM's super rules welcomed participation by industry associations, which often resist reforms embraced by their more proactive members. The process was impossibly slow and was made all the worse by voting rules that required unanimous consent. Rather than challenge the super rules governing ASTM (which are institutionalized in hundreds of ASTM committees worldwide), Gottfried and his fellow reformers created a new venue—the US Green Building Council—governed by different rulemaking procedures that facilitated the rapid development of new building standards and soon revolutionized the industry.

While some super rules decide who gets to participate, others determine whether and how the decisions that are made actually get implemented and lead to real change on the ground. To appreciate the importance of implementation rules, we need to briefly travel back to the origins of the modern environmental era in the United States.

MAKING IT MATTER

It was September 1970, and Maine Senator Edmund Muskie was moving fast. The American environmental movement was at the height of its influence, while confidence in the government had reached an all-time low during the fiasco of the Vietnam War. Amid the political pressure cooker of the 1960s and early 1970s, which opened so many social fissures within the American electorate, environmental protection was something that just about everyone could support. Muskie and his fellow rulemakers in Washington, DC, were under intense pressure to take action—to show that they could do something, anything, right by the people who elected them.

Senator Muskie already had considerable experience dealing with pollution and natural resource policy during his time as governor of Maine from 1955 to 1959. But the scope of the challenge before him in 1970—creating a comprehensive legal framework for protecting the environment—had never been attempted in the United States or, for that matter, in any country. As chair of the Senate Subcommittee on Air and Water Pollution, Muskie soon emerged as an eloquent spokesman for the environment. He joined forces with Senator Philip Hart of Michigan in a far-reaching effort to change the rules governing the relationship between the country's economy and its ecology. But the senators faced a problem. They understood that efforts to change industrial practices throughout the United States would not transpire magically once Congress signed some pieces of paper. They anticipated, rightly, that new anti-pollution laws would encounter stiff resistance from affected industries. The agencies charged with enforcement would lack the resources needed to aggressively tackle environmental problems on so many fronts at once. The senators' instincts were spot on; forty years later, a research review by economists Wayne Gray and Jay Shimshack reported that "the EPA and delegated states are responsible for overseeing more than forty-one million entities regulated under fifty-eight programs from fourteen key environmental statutes."[15]

"I think it is too much to presume," Muskie argued during the Senate debate over clean air legislation, "that, however well staffed or well intentioned these enforcement agencies, they will be able to monitor the potential violations."[16] "More tools are needed," he argued. Ultimately Senator Muskie's committee designed a tool so powerful that it would have lasting repercussions throughout the American political system. Section 304 of the 1970 Clean Air Act included a new kind of super rule known as the citizen suit. This rule empowered any US citizen to act as a "private attorney general" helping to enforce the law.[17] Unlike previous laws that allowed a person harmed by pollution to sue for personal damages, this new rule encouraged citizens to act on behalf of the general public, suing any person, corporation, or government agency that violated the new air rules. Language from the new super rule was subsequently copied and pasted into a dozen of the country's bedrock environmental laws, from the Clean Water Act to endangered species legislation.

It did not take environmental groups long to notice the revolutionary potential of the new super rule. In one of the first uses of citizen suits, the Natural Resources Defense Council successfully sued the Environmental Protection Agency for failing to include lead on its list of priority air pollutants. Over the coming decades, citizen suits were used by a wide range of groups to ensure that the new environmental laws were put into practice. Between 1995 and 2002, an average of 550 notices of intent to sue were filed each year by community organizations, corporations, states, and others. In a society where litigation is almost as popular as fast food, you might reasonably ask whether this super rule opened a Pandora's box of frivolous court proceedings. But the criteria for filing a citizen suit are quite strict. Plaintiffs must present an airtight case that demonstrates a pattern of repeated violations. In practice, by holding environmental agencies accountable, citizen suits have been a driving force in the development of environmental law in the United States. According to Jim May of Widener University School of Law, 75 percent of all judicial rulings in environmental law concern citizen suits. "The experiment worked," concludes May. "Citizen suits have secured compliance by myriad agencies and thousands of polluting facilities, diminished pounds of pollution produced by the billions, and protected hundreds of rare species and thousands of acres of ecologically important land."[18]

The United States is not the only country experimenting with super rules that change how environmental laws are enforced. In Brazil, the Ministério Público (Public Ministry) is an independent unit of the government that aggressively prosecutes polluting industries and even goes after other government agencies that fail to live up to their responsibilities. Employees at the ministry are selected through an extremely competitive process (less than 1 percent of applicants are accepted), and those who make the grade are well paid and granted sweeping powers under the Brazilian Constitution. In her book *Making Law Matter*, Lesley McAllister of the School of Law at the University of San Diego reports that, in a part of the world where there is often a yawning gap between the laws on the books and the reality on the ground, the Ministério Público has been a game changer.[19]

In South Africa, the National Environmental Management Act encourages communities to participate in enforcing environmental laws, giving them the right to sue polluters on behalf of the country, and encouraging government agencies to enter into cooperative agreements with communities to help with monitoring and enforcement. This super rule has provided crucial support for local groups in places like Durban, Africa's second busiest port, where citizens are trying to hold oil refineries accountable for the pollution they produce. Alex Aylett, a graduate student at the University of British Columbia, has been studying the results. He reports that government officials and local activists find themselves in an alliance born of necessity. A local government official confided, "You need the environmental movement, you know. The government can pass environmental regulations, but right now those are so weak. The big companies need to get *klapped* [smacked into line] and the community organizations keep them on their toes."[20] The essential point is that when confronting powerful polluters, local activists in South Africa are not acting alone; super rules issued by the South African government empower them to help enforce pollution laws.

MAINSTREAMING

A second innovation that bubbled up during the formative period of environmental policy was the need to mainstream environmental concerns, introducing these into decision making throughout the body politic rather than cordoning off the environment as a specialized concern with little relation to the whole. At the center of this pioneering approach was Lynton

Caldwell, a political scientist born in Montezuma, Iowa, in 1913, whose remarkable career straddled the worlds of research and action for over half a century. In 1963, Caldwell published a little-noticed academic article in *Public Administration Review* titled "Environment: A New Focus for Public Policy?" that would lay the foundations for modern environmental policy. In this surprisingly philosophical article, Caldwell emphasized the connectedness of environmental problems, taking a concept born of the ecological sciences (and with a lineage stretching back to numerous Eastern and indigenous cosmological traditions) and applying it to matters of public administration. "Segmental thinking, segmental decision making—the 'practical' approach to practical problems," wrote Caldwell, "has again and again produced some very impractical results." He argued that "means must be found for more effectively interrelating or integrating the tasks of the public agencies as they bear upon the environment."[21]

In 1968, Caldwell would have the opportunity to put his holistic perspective into practice, when he was hired by Senator Henry "Scoop" Jackson, chair of the Senate Committee on Interior and Insular Affairs, to help draft the National Environmental Policy Act. Caldwell proposed a new super rule called "environmental impact assessment." Any government action expected to significantly harm the environment would need to first undergo an evaluation to identify potential impacts and propose alternatives for decision makers to choose from. A government "action" was interpreted in the broadest sense, to include everything from approval for building permits to the construction of new highways. When the National Environmental Policy Act went into effect in 1970, suddenly everyone, from naval commanders to real estate developers, had to take seriously the environmental consequences of their decisions. Caldwell's idea was eventually adopted by more than fifty countries.

A number of countries, particularly in Europe, have gone even further than the United States in mainstreaming environmental concerns. The idea began in the late 1980s in the Netherlands, where Dutch leaders adopted a national environmental strategy that moved this holistic, multisectoral approach further upstream in the rulemaking process.[22] In the United States, environmental assessment takes place mostly at the level of specific projects (a highway extension, a new college campus), long after the big policy questions have been settled. US agricultural policy, for example, is made without a comprehensive and mandatory focus on concerns like biodiversity. It is only much later and further downstream that the Department of Agriculture prepares (to pick one recent example) an impact assessment for genetically engineered alfalfa, which includes consideration of its potential impacts on biodiversity. Indeed, America has no national biodiversity conservation strategy whatsoever, beyond the emergency room provisions of the Endangered Species Act, which kick in only when a species is already at risk. In contrast, strategic environmental planning mainstreams sustainability at the highest levels of decision making.

A LITTLE SUNSHINE

If you search the online directory of the Federal Register, the official record of government rulemaking in the United States, you will encounter an endless parade of laws and regulations, draft decisions, and requests for public comment. There is document 78 FR 48628, "proposing to allow, under certain conditions, the importation of commercial consignments of fresh papayas from Peru into the continental United States." If papayas don't stir your passions, there's always 78 FR 48608, the Coast Guard's Notice of Deviation from Drawbridge Regulations concerning the South Branch of the Elizabeth River, mile 7.1, at Chesapeake, Virginia. Although it's hard to resist poking fun at the minutiae of government rule-craft, expressed in the parched language of bureaucrat-ese, the consequences of these rules are not so minute if you happen to sail ships on the Elizabeth or support your family by importing tropical produce. Indeed, you will find in the Federal Register rules that shape the planet and the American populace in powerful ways, from highway safety to the protection of wetlands from industrial pollution. But as we wade through the river of rules that flow from Washington, DC, and pool in the Federal Register, it is easy to lose sight of the fact that there is a deeper rule at work here—a super rule—that requires the US government to publish its proposed decisions and to solicit public comments.

The Federal Register was initially created because the government was having a hard time keeping track of its own rules. (The problem reached such absurd proportions that in 1935, a legal challenge made it all the way to the Supreme Court before anyone realized that the regulation the plaintiff found so objectionable in fact no longer existed.[23]) But it was the Administrative Procedure Act of 1946—a powerful collection of super rules governing the conduct of American federal agencies—that elevated the importance of the Federal Register, requiring that all agencies use it to publicize their proposed rules and to hear what people think about them. Government transparency expanded over the coming decades through what came to be known as sunshine laws. These included the Freedom of Information Act of 1966, which required that the government hand over documentation about its decisions to any citizen who requests it. In 1976, the merrily named Government in the Sunshine Act required (with some exceptions) that "every portion of every meeting of an agency shall be open to public observation."[24]

We tend to take for granted the rules shaping our lives, rather than recognize these as the pliable products of political struggles. This is evident in our attitude toward government transparency. Nothing would seem more natural and appropriate than government agencies providing their citizens with a record of their decisions. Yet most governments do not; in fact, with the exception of Sweden, which adopted an information access law in 1766, there was not a government in the world that allowed citizens to scrutinize its information prior to the mid-20th century, when Finland and the United States passed laws to this effect. A handful of established democracies, including Australia, Canada, and the Netherlands, jumped on the transparency bandwagon in the 1970s and 1980s. Most European countries began to open up much more recently, with transparency rules going into effect in Germany and the United Kingdom only in 2005.[25]

Worldwide, transparency is now in vogue. Forty-six European countries have signed an international treaty on transparency, the Aarhus Convention on Access to Information, Public Participation in Decision-Making and Access to Justice in Environmental Matters. At the urging of advocacy groups like Transparency International, dozens of countries, from Mexico to Jamaica to Albania, have put in place new rules that allow citizens greater access than ever before to government information.[26] These efforts have achieved varying degrees of success, and it is easy to deride them where they fall short. But these reforms represent a radically new way of doing business in political settings where government decisions have always been made behind closed doors.[27] The results will become apparent on a timescale of decades, rather than weeks or months.

Although efforts to pry open national governments have attracted the most attention, you can have an impact in your immediate surroundings by advocating for greater transparency in your local government, workplace policies and practices, or college campus administration. Even in settings where those in power have elaborate justifications for limiting public participation, they are rarely able to muster a sound defense of refusing to share information about their activities. Of course, if absolutely every decision and discussion were made available for all to see, this could cripple organizational effectiveness. Complete transparency could stifle risk-taking and frank discussion of new ideas, for fear that they would be misconstrued by the public and prematurely snuffed out. But rare is the rulemaking process that is too open, too transparent. Moreover, sunshine laws typically exempt entire categories of information from public scrutiny, such as employee personnel decisions, matters involving national security, and proprietary information. The important point when designing transparency rules is to establish openness as the default position, placing the burden on decision makers to show convincingly to third-party arbiters (such as the courts or independent ombudsmen agencies) that a given piece of information should not be made public.

SYSTEM EFFECTS

You might expect that the most profound changes in rules accompany history's most dramatic political upheavals. But this is often not the case. A great social movement may produce little change if its demands are not encoded in rules and embedded in social structures that sustain its animating idea into the future. Conversely, a relatively small effort may produce a big bang. Consider that with a few strokes of the pen, in 2010 and again in 2014, the US Supreme Court put in place new rules that essentially legalized bribery in the United States, by equating campaign contributions with free speech.[28] It most countries, paying money for preferential access to rulemakers would be considered corruption. The word "corruption," however, implies corroding or warping the integrity of a system. In the United States, corruption increasingly *is* the system. It is a painful irony that in a country that has moved forward with such speed by tethering its stagecoach to the unruly beasts of democracy and capitalism, the latter should ultimately swallow the former by recourse to an argument cloaked in free speech. The quest for sustainability is made considerably more challenging when polluters and others with deep pockets can buy their way into a lawmaker's office as readily as they might purchase a corporate jet, and at a fraction of the price.[29]

Yet in this chapter we have seen that super rules—the rules that govern how other rules are made—can also be powerful forces for good. In 1993, two leading experts on the inner workings of social change, Thomas Rochon and Daniel Mazmanian, teamed up to take stock of the impact of social movements. When concerned citizens mobilize and try to change their world, does it really make any difference? To answer this question, they focused on two movements that inspired participation by many thousands of people throughout the 1980s. The first was an offshoot of the environmental movement: the push by local citizens' groups to control industrial hazardous waste in their communities. The second was the nuclear freeze movement which, amid the escalating rhetoric of the Cold War, called on the United States and the Soviets to halt the development of new nuclear weapons.

Both were mass social movements that enjoyed robust popular support. In the early 1980s, the nuclear freeze movement had between 1,400 and 2,000 local organizations. On June 12, 1982, an estimated 750,000 people gathered in support of the freeze in New York's Central Park—the largest single political protest in US history. But while the hazardous waste movement was successful in influencing industrial practices and government policies, the nuclear freeze had essentially zero impact on US foreign policy—and "the freeze movement soon melted without a trace."[30] What can account for the stark difference in impact between these two popular movements for change?

Rochon and Mazmanian argue that hazardous waste activists enjoyed greater success because they changed the *process* through which rules are made. They changed the super rules. Specifically, anti-toxics groups convinced officials to create local forums for citizen consultation and dispute resolution regarding the location and design of hazardous waste facilities. Over time, these local forums provided routine opportunities for ordinary people to participate in the decisions affecting their lives. The nuclear freeze movement, in contrast, focused its energies on a House of Representatives resolution in support of the freeze. In the end, the House passed a watered-down resolution that had little influence on Reagan-era policies to expand America's nuclear strike capabilities. Rochon and Mazmanian argue that activists could instead have tried to restore the constitutional role of the Senate in setting US foreign policy, providing balance to what were seen as shortsighted approaches taken by the White House.

By their nature, super rules ripple throughout a political system, affecting many decisions over long periods of time. In your toolkit for changing the world, super rules are the big hammer. While profound in their effects, super rules are not necessarily the hardest rules to change. When you advocate for a reform in how rules are made—like increasing participation and transparency—this creates opportunities to forge alliances with powerful constituencies who may care little about sustainability, but share the broader goal of democratic reform. Moreover, Rochon and Mazmanian observe that when decision makers are confronted with a public outcry for change, they are often more agreeable to expanding participation than they are to modifying this or that specific policy. In the short term, it may appear that rulemakers are merely coopting opponents by granting a token concession without making real changes. But those groups that have gained a foothold in a rulemaking process—establishing their concerns as a legitimate and ongoing part of the conversation—often see their influence grow over time, eventually resulting in concrete reforms.

11

Paper, Plastic, or Politics?

THE EMPEROR AND THE ACTIVIST

It would seem only fair that an author who places a question in the title of a book should be expected to answer it directly. So who *does* rule the earth? The answer is Napoleon. He did, after all, leave behind a legal code that, as we saw in chapter 1, continues to guide the behavior of billions of people throughout the world today. And then there's June Irwin, the country doctor who convinced the town of Hudson, Canada, to create rules banning nonessential pesticides and spawned a nationwide movement for change. And let's not forget her foes in the pesticide industry, who raced across America passing state preemption rules that deny local communities the right to regulate these poisons. The earth is ruled by the Roman Emperor Justinian, who created the legal precedent for public access to beaches (discussed in chapter 2), as well as the real estate developer David Gottfried (chapter 3), co-inventor of the rulemaking system for green buildings known as LEED. The list most certainly includes Edmund Muskie and Philip Hart, the senators who spearheaded passage of the US Clean Air Act and Clean Water Act in the early 1970s. But it also includes José Delfín Duarte, whose local water association is empowered to decide how water resources are managed in his small corner of Costa Rica—an effort that required revising rules at local and national levels. Whether famous figures like Jean Monnet, founder of the European Union, or tenacious groups of citizens like Portland's Bicycle Transportation Alliance, whether working at the level of empires or that of neighborhoods, the people who rule the earth are those who leave behind a legacy of rules that shape the actions and opportunities of generations to come. If we so choose—if we can put aside for a moment the "little things" we do for the earth, and think about the larger, lasting changes that result when people come together and rewrite the rules they live by—then the group of rulemakers also includes you and me. This, in turn, requires that we exercise our rights as citizens and move from passive observers to active participants in politics.

FLOWER POWER

How could planting a flower possibly be a political act? In a garden in the city of Long Beach, California, homeowner Jim Brophy learned just how political it can be. Brophy planned to replace his lawn with native plants—those species that have thrived in the parched climate of the American southwest for millennia by devising ingenious tricks for conserving water, while offering food to local bugs and birds who do the job of spreading pollen and seed. So Brophy tore up his lawn, planted a diverse group of natives including red-barked manzanita and fragrant purple-flowered sage, and waited. But it wasn't just the local hummingbirds that took an interest in his new eco-friendly landscape. He was contacted by officials from the local homeowners association, who demanded that he maintain a proper lawn on his property. Jim Brophy's act of botanic rebellion created what some neighbors considered an unruly island of foliage in their tidy sea of green lawns, whose uniformity was maintained by regular applications of pesticides and fertilizers. What's more, the homeowners association had the law on their side. Under the rules in place in California, the association was well within their rights in demanding that members carpet their landscape with thirsty grasses better suited to the East Coast than to the semi-arid landscape of southern California.[1]

To turn the situation around required changing the rules. The movement began at the state level, where in 2009, a year after Brophy's ordeal, the California state legislature introduced a new law forbidding homeowners associations from forcing their members to keep their lawns.[2] A separate law required all California cities to adopt water-efficient landscape ordinances by 2010. Officials in the Long Beach Water Department were already on board with the concept. They needed to find a long-term solution to the brutal mathematics of limited water supply (about 12 inches of rain per year, on average) plus a growing economy and expanding population in one of the world's major port cities. In 2010, the Water Department introduced a new Lawn-to-Garden Incentive Program. Under the new rules, residents received a rebate of $2.50 per square foot to remove all grass turf and replace it with water-wise plants and permeable groundcover materials that allow water to percolate into the soil rather than flow into the city sewer. Five hundred homeowners took advantage of the program in the first seven months alone. The water agency proudly declared their Lawn-to-Garden initiative "the most successful turf-removal program in California."[3]

If you dig down and examine the roots of the environmental successes and failures that populate our landscapes, you will find that politics is never far below the surface. I experienced the immediacy of politics a couple of years ago while purchasing a coffee at a Starbucks near my office at Harvey Mudd College. It had always struck me as odd that there were no recycling bins in the store. Why would a company that has so much public exposure, and proudly boasts of partnerships with the likes of Conservation International, not make recycling bins available for its customers?

The employees at this particular store know me well, so while the young barista rang up the order, I nonchalantly posed my question to her. She cast a furtive glance in the direction of the back office, leaned forward, and in hush tones confided, "It's crazy, isn't it? We wish we had recycling too." I learned that she and her fellow baristas had actually taken to filling their cars with empty milk containers at the end of their shifts and were carting these by the dozens to recycling bins at their homes. I asked why this was necessary, and she replied that the store manager forbids recycling. Increasingly interested now, I asked to see the manager, who emerged from the back office and politely explained that she would love nothing more than to offer recycling, but the owner of the building leased by this particular store doesn't allow it. Following the trail, I asked for the owner's phone number and called him. The building owner explained that he would love nothing more than to offer recycling, but the city of Claremont prevents it, because his building is too small to qualify for a large recycling dumpster. He was telling me, in effect, that recycling would require a political push to modify the city's rules. (Soon after my little investigation, a city recycling dumpster mysteriously appeared behind the store.)

What is it about efforts to protect the environment that so quickly lead us to politics? It's not enough to breezily declare that "politics affects everything." There are plenty of things we do, such as exercise, that improve the quality of our lives without requiring that we convene community meetings and shout through a bullhorn on the capitol steps. But our physical environment—derived from the ancient French word *environs*, or that which surrounds us—is like a vast web that binds us together both physically and socially. One person's decisions (what to plant, how to dispose of waste) have consequences for others, and so the pursuit of individual health and prosperity is often impossible without some sort of coordinated action like pollution control. This is why sustainability is a not merely an individual choice, but a social choice, and frequently a political one, requiring that people work together and press politicians and others in positions of influence to promote the public good. And all forms of social organization, from the European Court of Justice to your favorite restaurant, operate through that other mode of human connectedness: social rules.

It is a wonder, and a formidable challenge to any researcher trying to explain it, that something as powerful as a social rule could be completely invisible—a mere concept that we carry in our minds, and jot down in symbols in legal texts—yet can literally move mountains, as we saw in chapter 4 with the mining rules promoting mountaintop removal throughout Appalachia. If I have had any success in pressing my point, you will now find that the world looks somewhat different than before. (Or as one of my students complained in a course evaluation, "You have ruined me. Now I see social rules everywhere.") It seems to me that the task before us is to reveal and repair—to bring to light the social rules that shape our world and to have a hand in making them. To help with the reveal, my students have created a variety of multimedia educational resources, available for free at www.rulechangers.org. The second challenge is to repair—to do the work of democratically tinkering with the rules that shape our lives, to make sure they promote the common good over the long term. This is where the author's pen stops and the creative intelligence of engaged citizens takes over. I do not presume any particular level or type of political engagement on your part, nor do I hold myself up as a model to emulate. Whether politics is something you do every day, or at a monthly meeting, or in a once-in-a-lifetime burst of activity around an issue close to your heart, is entirely your call. But I would like to suggest, and I hope to have shown, that if we pursue the goal of sustainability while ignoring its political dimensions, we will simply never get there.

To move effectively through the world requires a precarious balance of confidence and humility; each must be kept in check to avoid the extremes of hubris or self-doubt. As I turn to recommendations for action, I find that my humility dial needs to be turned way up. The reason is that social science deals in generalizations, while the prerequisites for social change are very context-specific. A chemist can with great confidence proclaim that atmospheric carbon dioxide traps heat, whether it is drifting above Chicago or Shanghai. But the appropriate course of action for citizens hoping to reduce carbon emissions in those two settings is an entirely different matter, requiring a solid grasp of the particular place. The challenge is a little like that facing the author of a gardening book, who offers landscaping recommendations for an audience scattered across diverse ecological zones. Plants that will flourish in some environments will wither in others; within a particular zone, gardeners must use their knowledge of microclimates and site-specific soil conditions. Even then, it takes a lot of trial and error to figure out what works. There is a time dimension too. The appropriate course of action in year twenty, when a maturing oak dominates the landscape, is quite different from the options available at the outset; so too in the process of social change, reformers must take account of the established ecosystem of interests and organizations. Like the skilled gardener with books on the shelf and hands in the soil, often it is the action-oriented thinker with some distance from academia who is in the best position to combine general theoretical knowledge with the site-specific skills and know-how required to produce real results.

With these qualifiers in mind, here are eight principles for action that are general enough in their prescriptions that they can be applied in diverse political settings, yet specific enough that they lend a practical flavor to some of the more general themes discussed throughout this book.

1. RULE THE EARTH

In practice, to be a rule changer does not imply a fanatical focus on creating new regulations at every turn. Nor does it suggest that you must adopt the arrogance of an emperor issuing edicts from on high. It could take the form of encouraging your bicycling club to ask some searching questions about why there are so few bike lanes. You might nudge your professional association to open a dialogue and adopt positions on environmental policy issues affecting your profession's long-term interests, providing a counterweight to those industry actors who reflexively oppose environmental regulations at every turn. After identifying one or more priority concerns, undertake a diagnostic exercise like the one I used to identify barriers to recycling at my local Starbucks. Arm yourself with a toddler's tactic of relentlessly asking *Why?* until you arrive at the root causes of outcomes like poor public transit or overfishing.

For groups that are already involved in advocacy, the perspective that emerges from this book is first and foremost to think long term. Today's well-attended community meeting, all aglow with a spirit of volunteerism and overflowing with ideas, may amount to little down the road if you fail to institutionalize new practices and perspectives. So drop an anchor. Rewrite the purchasing contracts. Revisit the rules governing how energy is bought and sold. Scrutinize the technical standards that guide professional best practices. Design rules to be flexible to take account of changing needs, but don't make them too flexible. As we saw with the US Forest Service, discussed in chapter 9, the trick is to balance durability—which is a fundamental part of social change—with the recognition that we never have all the answers.

In this and other aspects of rulemaking, pay special attention to super rules—the rules that govern the rulemaking process itself. These are the foundational by-laws and constitutions, electoral systems and voting requirements, standards of evidence and decision criteria that determine what it takes to modify the rules we live by. It is one thing to catch the ear of a sympathetic city official, who helps you secure a permit to launch a community garden. It is quite another to introduce into city planning guidelines a new emphasis on preserving local food production systems, which affects all sorts of rules long into the future. Notice the super rules that govern participation. Who gets to decide what our children eat in their school cafeterias—a distant bureaucrat, or a local committee with input from parents, nutrition experts, chefs, local growers, and (gasp) the children themselves? And remember that more rules are not necessarily better rules. Sometimes what we need most is to toss in the compost pile an old rule that may have made perfect sense in its time, but is now preventing people from exercising their creative discretion and doing the right thing.

2. BRIDGE RESEARCH AND ACTION

Intelligence is a social activity. We like to celebrate the achievements of brainy individuals, but in reality, major advances like space travel or carbon-neutral buildings result from rich aggregations of knowledge, shared among large numbers of people with diverse skills and expertise. Unfortunately, the flow of knowledge about the root causes of environmental problems is blocked by a mile-high partition separating the producers from would-be consumers of research. Those of us who study the institutional underpinnings of society, including the hundreds of researchers I have cited in the notes, are a bit like the microscopic village in Dr. Seuss's book *Horton Hears a Who!*, shouting "We're here! We're here! We're here!" (The analogy is not so far-fetched; in a recent book explaining why some nations are rich and others poor, a prominent group of researchers subtitled a chapter "Institutions, Institutions, Institutions."[4]) For social scientists who spend their days trying to understand the forces that move the world, there is often the will, but less often an obvious way, to share research findings and to collaborate with people from all walks of life on issues of shared concern. While researchers are hidden from public view, people who care deeply about sustainability, and are eager to learn more, have been denied access to the best research. Citizens scouring the Internet for insights have had to pay to read the professional research journals, unless they have access to a university library account; meanwhile, all manner of nonsense has been abundant and accessible for free.

This is now beginning to change, as a result of two trends—neither of which is a reduction in the quantity of nonsense. The first is Google Scholar, which provides an easy way to search for research articles and books on issues that you care about. The second trend is the movement toward open-access publications that are freely available to all. Open access is still the exception rather than the rule, but high quality research on the social dimensions of environmental issues is quickly becoming available to the public as never before.[5] Empowering yourself with this knowledge will give you a leg up when participating in public debates and choosing an appropriate course of action. In the Internet age, the intelligence that you bring to bear on social questions stems less from your innate IQ and more from your acquired QI, or quality information. It remains to be seen whether greater public access to research will inspire researchers to communicate our results differently—writing abstracts in jargon-free language, for example, or including appendices or video links for public consumption.

Beyond reading the research, we need more routine (dare I say, institutionalized) opportunities for meaningful collaboration between researchers and agents of change. Despite the considerable resources and physical grandeur of the college campuses where we work, professors typically operate as "lone wolves" without substantial staff support or other organizational capacities. As a result, we're poorly positioned to compete in the crowded marketplace of ideas unleashed by the Internet, unable to create even a compelling website, let alone an organization devoted to putting new ideas into practice. Collaborations between researchers and practitioners carry their own challenges, but can be deeply rewarding.[6] I suspect that many researchers, who devote their lives to studying the hows and whys of the world, would jump at the chance to partner with a community group or other organization in need of information.

3. BUILD UNCONVENTIONAL COALITIONS

We know from the research on political participation that those who are most active in politics, and who energetically engage in the civic life of their communities, are also more likely to attach themselves firmly to one or another political party.[7] This is understandable, to a degree; if you care, you commit. Yet political partisanship also carries risks for any effort to change the rules that govern the earth. For social rules to matter, they must endure long enough to make a difference. Convincing farmers to change their land use practices, modifying industrial processes to eliminate toxic waste, creating a pedestrian-friendly metropolitan area—these are long-term undertakings that span multiple election cycles and periods of economic growth and recession. To go the distance, new rules must have support from multiple political parties and diverse social groups who can defend the new arrangement and prevent reversals during times of political and economic change.

In the United States, political discourse has become increasingly divisive, incited by pundits who rally their supporters (and boost their media ratings) by demonizing opponents. This works against the spirit of coalition building required for lasting change. Myopic partisanship is by no means limited to the United States. I encountered a South African version during an interview in 2001 with one of the country's top environmental policymakers.

In the elated atmosphere of newly democratic South Africa, he and his colleagues were young, smart, and eager to replace apartheid-era structures with forward-looking social policies. My interview subject was especially excited about a new committee created by President Mbeki, which regularly brought together a wide range of agency leaders to coordinate policy on sustainable development and foreign affairs. At the end of our interview, I asked him about the likelihood that this new environmental initiative would last, particularly if the African National Congress were voted out of power. "I haven't even thought about the possibility of it not continuing to exist," he confessed. His upbeat assessment was that the structure would remain in place because the African National Congress would likely remain in power for the foreseeable future. This sort of here-and-now-ism dominates the thinking of many an impatient change maker. Yet as we saw in chapter 6, innovative rules for promoting sustainability are often swept away with a change in agency leaders or ruling coalitions. If rules are to last, rather than be jettisoned at the first sign of waning support, the new rules must enjoy the support of diverse constituencies.

One way to diagnose the durability of new institutional innovations is to envision future scenarios, posing a series of "what if" questions like the ones we considered in chapter 9 with respect to the green building industry.[8] What if the community group spearheading the effort were no longer around? What if another political party occupied city hall? What if an economic recession hit, or federal funding for this sort of initiative were reduced? Once you identify vulnerabilities, you can enlist participation by those who are in a position to help your initiative weather the inevitable storms. Cultivating these diverse perspectives also makes it more likely that the rules themselves will be well designed, taking account of the big picture, instead of imposing social costs that could have been avoided with more foresight, or advancing one environmental goal at the expense of another. To build coalitions for sustainability requires reaching out beyond our comfortable social circles and networks of like-minded folk, to include others with different perspectives and priorities.

4. CREATE PUBLIC VALUE

In chapter 3 we considered why it is that we do not live in the best of all feasible worlds. The information that we need to make good decisions is often costly to find; most businesses are unaware of how they use energy, let alone where to find proven technologies that could save them money while helping the environment. Free-riding behavior is pervasive, leading people to shun causes that produce shared benefits, even when cooperation would make everyone better off. Organizations are heavy with commitments from years past; deeply ingrained choices about staff hires, geographic location, and signature strategies hamper the ability of organizations to tackle new problems.

This all sounds rather pathetic. But for anyone who has ever wondered whether they can make a difference in the world, it is actually cause for celebration. There are opportunities for enterprising individuals to share the information, to bring together the buyers and sellers, to provide members-only benefits (throw a party!) that grease the skids of cooperation, and to create new organizations free of the trappings of those created long ago. The posture is that of the entrepreneur, who in the words of the 18th-century economist Jean-Baptiste Say, "shifts economic resources out of an area of lower and into an area of higher productivity"—but does so to advance public well-being rather than private profit.

Sometimes value creation brings benefits to our economy and our ecology at the same time. Through its Pollution Prevention Pays initiative, 3M Corporation has reduced pollution emissions by 3.8 billion pounds, at a cost savings of almost $2 billion. We saw in chapter 5 that when cap-and-trade rules were introduced into the Clean Air Act in 1990, sulfur dioxide emissions in the United States fell by two-thirds, while industry saved about $1 billion compared to the cost of traditional one-size-fits-all regulation. Similarly, when communities adopt rules that make it easier to conserve open space, this protects wildlife, enhances recreation opportunities, improves public health, and increases property values for homeowners. Costa Rica's Payment for Ecosystem Services program increases forest habitat, generates income for farmers who grow trees, pulls carbon dioxide out of the air, and offers investors from industrialized countries a cheaper way to tackle global warming.

Of course, not every social change, however visionary, can promise a win-win solution and a hands-around-the-campfire "Kumbaya." Sometimes change requires dislodging those who are entrenched in positions of power, whether a brutal dictator or an intransigent school board. When Martin Luther King Jr. said "I'm interested in power that is moral," he was not only calling for a more humanistic approach to the exercise of power.

King advocated a power-based approach to advancing human rights, recognizing that power shifts are necessary for improving the human condition. Nor is every win for the environment a win for the economy. While researchers have shown that environmental regulations frequently increase competitiveness and bolster innovation, this is not always the case. Double green solutions—those helping both your planet and your pocketbook—are more likely when our rules include clear and ambitious standards, a predictable regulatory environment, incentives for businesses to go green, phase-in periods that allow firms time to make the transition, and flexibility in the methods used to achieve environmental goals.[9]

Value creation is at the heart of efforts to reconcile biodiversity conservation and economic development in poor countries. In some countries, national parks are governed by rules that explicitly protect the rights of local communities, balancing these against the larger aims of conservation. Bolivia's Kaa-Iya del Gran Chaco National Park, for example, protects the land rights of the indigenous Guaraní people while providing habitat for what scientists estimate to be upward of 1,000 jaguars.[10] In other cases, parks have been imposed by central governments with little consideration for human rights and local livelihoods.[11] Economic development and the environment are at best overlapping circles. It would be naive to assume that protecting the environment always promotes prosperity, or that economic growth necessarily benefits the environment. The core challenge of sustainability is to push those two circles closer together. That's what value creation is all about.

5. BEG, BORROW, OR STEAL

Rulemaking experiments are underway all around the globe: New corporate codes of conduct governing mining practices in Africa. Tax incentives to turn abandoned lots into community gardens in Philadelphia. Urban "dark skies" initiatives to preserve the starscape at night in places like Flagstaff, Arizona. Multinational agreements for the conservation of shared water resources in the Great Lakes. My students have identified over 1,000 distinct policy tools that have been used for the protection of biodiversity. With so many experiences to draw on, a smart way to approach the task of reworking the rules that shape the planet is to troll for ideas that have been tried elsewhere, learning from the successes and failures of others.

Richard Rose, a political scientist at the University of Strathclyde in Glasgow, observes that "problems that are unique to one country, such as German reunification, are abnormal. The concerns for which ordinary people turn to government—education, social security, health care, safety on the streets, a clean environment, and a buoyant economy—are common on many continents."[12] Borrowing and adapting rules that were first developed elsewhere brings a number of practical advantages. For one thing, the causes of environmental outcomes are often exceedingly complex, involving a multitude of actors and interests, unpredictable market trends, and unanticipated technological breakthroughs. Add to this the uncertainties regarding the natural systems themselves, which are vast and only partially understood, and it can be difficult to anticipate on the drawing board which approaches will work and which ones will flop. There is a lot of trial and error involved in institutional innovation, and good reason to learn from the experiences of others.

Borrowing ideas from other places also marshals the power of demonstration effects, which can disarm naysayers and provide inspiration for would-be adopters of innovations. It is one thing to tell a mayor that she can revitalize her city's downtown area through innovative urban planning, and quite another to invite her for a stroll along a charming promenade in a city that has already done just that. Beyond offering a proof of concept, these earlier innovations are associated with cadres of professionals who had to figure out how to get the job done—law firms with sample contracts, arborists who understand which tree species are appropriate for public walkways, and engineering firms and building managers who learned the hard way that this cooling system works best in that type of building.

Importing ideas allows you to take advantage of peer-to-peer learning. Everett Rogers, whose pathbreaking work launched the field of communications theory, shows in his book *Diffusion of Innovations* that people are more likely to adopt new practices from others who share their social background.[13] A senior manager at a water treatment plant may be unimpressed with your group's vision for change (which might include the use of recycled "gray water" on public landscapes), dismissing it as romantic and impractical. If you introduce that person to a senior manager from another treatment plant who has already implemented such a system, a very different type of conversation follows—one that can open up opportunities for meaningful change.

6. CULTIVATE PROCESS EXPERTISE

If you wish to change the rules that rule your world, you need to involve people with process expertise—an intimate understanding of what it takes to bring about change in a specific social setting. This type of knowledge is rarely written down, but resides in the minds of people who have spent many years observing and participating in rulemaking in a given organization or political venue.[14] Seek their counsel early and often to gain an understanding of "how things work around here." Where does the decision-making power reside, both in terms of official authority, and influential stakeholders who have the power to torpedo proposals they find objectionable? What are the coalitions and histories of conflict among key players? Who controls the resources (budgets, land, staff, technical expertise) needed to bring about change? What types of arguments are most likely to find traction? Which constituencies should be consulted, and in what order? Who within a given group is most likely to entertain your new idea, and who should approach them? What has been tried in the past, and what is the track record of success? Is the timing right?

Process expertise receives little public praise or recognition, compared to more widely lauded forms of knowledge like marine science or nanotechnology. There are at least two reasons for this. First, unlike most experts, people deploying political savvy often prefer to apply their skills in a quiet manner. You won't see a professor distribute a paper titled "How I worked the system and convinced the university administration to cover the cost of my new laboratory." Nor will a mayor proudly announce to the media the political promises she had to make behind the scenes to win support for her plan. A second reason why process expertise receives less attention than it deserves is that it changes from one social setting to the next. What works in one place simply doesn't cut it in another. Scientists (social or otherwise) love findings that can be applied broadly; we tend to spurn those portions of reality that vary from place to place and, heaven forbid, change rapidly over time. It's enough to give any self-respecting researcher heartburn.

Consider the disparate political geography traversed by the endangered cerulean warbler, discussed in chapter 4, as it migrates along the foothills of the Andes in Peru, up through Central America and the Caribbean, to its seasonal home in the eastern United States and Canada. Forest habitats are rapidly disappearing along the bird's migratory route, requiring a coordinated response among conservation groups across the Americas. Yet there is not a person in the world with an in-depth understanding of how to change the rules affecting forests in each of the cerulean's landing spots, let alone the local credibility required to bring these changes about. The anthropologists and indigenous leaders who understand the rulemaking norms of the Kogi Indians on the mountain slopes of northern Colombia wouldn't know the first thing about how to change the rules governing land use in West Virginia. The Guatemalan lawmaker who has spent decades cultivating political connections throughout her country would be at a loss effecting change in neighboring Mexico, and would run afoul of fierce nationalist sentiment if she tried. Efforts to conserve migratory birds suggest a unity of purpose that transcends national borders. But any coordinated effort to save the cerulean from extinction will require tapping the parochial process expertise of hundreds of individuals who understand how to bring about change in particular places.

Faced with this complexity, we can take solace in the fact that social questions (like the ones at the beginning of this section) often travel better than do their answers. That is, we can apply a similar strategy of inquiry across many settings, even if the resulting conclusions differ from one place to the next. So where does one find the process experts capable of providing reliable answers in a given locale? If you wish to bring about change on a college campus, confer with those members of the faculty and administration who have a finger on the pulse of the place, and who know how to get things done. To reform city practices, seek out seasoned journalists, former elected officials (who may speak with greater candor than those currently in office), and others who have stewed in the broth of local politics and decision making over many years. And keep in mind, the individuals you rely on to explain how things work in their "tribe" (be it a labor union, a government agency, or a corporate sales division) may impart only a partial picture, and could have agendas of their own. Be sure to diversify your sources.

7. THINK VERTICALLY

The habit of vertical thinking does not come easily. We often assume that if you wish to improve the energy efficiency of cities, you need to work for change at the city level, right? Wrong. It may be that you should focus your efforts at the level of states and provinces, convincing rulemakers there to provide incentives to cities, perhaps through a grant program, or by tying their infrastructure funding to new requirements for energy planning. Vertical thinking also works in the opposite direction. If you want to change climate policy in your country, it might seem that the practical thing to do is put on a power suit, grab a briefcase, and head to the nation's capital. Yet it may be that what's needed most are successful demonstration programs at local levels, to convince national policymakers that the proposed change is feasible and needs scaling up.

Thus the first principle of vertical thinking is that the location of a problem, and the site of maximum leverage over that problem, are not one and the same. We saw in chapter 8 that the global movement for decentralization, which is increasing the rulemaking power of local governments and communities, was from the start a multi-tiered affair, initiated not by a groundswell from below but by national-level policymakers, and later promoted at the international level by organizations like the European Union. Similarly, progress in protecting the land rights of indigenous peoples came about when leaders of these groups worked vertically, using what Margaret Keck and Kathryn Sikkink call the "boomerang effect." Faced with intransigent national leaders, indigenous groups and their allies in the Western world turned to international organizations for help. They convinced the International Labor Organization and other UN bodies to apply pressure on national governments, which introduced major reforms in the 1990s and 2000s to recognize indigenous citizenship and traditional land rights.[15] This precipitated a major change in who rules the earth, including the return of millions of acres of land to indigenous control for the first time since Spanish conquistadors imposed new property rules five hundred years ago.

The second principle of vertical thinking is what we might call the power-sharing paradox. The power-sharing paradox runs as follows: If you want to change the balance of power between two levels of governance (such as a town and the larger county or state in which it is situated), the authority to make such a change resides at the level that already has most of the power. The paradox is that those who possess power may be loath to let it go. Yet we have seen that the effects of the power-sharing paradox are not absolute. Government entities do not seek to maximize control over decision making at any cost; often the cost is too high, and they are eager to dispatch responsibilities in order to shore up public support.[16]

Frequently rulemakers hand over responsibility for environmental management to another level of government without conferring the authority and financial resources needed to fulfill those responsibilities. Therein lies the political struggle. This is true of national governments in their posture toward scaling down (sharing power with local communities) and scaling up (deferring to international organizations). Yet here again, the power-sharing paradox is not insurmountable. National rulemakers often miscalculate, imagining that they can introduce token reforms while preventing more meaningful change. Often they are surprised when newly empowered communities demand more, or when a new international organization with limited authority (like the European Union during its formative stage) develops into a major force for change.

8. AND YES, KEEP RECYCLING

If you embrace a new outlook on environmental issues, one that reveals the deeper social institutions shaping the outcomes you care about, this in no way suggests that you should stop doing the little things like recycling, growing your own food, or purchasing compostable cups for the company picnic. It is a matter of balance, complementing these everyday acts of individual conscience with larger actions that promote social change.

It seems plausible that the retreat into individualism that characterizes so much of environmentalism today stems from a sort of aesthetic longing— a desire to make elementary connections with the soil and sky, offering rewards as tangible and immediate as that first bite of a juicy organic peach. The physicality of the natural world makes environmentalism unique among political causes, offering to those who open their senses a powerful elixir of body, mind, and spirit, all triggered by actions as simple as sitting quietly in a wooded field at sunset. As the poet Wendell Berry put it, "What we need is here." Social rules, on the other hand, offer few opportunities to quench this aesthetic thirst. They occupy a sort of shadow world, the ghosts of politics past, nudging us this way and that but ultimately untouchable and even unnoticed.

Caring for a community garden provides a very different sort of engagement with the earth than, say, learning that a local factory is raining mercury emissions down on your garden and marching over to city hall to demand a change. Yet political action can be deeply satisfying in its own right, fulfilling another primordial human need: a meaningful connection with others. When we make common cause by cultivating human relationships and building communities of shared concern—when we feel the rush of excitement at realizing that we are on to something, and that our efforts may well lead to change—then we are tending not only to the health and well-being of our physical surroundings, but to that of our inner selves. Connecting with the planet, and connecting with one another, are two important and complementary activities for the modern citizen looking for creative ways to take part in ruling the earth.

Notes

CHAPTER 1

1. Michael F. Maniates (2001) Individualization: Plant a Tree, Buy a Bike, Save the World?, *Global Environmental Politics* 1(3):31–52. Quote p. 33.
2. Chris Wilkins, Hudson Town Councilor, 1991, quoted in the film *A Chemical Reaction* (2009), directed by Brett Plymale.
3. This history of local pesticide bans in North America draws primarily on Sarah B. Pralle (2006) Timing and Sequence in Agenda-setting and Policy Change: A Comparative Study of Lawn Care Pesticide Politics in Canada and the US, *Journal of European Public Policy* 13(7):987–1005; and Plymale, op. cit.
4. Research on the susceptibility of children to pesticides and other synthetic chemicals is summarized in Philippe Grandjean and Philip J. Landrigan (2006) Developmental Neurotoxicity of Industrial Chemicals, *The Lancet* 368:2167–78.
5. See Aaron K. Todd, *Changes in Urban Stream Water Pesticide Concentrations One Year after a Cosmetic Pesticides Ban*, Environmental Monitoring and Reporting Branch, Ontario Ministry of the Environment, November 2010. For a closer look at the impact of the municipal pesticide bans on lawncare practices, see Donald C. Cole et al. (2011) Municipal Bylaw to Reduce Cosmetic/Non-essential Pesticide Use on Household Lawns—A Policy Implementation Evaluation, *Environmental Health* 10(1):1–17.
6. Mike Christie, *Private Property Pesticide By-laws in Canada: Population Statistics by Municipality*, Ottawa, December 31, 2010.
7. Marty Whitford, It's in da BAG, *Landscape Management*, September 16, 2008.
8. Five additional states (Ohio, West Virginia, Utah, Wyoming, and South Carolina) have preemption laws, but the year of adoption is unavailable

from state legislative records for inclusion in Figure 1.1. The data on preemption laws were compiled from the National Association of State Departments of Agriculture, *Federal Pesticide Regulation Preemption*, Washington, DC, June 22, 1993; Pralle, op. cit.; Elena S. Rutrick (1993) Local Pesticide Regulation Since *Wisconsin Public Intervenor v. Mortier*, Boston College Environmental Affairs Law Review 20(1):65–97; the Lexis-Nexis State Capital Database; and the websites of state agencies.

9. Data on the quantity of pesticides used on lawns throughout the United States are from Table 5.8 of US Environmental Protection Agency, *Pesticides Industry Sales and Usage 2006 and 2007—Market Estimates*, Washington, DC, 2011. This figure underreports actual usage on lawns because it includes only home use (omitting public spaces like parks and highway median strips) and excludes all professional applications in homes. The EPA data also only report the pounds of "active ingredient" used, excluding other chemicals (such as solvents and stabilizers) included in pesticide formulas.

10. Throughout the book I will use the terms "social rules" and "institutions" interchangeably, but I prefer the former for two reasons. First, when most people encounter the word "institutions" they think of organizations, like Chevron or the Environmental Protection Agency. Although organizations create and contain social rules, they are not the same thing. Second, researchers use the term "institutions" differently across disciplines. For economists, institutions are social rules, pure and simple. Within political science, the "institutions" label is commonly applied to both rules and organizations. When sociologists use the term, they mean all enduring social structures, including family, caste, religion, and social class. So rather than wade through the definitional muck surrounding institutions, I have chosen to use the vocabulary of social rules, which play a central role in every approach to institutions across the social sciences. For an overview of research traditions in institutional analysis, see Peter A. Hall and Rosemary C. R. Taylor (1996) Political Science and the Three New Institutionalisms, *Political Studies* 44(5):936–57.

11. Barnett and Finnemore point out that the power of rulemaking bodies stems not only from the sanctions and rewards associated with following their rules, but from their ability to define, through their actions and pronouncements, what is normal and right and appropriate. See Michael Barnett and Martha Finnemore, *Rules for the World: International Organizations in Global Politics*, Cornell University Press, Ithaca, NY, 2004.

12. The pace of tropical forest loss is from data collected between 2000 and 2005, reported on page 19 of UN Food and Agriculture Organization, *Global Forest Land Use Change 1990–2005*, FAO Forestry Paper 169, Rome, 2012.

13. Elinor Ostrom et al., *The Drama of the Commons*, National Academies Press, Washington, DC, 2002.

14. Tanja A. Borzel (2000) Why There is No "Southern Problem": On Environmental Leaders and Laggards in the European Union, *Journal of European Public Policy* 7(1): 141–62.

15. Examples of research on property rights as they affect environmental quality include Daniel H. Cole, *Pollution and Property: Comparing Ownership Institutions for Environmental Protection*, Cambridge University Press, New York, 2002; and Claudio Araujoa et al. (2009) Property Rights and Deforestation in the Brazilian Amazon, *Ecological Economics* 68(8–9):2461–68.

16. Kate O'Neill, The Comparative Study of Environmental Movements, pp. 115–42 in Paul F. Steinberg and Stacy D. VanDeveer, *Comparative Environmental Politics: Theory, Practice, and Prospects*, MIT Press, Cambridge, MA, 2012.

17. Melinda Herrold (2001) Which Truth? Cultural Politics and Vodka in Rural Russia, *Geographical Review* 91(1–2):295–303.

18. In 1991, Sagan participated in a remarkable debate, published in the journal *Science,* on the relation between research and action. Sagan wrote a letter to the editor about his discovery that a well-known science journalist had publicly mocked Sagan and other scientists who engaged in advocacy. "Suppose you had found that the global consequences of nuclear war were much worse than had been generally understood and that military establishments worldwide had overlooked those consequences," he wrote. "Would you think it your responsibility to keep quiet about this because the results were not absolutely certain, or because the full-scale experimental verification had not yet been obtained? Or would you consider it your obligation to your children and the children of everyone else to speak up? Keeping quiet under such circumstances seems bizarre and reprehensible to me." Another target of the journalist's attack was Harvard biologist E. O. Wilson, who has done more than anyone to publicize the problem of global species extinction. "It is reasonable," Wilson wrote in a letter that appeared alongside Sagan's, "to ask what scientists are expected to do when they hit upon a serious environmental problem. Whisper in the ear of a journalist? Entirely and chastely refrain from publishing outside technical journals, hoping the results will be discovered by nonscientists?" (Carl Sagan, Edward O. Wilson, and Daniel E. Koshland, Jr. (1993) Speaking Out, *Science* 260(5116):1861).

CHAPTER 2

1. For a critical overview of the Scottish land reform initiative, see John Bryden and Charles Geisler (2007) Community-based Land Reform: Lessons from Scotland, *Land Use Policy* 24:24–34. Cross-national differences in the rules governing coastal access are described in Peter Scott, ed., *Coastal Access in Selected European Countries: Report Prepared for The Countryside Agency*, Peter Scott Planning Services Ltd., Edinburgh, 2006.

2. Rachel Carson, *The Sea Around Us,* Oxford University Press, New York, 1951.

3. Daniel Summerlin (1996) Improving Public Access to Coastal Beaches: The Effect of Statutory Management and the Public Trust Doctrine, *William and Mary Environmental Law & Policy Review* 20:425–44; and Katherine Niven (1978) Beach Access: An Historical Overview, *New York Sea Grant Law and Policy Journal* 2:161–99.

4. Illinois Central Railroad Co. v. State of Illinois. 146 U.S. 387 (1892). The New Jersey decision quoted by the Supreme Court is *Robert Arnold v. Benajah Mundy*. Supreme Court of New Jersey. 6 N.J.L. 1 (1821).

5. Pamela Pogue and Virginia Lee (1999) Providing Public Access to the Shore: The Role of Coastal Zone Management Programs, *Coastal Management* 27:219–37.

6. For examples of the coastal access movement, see Marc R. Poirier (1996) Environmental Justice and the Beach Access Movements of the 1970s in Connecticut and New Jersey: Stories of Property and Civil Rights, *Connecticut Law Review* 28:719–812.

7. Surfrider Foundation is at the forefront of efforts to prevent private landowners from illegally blocking access to the US shoreline. The city of Malibu, California, is among the holdouts.

8. For a personal account of the effort to desegregate beaches, see Gilbert R. Mason with James Patterson Smith, *Beaches, Blood, and Ballots: A Black Doctor's Civil Rights Struggle*, University Press of Mississippi, Jackson, 2000.

9. Ronald B. Mitchell, *Intentional Oil Pollution at Sea: Environmental Policy and Treaty Compliance*, MIT Press, Cambridge, MA, 1994.

10. For an overview of the state-by-state battle to clean the air, see Scott H. Dewey, *Don't Breathe the Air: Air Pollution and U.S. Environmental Politics, 1945–1970*, Texas A&M University Press, College Station, 2000.

11. James M. Acheson, *Capturing the Commons: Devising Institutions to Manage the Maine Lobster Industry*, University Press of New England, Lebanon, NH, 2003.

12. Jonathan Roughgarden and Fraser Smith (1996) Why Fisheries Collapse and What to Do About It, *Proceedings of the National Academy of Sciences* 93(10):5078–83.

13. Edward A. Parson, *Protecting the Ozone Layer: Science and Strategy*, Oxford University Press, New York, 2003.

14. James J. Corbett et al. (2007) Mortality from Ship Emissions: A Global Assessment, *Environmental Science & Technology* 41(24):8512–18.

15. In the original French, Napoleon was quoted by his assistant as saying "Ma vraie gloire n'est pas d'avoir gagné quarante batailles; Waterloo effacera le souvenir de tant de victoires; ce que rien n'effacera, ce qui vivra éternellement, c'est mon Code Civil."

CHAPTER 3

1. Christopher P. Hood (2006) From Polling Station to Political Station? Politics and the Shinkansen, *Japan Forum* 18(1):45–63; Henrik Selin and Stacy D. VanDeveer (2006) Raising Global Standards: Hazardous Substances and E-Waste Management in the European Union, *Environment* 48(10):7–18; and Martin E. Halstuk and Bill F. Chamberlin (2006) The Freedom of Information Act 1966–2006: A Retrospective on the Rise of Privacy Protection Over the Public Interest in Knowing What the Government's Up To, *Communication Law and Policy* 11(4):511–64.

2. David Gottfried, *Greed to Green: The Transformation of an Industry and a Life*, WorldBuild Publishing, Berkeley, 2004. Quote p. 5.

3. There is a large research literature on certification rulemaking systems that evaluate and publicize the environmental impacts of goods and services. See, for example, Benjamin Cashore, Graeme Auld, and Deanna Newsom, *Governing Through Markets: Forest Certification and the Emergence of Non-state Authority*, Yale University Press, New Haven, CT, 2004; Matthew Potoski and Aseem Prakash (2005) Green Clubs and Voluntary Governance: ISO 14001 and Firms' Regulatory Compliance, *American Journal of Political Science* 49(2):235–48; and Mrill Ingram and Helen Ingram, Creating Credible Edibles: The Organic Agriculture Movement and the Emergence of U.S. Federal Organic Standards, pp. 121–48 in David S. Meyer, Valerie Jenness, and Helen Ingram (eds.), *Routing the Opposition: Social Movements, Public Policy, and Democracy*, University of Minnesota Press, Minneapolis, 2005.

4. Data on the expansion of LEED are from the US Green Building Council. Additional industry estimates are from McGraw-Hill Construction, *Green Outlook 2013*, summarized at http://www.construction.com/about-us/press/green-building-outlook-strong-for-both-non-residential-and-residential.asp

5. Daniel C. Matisoff, Douglas S. Noonan, and Anna M. Mazzolini (2014) Performance or Marketing Benefits? The Case of LEED Certification, *Environmental Science & Technology* 48(3):2001–07.

6. H. L. Jelks et al. (2008) Conservation Status of Imperiled North American Freshwater and Diadromous Fishes, *Fisheries* 33(8):372–407.

7. André Nel (2005) Air Pollution-Related Illness: Effects of Particles, *Science* 308:804–06. On the health impact of particulates in the United States, see also C. Arden Pope III, Majid Ezzati, and Douglas W. Dockery (2009) Fine-Particulate Air Pollution and Life Expectancy in the United States, *New England Journal of Medicine* 360:376–86. War fatality data are from Ziad Obermeyer, Christopher J. L. Murray, and Emmanuela Gakidou (2008) Fifty Years of Violent War Deaths from Vietnam to Bosnia: Analysis of Data from the World Health Survey Programme, *British Medical Journal* 336(7659):1482–86.

8. Jos G. J. Oliver et al., *Long-term Trends in Global CO_2 Emissions: 2011 Report*. PBL Netherlands Environmental Assessment Agency, The Hague, 2011.

9. Kathleen Reytar, Mark Spalding, and Allison Perry, *Reefs at Risk Revisited*, World Resources Institute, Washington, DC, 2011.

10. World Health Organization and UNICEF, *Progress on Drinking Water and Sanitation: 2013 Update*, Geneva, 2013.

11. The potential for fear-based strategies to promote or hinder environmental behavior is discussed on p. 526 of Paul C. Stern (2000) Psychology and the Science of Human-Environment Interactions, *American Psychologist* 55(5):523–30.

12. For an insightful overview of research on the limits of human decision making, see Richard H. Thaler and Cass R. Sunstein, *Nudge: Improving Decisions about Health, Wealth, and Happiness*, Yale University Press, New Haven, CT, 2008. Where I part ways with these authors is in their insistence that nonbinding guidelines are inherently superior to social rules such as laws and policies. Historically, great advances in human rights, environmental protection, and other areas have never come about from a nudge. The founders of the United States did not nudge the British to grant them independence. Nor was the civil rights era a polite nudge. Our air is cleaner now not because companies were asked to please consider changing their production processes, or drivers to kindly contemplate using catalytic converters on their cars. Indeed, the ability of scholars to publicly disagree on the importance of binding social rules is itself guaranteed by a binding social rule—the First Amendment to the US Constitution.

13. Herbert A. Simon (1990) Invariants of Human Behavior, *Annual Review of Psychology* 41:1–19. Quotes are from page 17.

14. Gottfried 2004, op. cit., p. 69.

15. Mancur Olson, *The Logic of Collective Action: Public Goods and the Theory of Groups*, Schocken Books, New York, revised edition, 1971. Quotes are from page 2.

16. These members-only benefits take many forms. The young civil rights activists who endured humiliation and violence by sitting at whites-only lunch counters throughout the American South in the 1960s were not just contributing to a greater cause. After all, why not let someone else risk their neck and then enjoy the shared benefits of progress in civil rights? Civil rights protesters were motivated by an intense psychological gratification arising from feelings of solidarity and mutual obligation with the small groups of protesters with whom they trained, marched, sang, and shared jail cells. For an overview of research on solidarity in social movements, see Scott A. Hunt and Robert D. Benford, Collective Identity, Solidarity, and Commitment, pp. 433–57 in David Snow, Sarah A. Soule, and Hanspeter Kriesi (eds.), *The Blackwell Companion to Social Movements*, John Wiley & Sons, Hoboken, NJ, 2008.

17. Michael D. Cohen, James G. March, and Johan P. Olsen (1972) A Garbage Can Model of Organizational Choice, *Administrative Science Quarterly* 17(1):1–25.

18. David Osborne and Ted Gaebler, *Reinventing Government: How the Entrepreneurial Spirit Is Transforming the Public Sector*, Addison-Wesley, Reading, MA, 1992.

19. James G. March and Johan P. Olsen, Elaborating the "New Institutionalism," pp. 3–20 in R. A. W. Rhodes, Sarah A. Binder, and Bert A. Rockman, *The Oxford Handbook of Political Institutions*, Oxford University Press, New York, 2006. Quote p. 15.

20. For a contemporary example of a realist perspective on foreign policy, see Charles Krauthammer, *Democratic Realism: An American Foreign Policy for a Unipolar World*, The Irving Kristol Lecture, American Enterprise Institute Press, Washington, DC, 2004. For an alternative view see Joseph Nye, Jr., *The Future of Power*, Public Affairs, New York, 2011.

21. Reinhold Niebuhr, *Moral Man and Immoral Society: A Study of Ethics and Politics*, Westminster John Knox Press, Louisville, KY, 2002 (orig. 1932). Quotes pp. xxv–xxvi.

22. Niebuhr, op. cit., pp. 26 and 28. Niebuhr discusses at length the need to distinguish between immoral and morally justifiable uses of coercive power. In ways that Niebuhr does not fully acknowledge, this process of weighing the ethics of coercive action requires the very type of moral reasoning and deliberation that Niebuhr downplays as a vehicle for social progress.

23. John Gaventa, *Power and Powerlessness: Quiescence and Rebellion in an Appalachian Valley*, University of Illinois Press, Urbana, 1980.

24. The distinction among these dimensions of power was originally developed by Steven Lukes in *Power: A Radical View*, Macmillan, New York, 1974. See also Peter Bachrach and Morton S. Baratz (1962) Two Faces of Power, *American Political Science Review* 56(4):947–52.

25. Center for Responsive Politics, 2013.

26. In recent years the concept of social entrepreneurship has gained popularity in the nonprofit sector and has found a foothold in business school curricula. See J. Gregory Dees, The Meaning of "Social Entrepreneurship," unpublished manuscript, Graduate School of Business, Stanford University, May 30, 2001; and David Bornstein, *How to Change the World: Social Entrepreneurs and the Power of New Ideas*, Oxford University Press, New York, 2004.

27. Roger Fisher and William Ury, *Getting to Yes: Negotiating Agreement without Giving In*, Houghton Mifflin, Boston, 1981.

28. In 1993, the economist Joel Waldfogel published a paper calculating the economic loss from gift giving during the holiday season. He surveyed students, asking them to calculate two figures regarding their gifts: how much each gift was worth to them personally (what they would be willing to pay for it in a store), and their best guess as to the actual purchase price. On this basis, Waldfogel estimated that holiday gift giving results in a $4 billion annual loss in value in the United States. Of course, this research did not measure the inherent value that people place on the process of selecting and exchanging gifts. (It's the thought that counts, right?) But the Waldfogel study is a classic demonstration of how social value can be created or destroyed. See Anon., Is Santa a Deadweight Loss?, *The Economist*, December 20, 2001; and Joel Waldfogel (1993) The Deadweight Loss of Christmas, *The American Economic Review* 83(5):1328–36.

29. Eugene Bardach, *A Practical Guide for Policy Analysis: The Eightfold Path to More Effective Problem Solving*, Chatham House, New York, 2000.

CHAPTER 4

1. Michael C. Blumm and Lucas Ritchie (2005) The Pioneer Spirit and the Public Trust: The American Rule of Capture and State Ownership of Wildlife, *Environmental Law* 35(4):101–47.

2. Missouri v. Holland, 252 U.S. 416 (1920).

3. William Blackstone, *Commentaries on the Laws of England*, Vol. 2., University of Chicago Press, Chicago, 1979 (1766), p. 2, as quoted in Bruce G. Carruthers and Laura Ariovich (2004) The Sociology of Property Rights, *Annual Review of Sociology* 30:23–46. Quote on p. 23.

4. Adam Smith, *An Inquiry into the Nature and Causes of the Wealth of Nations*, Pennsylvania State Electronic Classics Series Publication, 2005 [1776]. Quote p. 580.

5. Smith, op. cit., p. 337.

6. Carol Rose, *Property and Persuasion: Essays on the History, Theory and Rhetoric of Ownership*, Westview Press, Boulder, CO, 1994.

7. The science of climate change is not nearly as uncertain as some political pundits would have it. See William R. L. Anderegg et al. (2010) Expert Credibility in Climate Change, *Proceedings of the National Academy of Sciences* 107(27):12107–09.

8. Joseph Singer, *Entitlement: The Paradoxes of Property*, Yale University Press, New Haven, CT, 2000. Quote p. 9.

9. The BBS data reported here are from North American Bird Conservation Initiative, U.S. Committee, *The State of the Birds: United States of America, 2009*, U.S. Department of Interior, Washington, DC, 2009.

10. Birdlife International, *State of the World's Birds: Indicators for Our Changing World*, Birdlife International, Cambridge, UK, 2008.

11. North American Bird Conservation Initiative, op. cit.; and Birdlife International, op. cit.

12. Information on the migratory route and potential landing spots of the cerulean warbler draws on Theodore A. Parker III (1994) Habitat, Behavior, and Spring Migration of Cerulean Warbler in Belize, *American Birds* 48(1):70–75; various publications of the Cerulean Warbler Technical Group, including Paul B. Hamel, Deanna K. Dawson, and Patrick D. Keyser (2004) How We Can Learn More about the Cerulean Warbler (*Dendroica cerulea*), *The Auk* 121(1):7–14; S. Barker et al., *Modeling the South American Range of the Cerulean Warbler*, ESRI International User Conference Papers, San Diego, August 2006; and dozens of additional articles documenting the bird's presence in particular sites.

The landing spots highlighted in this chapter were chosen on the basis of two criteria. The first is documented or plausible landing sites for cerulean populations. I judged a site to be a plausible stopover if its landscape is characterized by the cerulean's preferred habitat—forested mountain slopes at low- to mid-elevations—and is located in a subnational region with documented sightings of ceruleans. My second criterion was that the site must be the focus of high-quality social science research on the subject of property rights. Often this meant bypassing a region known to be a major stopover for ceruleans in favor of another with a thinner biological record.

13. Russell A. Mittermeier and Timothy B. Werner (1990) Wealth of Plants and Animals Unites "Megadiversity" Countries, *Tropicus* 4(1):1, 4–5.

14. The cultural history of Callanga is described in Glenn H. Shepard, Jr., Klaus Rummenhoeller, Julia Ohl-Schacherer, and Douglas W. Yu (2010) Trouble in Paradise: Indigenous Populations, Anthropological Policies, and Biodiversity Conservation in Manu National Park, Peru, *Journal of Sustainable Forestry* 29:252–301.

15. The vertical archipelago of Andean-to-Amazon trade was originally described by John Murra. See John Victor Murra, Andean Societies Before 1532, pp. 59–90 in Leslie Bethell (ed.), *The Cambridge History of Latin America, Vol. 1: Colonial Latin America*, Cambridge University Press, 1984.

16. Catherine J. Julian (1988) How the Inca Decimal Administration Worked, *Ethnohistory* 35(3):257-79. The quipu system and its numeric and linguistic meaning have yet to be fully deciphered and are the subject of ongoing research and debate. See Charles Mann (2005) Unraveling Khipu's Secrets, *Science* 309(5737):1008-09.

17. Friar Diego de Landa, *Yucatan Before and After the Conquest*, translated by William Gates, Dover Publications, New York, 1978. Quote pp. 159-60.

18. Gates, op. cit., p. vi.

19. Letter from Christopher Columbus to Ferdinand and Isabella, January 20, 1494, as quoted on p. 223 of Nicolás Wey Gómez, *The Tropics of Empire: Why Columbus Sailed South to the Indies*, MIT Press, Cambridge, MA, 2008.

20. Timothy J. Yeager (1995) Encomienda or Slavery? The Spanish Crown's Choice of Labor Organization in Sixteenth-Century Spanish America, *The Journal of Economic History* 55(4):842-59.

21. The most reliable estimate of silver and gold exports comes from Clarence Haring, a historian who in 1915 made a systematic review of the original ledgers of the royal treasurers of the Spanish colonial government, housed in library archives in Seville. Based on meticulous calculations, Haring arrived at the estimate of 1.6 million pounds of silver and gold exported from Peru to Spain between 1533 and 1560. Clarence H. Haring (1915) Gold and Silver Production in the First Half of the Sixteenth Century, *Quarterly Journal of Economics* 29(3):433-79.

22. James Lockhart (1969) Encomienda and Hacienda: The Evolution of the Great Estate in the Spanish Indies, *The Hispanic American Historical Review* 49(3):411-29; Kay, op. cit., p. 189.

23. Mary L. Barker (1980) National Parks, Conservation, and Agrarian Reform in Peru, *The Geographical Review* 70(1):1-18.

24. Cristóbal Kay, The Agrarian Reform in Peru: An Assessment, pp. 185-239 in A. K. Ghose (ed.), *Agrarian Reform in Contemporary Developing Countries*, Croom Helm, London, and St. Martin's Press, New York, 1983.

25. Forest ownership data are from Arun Agrawal, Local Institutions and the Governance of Forest Commons, pp. 313-40 in Paul F. Steinberg and Stacy D. VanDeveer (eds.), *Comparative Environmental Politics: Theory, Practice, and Prospects*, MIT Press, Cambridge, MA, 2012; and William D. Sunderlin, Jeffrey Hatcher, and Megan Liddle, *From Exclusion to Ownership? Challenges and Opportunities in Advancing Forest Tenure Reform*, The Rights and Resources Initiative, Washington, DC, 2008.

26. Jessica Hidalgo and Carlos Chirinos, *Manual de Normas Legales sobre Tala Ilegal* (Manual of Legal Norms Concerning Illegal Logging), Sociedad Peruana de Derecho Ambiental (Peruvian Society for Environmental Law) and International Resources Group, Lima, 2005, as cited by Robin R. Sears and Miguel Pinedo-Vasquez (2011) Forest Policy Reform and the Organization of Logging in Peruvian Amazonia, *Development and Change* 42:609-31.

27. Sears and Pinedo-Vasquez, op. cit., p. 613.

28. Paulo J. C. Oliveira et al. (2007) Land-Use Allocation Protects the Peruvian Amazon, *Science* 317(5842):1233-36.

29. Sears and Pinedo-Vasquez, op. cit. Quote p. 617.

30. Matt Finer et al. (2008) Oil and Gas Projects in the Western Amazon: Threats to Wilderness, Biodiversity, and Indigenous Peoples, *PLOS ONE* 3(8):e2932.

31. Fabio Sánchez, María del Pilar López-Uribe and Antonella Fazio (2010) Land Conflicts, Property Rights, and the Rise of the Export Economy in Colombia, 1850-1925, *The Journal of Economic History* 70(2):378-99.

32. Andrés Guhl, Coffee Production Intensification and Landscape Change in Colombia, 1970–2002, pp. 93–116 in Wendy Jepson and Andrew Millington (eds.), *Land Change Science in the Tropics: Changing Agricultural Landscapes*, Springer Verlag, 2008.

33. Guhl, op. cit.

34. The amount of forest cover maintained under shade-grown coffee systems is discussed in Patricia Moguel and Victor M. Toledo (1999) Biodiversity Conservation in Traditional Coffee Systems of Mexico, *Conservation Biology* 13(1):11–21.

35. On the biodiversity benefits of wildlands versus plantations, see Jos Barlow, Luiz A. M. Mestre, Toby A. Gardner, and Carlos A. Peres (2007) The Value of Primary, Secondary and Plantation Forests for Amazonian Birds, *Biological Conservation* 136:212–31.

36. http://www.rainforest-alliance.org/multimedia/migratory-bird-day.

37. Camilo Montes et al. (2010) Clockwise Rotation of the Santa Marta Massif and Simultaneous Paleogene to Neogene Deformation of the Plato-San Jorge and Cesar-Ranchería Basins, *Journal of South American Earth Sciences* 29(4):832–48.

38. Ralf Strewe and Cristobal Navarro (2004) New and Noteworthy Records of Birds from the Sierra Nevada de Santa Marta Region, *Bulletin of the British Ornithologists' Club* 124(1):38–51.

39. A fourth group, the Kankuamo, share a common lineage with these three but have assimilated into mainstream Colombian society. I provide less background on the Kankuamo because they do not share the same perspective on property and land management.

40. Astrid Ulloa (2009) Indigenous Peoples of the Sierra Nevada de Santa Marta-Colombia: Local Ways of Thinking [sic] Climate Change, *Institute of Physics (IOP) Conference Series: Earth and Environmental Science* 6:1–2; G. Reichel-Dolmatoff (1982) Cultural Change and Environmental Awareness: A Case Study of the Sierra Nevada de Santa Marta, Colombia, *Mountain Research and Development* 2(3):289–98; and Guillermo E. Rodríguez-Navarro (2000) Indigenous Knowledge as an Innovative Contribution to the Sustainable Development of the Sierra Nevada of Santa Marta, Colombia, *Ambio* 29(7):455–58.

41. *The Heart of the World: Elder Brothers' Warning* (1990), documentary film directed by Alan Ereira, British Broadcasting Corporation.

42. Reichel-Dolmatoff, op. cit., p. 293.

43. Ibid.

44. The term "indigenous peoples" is a broad category denoting descendants of cultures that were in place before the arrival of Europeans, whom the Spaniards referred to as Indians. Most Latin Americans are either directly descended from the Spanish, or have mixed Spanish-indigenous roots ("mestizo"), or are indigenous. In Colombia, Brazil, and elsewhere, the descendants of African slaves also constitute an important and culturally distinct part of the population.

45. For an overview of the domestic legislative accomplishments of the indigenous rights movement, see Roque Roldán Ortiga, *Models for Recognizing Indigenous Land Rights in Latin America*, The World Bank, Washington, DC, 2004.

46. Website of the Gonawindua Tayrona Organization, www.tairona.org, accessed April 10, 2012.

47. J. V. Remsen, Jr. and Thomas S. Schulenberg (1997) The Pervasive Influence of Ted Parker on Neotropical Field Ornithology, *Studies in Neotropical Ornithology* 48:7–19.

48. Noel Maurer and Carlos Yu, *The Big Ditch: How America Took, Built, Ran, and Ultimately Gave Away the Panama Canal*, Princeton University Press, Princeton, NJ, 2011; William H. Chaloner (1959) The Birth of the Panama Canal, 1869–1914, *History Today* 9(7):482–92; and John M. Thompson (2011) "Panic-Struck Senators, Businessmen and Everybody Else": Theodore Roosevelt, Public Opinion, and the Intervention in Panama, *Theodore Roosevelt Association Journal* 32(1/2):7–28.

49. Zachary Langford et al., *Socio-Environmental Impacts of Land Cover Change in the Panama Canal Watershed*, paper prepared for the 62nd International Astronautical Congress, Cape Town, South Africa, 2010.

50. Mike Fotos, Quint Newcomer, and Radha Kuppalli (2007) Policy Alternatives to Improve Demand for Water-Related Ecosystem Services in the Panama Canal Watershed, *Journal of Sustainable Forestry* 25(1–2):195–216.

51. Rodrigo A. Arriagada et al. (2012) Do Payments for Environmental Services Affect Forest Cover? A Farm-Level Evaluation from Costa Rica, *Land Economics* 88:382–99; and Ina Porras, *Fair and Green? Social Impacts of Payments for Environmental Services in Costa Rica*, International Institute for Environment and Development, London, December 2010.

52. The authors evaluated a 3,000-square-kilometer region of Costa Rica called Sarapiquí. They point out that the Payment for Ecosystem Services program in this region is well run, and may or may not be representative of the impact of the program in the rest of the country.

53. David Barton Bray and Peter Klepeis (2005) Deforestation, Forest Transitions, and Institutions for Sustainability in Southeastern Mexico, 1900–2000, *Environment and History* 11:195–223; and Antonio García de León (2005) From Revolution to Transition: The Chiapas Rebellion and the Path to Democracy in Mexico, *The Journal of Peasant Studies* 32(3/4):508–27.

54. Camille Antinori and David Barton Bray (2005) Community Forest Enterprises as Entrepreneurial Firms: Economic and Institutional Perspectives from Mexico, *World Development* 33(9):1529–43.

55. Reforms introduced in 1992 loosened legal restrictions on the use of ejido property, allowing the buying and selling of lots within ejidos and the right to devote some lands to private property, following a two-thirds vote of approval by the community. See Grenville Barnes (2009) The Evolution and Resilience of Community-based Land Tenure in Rural Mexico, *Land Use Policy* 26(2):393–400. Figures on the current number of ejidos in Mexico are from the most recent ejido census: Instituto Nacional de Estadística, Geografía e Informática, *Resultados Preliminares del IX Censo Ejidal* (Preliminary Results of the 9th Ejido Census), Comunicado Número 069/08, Aguascalientes, Mexico, April 11, 2008.

56. The description of the Community Forest Enterprise in Ixtepeji draws on Ross E. Mitchell (2006) Environmental Governance in Mexico: Two Case Studies of Oaxaca's Community Forest Sector, *Journal of Latin American Studies* 38:519–48; and Salvador Anta Fonseca, *Forest Management in the Community Enterprise of Santa Catarina Ixtepeji, Oaxaca, Mexico*, The Rights and Resources Initiative, Washington, DC, 2007.

57. Peter T. Leeson (2007) An-arrgh-chy: The Law and Economics of Pirate Organization, *Journal of Political Economy* 115(6):1049–94. Quote p. 1072.

58. A concise overview of the history of Law of the Sea negotiations is provided in David D. Caron, Negotiating Our Future with the Oceans, pp. 25–34 in Laurence Tubiana, Pierre Jacquet, and Rajendra K. Pachauri (eds.), *A Planet for Life 2011—Oceans:*

The New Frontier, Institute for Sustainable Development and International Relations, and the French Development Agency, Paris, 2011.

59. Arvid Pardo, Maltese Ambassador to the United Nations, UN General Assembly, 22nd Session, general debate with respect to Agenda Item 92, New York, November 1, 1967.

60. The baseline used to calculate the edge of a country's shore is a bit more complex than this, but close enough for our purposes. For further details see Caron, op. cit.

61. Juliet Eilperin, U.S. Oil Drilling Regulator Ignored Experts' Red Flags on Environmental Risks, *Washington Post*, May 25, 2010, p. A1; and Ian Urbina, Inspector General's Inquiry Faults Regulators, *New York Times*, May 24, 2010, p. A16.

62. Ronald M. Atlas and Terry C. Hazen (2011) Oil Biodegradation and Bioremediation: A Tale of the Two Worst Spills in U.S. History, *Environmental Science and Technology* 45(16):6709–15; and David B. Irons (2000) Nine Years after the *Exxon Valdez* Oil Spill: Effects on Marine Bird Populations in Prince William Sound, Alaska, *The Condor* 241:723–37.

63. Bruce A. Stein, Lynn S. Kutner, and Jonathan S. Adams (eds.), *Precious Heritage: The Status of Biodiversity in the United States*, Oxford University Press, New York, 2000.

64. The extent of the environmental destruction wrought by mountaintop removal is documented in Brian D. Lutz, Emily S. Bernhardt, and William H. Schlesinger (2013) The Environmental Price Tag on a Ton of Mountaintop Removal Coal, *PLOS ONE* 8(9):e73203; and US Environmental Protection Agency, *The Effects of Mountaintop Mines and Valley Fills on Aquatic Ecosystems of the Central Appalachian Coalfields*, Washington, DC, March 2011, EPA/600/R-09/138F.

65. On the distinction between property and control over resources, see Jesse C. Ribot and Nancy Lee Peluso (2003) A Theory of Access, *Rural Sociology* 68(2):153–81. See also Aileen McHarg, Barry Barton, Adrian Bradbrook, and Lee Godden (eds.), *Property and the Law in Energy and Natural Resources*, Oxford University Press, New York, 2010.

66. Jeff Goodell, *Big Coal: The Dirty Secret Behind America's Energy Future*, Houghton Mifflin Harcourt, Boston, 2007.

67. Goodell, op. cit., p. 23 and Federal Election Committee data available at www.fec.gov.

68. Land Trust Alliance, *2010 National Land Trust Census Report*, Washington, DC, 2011. The National Conservation Easement Database provides a superbly fine-grained analysis of the use of this new property rule throughout the country. See http://www.conservationeasement.us

69. Interview with Dr. Patrick Angel, Senior Forester/Soil Scientist, US Department of the Interior, Office of Surface Mining Reclamation and Enforcement, Appalachian Regional Office, March 22, 2012.

CHAPTER 5

1. US Treasury Department, Public Health Service, Proceedings of a Conference to Determine Whether or Not There Is a Public Health Question in the Manufacture, Distribution, or Use of Tetraethyl Leaded Gasoline, *Public Health Bulletin* 158, Government Printing Office, Washington, DC, 1925. Quotes from p. 98. See also Alice Hamilton, Paul Reznikoff, and Grace M. Burnham (1925) Tetraethyl Lead, *Journal of the American Medical Association* 84(20):1481–86; and Alice Hamilton, *Exploring the Dangerous Trades: The Autobiography of Alice Hamilton*, M.D., OEM Press, Beverly, MA, 1995 (orig. 1943).

2. For a harrowing and insightful account of the health effects of lead use over the centuries, see Richard P. Wedeen, *Poison in the Pot: The Legacy of Lead*, Southern Illinois University Press, Carbondale, 1984. Equally valuable is Christian Warren, *Brush with Death: A Social History of Lead Poisoning*, Johns Hopkins University Press, Baltimore, MD, 2000.

3. Letter from Benjamin Franklin to Benjamin Vaughan, July 31, 1786. Quoted in Carey P. McCord (1953) Lead and Lead Poisoning in Early America; Benjamin Franklin and Lead Poisoning, *Industrial Medicine & Surgery* 22(9):393-99. Quote pp. 398-99.

4. Table 2 of James L. Pirkle et al. (1994) The Decline in Blood Lead Levels in the United States: The National Health and Nutrition Examination Surveys, *Journal of the American Medical Association* 272(4):284-91. On the relative contributions of leaded gasoline and interior household paints to elevated blood lead levels, see Howard W. Mielke and Patrick L. Reagan (1998) Soil Is an Important Pathway of Human Lead Exposure, *Environmental Health Perspectives* 106(Supplement 1):217-29.

5. In Europe, the removal of lead from gasoline followed a combination of an international treaty, national requirements, and market-based regulation, imposing a higher tax rate on ethyl gas to promote the switch. Germany was the exception to the rule of Europe lagging behind the United States, having introduced policies to promote unleaded gas in 1972. Hammar and Löfgren argue that tradable permits would have made less sense in Europe because there were relatively few refineries (making for a smaller, more homogenous market) and the program would have been difficult to administer given the large number of currencies in circulation prior to European unification. See Henrik Hammar and Åsa Löfgren, Leaded Gasoline in Europe: Differences in Timing and Taxes, pp. 192-205 in Winston Harrington, Richard D. Morgenstern, and Thomas Sterner (eds.), *Choosing Environmental Policy: Comparing Instruments and Outcomes in the United States and Europe*, Resources for the Future, Washington, DC, 2004; and Hans von Storch et al. (2003) Four Decades of Gasoline Lead Emissions and Control Policies in Europe: A Retrospective Assessment, *Science of the Total Environment* 311:151-76.

6. "Elevated" blood lead levels refers to 10 micrograms of lead per deciliter of blood, the amount that the US Centers for Disease Control consider sufficiently dangerous to warrant local mitigation measures. These data were collected in the NHANES survey, an ambitious national health screening program combining household interviews with blood samples drawn in a mobile medical unit. The level dropped from 88.2 percent of children ages one to five in the first period surveyed (1976-80) to 8.9 percent in the second period (1988-91). See Table 2 of Pirkle et al., op. cit. The sources of remaining exposures are discussed in Mielke and Reagan 1998, op. cit.

7. *Arctic ice*: Units are picograms of lead per gram of snow or ice. Source: Jean-Pierre Candelone et al. (1995) Post-Industrial Revolution Changes in Large-scale Atmospheric Pollution of the Northern Hemisphere by Heavy Metals as Documented in Central Greenland Snow and Ice, *Journal of Geophysical Research: Atmospheres* 100(8):16605-16. *California coastal sediments*: Units are parts per billion of lead, normalized to background aluminum levels. Source: Bruce P. Finney and Chih An Huh (1989) History of Metal Pollution in the Southern California Bight: An Update, *Environmental Science & Technology* 23(3):294-303. *Swedish lake sediments*: Units are micrograms of lead per gram of dry sediment. Source: Ingemar Renberg, Maja-Lena Brännvall, Richard Bindler, and Ove Emteryd (2000) Atmospheric Lead Pollution History during Four Millennia (2000 BC to 2000 AD) in Sweden, *Ambio* 29(3):150-56. *Blood lead levels*: Units are micrograms of lead per deciliter of blood. Sources: Robert L. Jones et al. (2009) Trends

in Blood Lead Levels and Blood Lead Testing among US Children Aged 1 to 5 Years, 1988–2004, *Pediatrics* 123(3):e376–85; and Hans Von Storch et al. (2003) Four Decades of Gasoline Lead Emissions and Control Policies in Europe: A Retrospective Assessment, *Science of the Total Environment* 311(1):151–76.

8. Eiliv Steinnes et al. (1994) Atmospheric Deposition of Trace Elements in Norway: Temporal and Spatial Trends Studied by Moss Analysis, *Water, Air, and Soil Pollution* 74(1–2):121–40; Hans Von Storch et al., op. cit.; Ylva Lind, Anders Bignert, and Tjelvar Odsjö (2006) Decreasing Lead Levels in Swedish Biota Revealed by 36 Years (1969–2004) of Environmental Monitoring, *Journal of Environmental Monitoring* 8(8):824–34; Roberto Bono et al. (1995) Updating about Reductions of Air and Blood Lead Concentrations in Turin, Italy, Following Reductions in the Lead Content of Gasoline, *Environmental Research* 70(1):30–34; and Candelone et al., op. cit.

9. We lack detailed data on the price of permits during this early experiment, so a precise figure for cost savings is unavailable. Based on the volume of permits traded, environmental economists Robert Hahn and Gordon Hester estimate the figure to run roughly in the hundreds of millions. See Robert W. Hahn and Gordon L. Hester (1989) Marketable Permits: Lessons for Theory and Practice, *Ecology Law Quarterly* 16:361–406.

10. James Q. Wilson, *Bureaucracy: What Government Agencies Do and Why They Do It*, BasicBooks, New York, 1989. Quote p. 114.

11. Edward D. Andrews, *The People Called Shakers: A Search for the Perfect Society*, Dover Publications, New York, 1963. Quote p. 257.

12. Michael E. Porter and Mark R. Kramer (2006) Strategy and Society: The Link Between Competitive Advantage and Corporate Social Responsibility, *Harvard Business Review* 84(12):78–92.

13. Tiffany & Co.'s rules to prevent commerce in conflict diamonds are discussed in Matthew Schuerman (2004) Behind the Glitter: Tiffany and Co. Moves to Get African "Conflict Diamonds" Out of Its Stores, *Stanford Social Innovation Review* 2(2).

14. See Benjamin Cashore, Graem Auld, and Deanna Newsom, *Governing through Markets: Forest Certification and the Emergence of Non-State Authority*, Yale University Press, New Haven, CT, 2004.

15. For an overview of corporate reforms designed to promote the public interest, see David Vogel, *The Market for Virtue: The Potential and Limits of Corporate Social Responsibility*, Brookings Institution, Washington, DC, 2005.

16. Douglass North, *Institutions, Institutional Change, and Economic Performance*, Cambridge University Press, New York, 1990.

17. See Peter Evans and James E. Rauch (1999) Bureaucracy and Growth: A Cross-National Analysis of the Effects of "Weberian" State Structures on Economic Growth, *American Sociological Review* 64(5):748–65; Peter Evans, *Embedded Autonomy: States and Industrial Transformation*, Princeton University Press, Princeton, NJ, 1995; and Stephan Haggard (2004) Institutions and Growth in East Asia, *Studies in Comparative International Development* 38(4):53–81.

18. Dani Rodrik (2000) Institutions for High-Quality Growth: What They Are and How to Acquire Them, *Studies in Comparative International Development* 35(3):3–31. Quote p. 4.

19. Daron Acemoglu and James A. Robinson, *Why Nations Fail: The Origins of Power, Prosperity and Poverty*, Crown Publishers, New York, 2012.

20. Arthur C. Pigou, *The Economics of Welfare: Volume 1*, Transaction Publishers, 2009 [orig. 1920], New Brunswick, NJ. Quote p. 184.

21. This estimate takes into account government expenses as well as the war's broader impact on the US economy. Joseph E. Stiglitz and Linda J. Bilmes, The True Cost of the Iraq War: $3 Trillion and Beyond, *Washington Post*, September 5, 2010.

22. In 2013, 134.51 billion gallons of gas were consumed in the United States according to the US Energy Information Administration.

23. Anon., Global Warming Will Have Significant Economic Impacts on Florida Coasts, Reports State, *Science Daily*, October 1, 2008.

24. Fred Krupp and Miriam Horn, *Earth: The Sequel*, W.W. Norton & Co., New York. Quote p. 11.

25. Jason Thompson, New Emissions and Fuel Efficiency Standards, *Diesel Magazine*, February 1, 2011.

26. These data are from Taylor et al. (2005) Regulation as the Mother of Innovation: The Case of SO_2 Control, *Law & Policy* 27(2):348–78.

27. Adam B. Jaffe, Steven R. Peterson, Paul R. Portney, and Robert N. Stavins (1995) Environmental Regulation and the Competitiveness of U.S. Manufacturing: What Does the Evidence Tell Us?, *Journal of Economic Literature* 33(1):132–63.

28. White House Office of Management and Budget, Office of Information and Regulatory Affairs, *2011 Report to Congress on the Benefits and Costs of Federal Regulations and Unfunded Mandates on State, Local, and Tribal Entities*, Washington, DC, 2011.

29. Michael Porter and Class van der Linde (1995) Green and Competitive: Ending the Stalemate, *Harvard Business Review* 73(5):120–34.

30. Stefan Ambec, Mark A. Cohen, Stewart Elgie, and Paul Lanoie, *The Porter Hypothesis at 20: Can Environmental Regulation Enhance Innovation and Competitiveness?*, Discussion Paper 11-01, Resources for the Future, Washington, DC, January 2011.

31. See Tom Tietenberg (2010) Cap-and-Trade: The Evolution of An Economic Idea, *Agricultural and Resource Economics Review* 39(3):359–67.

32. Ronald Coase (1960) The Problem of Social Cost, *Journal of Law and Economics* 3(1):1–44.

33. Coase did not explicitly use the language of pollution permits, which arose in later years as one method for implementing his proposal. Coase applied his logic of tradable property rights to a variety of socially harmful activities, from air pollution to cattle wandering onto a neighbor's property. He recognized that his proposal only works if the companies can strike a deal in a relatively low-cost manner. This can be challenging when dealing with large numbers of firms in a global economy. Here again, social rules are needed to reduce these "transaction costs" by creating common technical standards, ensuring credible enforcement so that firms don't exceed their allotted pollution permits (which would give cheaters a competitive advantage), and establishing a transparent mechanism for trade. Typically this rulemaking function is played by governments. For a discussion, see Robert N. Stavins (1995) Transaction Costs and Tradeable Permits, *Journal of Environmental Economics and Management* 29(2):133–48.

34. W. David Montgomery (1972) Markets in Licenses and Efficient Pollution Control Programs, *Journal of Economic Theory* 5:395–418; and John H. Dales, *Pollution, Property and Prices*, University of Toronto Press, Toronto, 1968.

35. Hugh S. Gorman and Barry D. Solomon (2002) The Origins and Practice of Emissions Trading, *Journal of Policy History* 14(3):293–320.

36. Eventually most major environmental groups embraced the idea of tradable pollution rights. In 2011, the Natural Resources Defense Council—the same group that

sued the EPA in the early years to halt the program—released a statement arguing, "A well-designed cap and trade program is critical to ensuring pollution reductions and spurring innovation." Kristin Eberhard, Natural Resources Defense Council, Air Board Should Move Ahead with AB 32 Scoping Plan: California's Blueprint for Transitioning to a Clean Energy Economy, Posted August 24, 2011. http://switchboard.nrdc.org/blogs/kgrenfell/air_board_should_move_ahead_wi.html.

37. Gorman and Solomon, 2002, op. cit., p. 308.

38. The EPA data are reported in *The New York Times* at http://www.nytimes.com/gwire/2011/03/31/31greenwire-has-emissions-cap-and-trade-created-toxic-hotsp-4746.html.

39. Robert N. Stavins (1998) What Can We Learn from the Grand Policy Experiment? Lessons from SO_2 Allowance Trading, *Journal of Economic Perspectives* 12(3): 69–88.

40. For a history of the environmental justice movement, see Robert D. Bullard (ed.), *The Quest for Environmental Justice: Human Rights and the Politics of Pollution*, Sierra Club Books, San Francisco, 2005; and Andrew Szasz, *Ecopopulism: Toxic Waste and the Movement for Environmental Justice*, University of Minnesota Press, Minneapolis, 1994.

41. Evan J. Ringquist (2011) Trading Equity for Efficiency in Environmental Protection? Environmental Justice Effects from the SO_2 Allowance Trading Program, *Social Science Quarterly* 92(2):297–323.

42. The evidence documenting unequal distribution of industrial pollution in America is summarized in Paul Mohai and Robin Saha (2007) Racial Inequity and the Distribution of Hazardous Waste: A National Level Reassessment, *Social Problems* 54(3):343–70; and Evan Ringquist (2005) Assessing the Evidence Regarding Environmental Inequities: A Meta-Analysis, *Journal of Policy Analysis and Management* 24(2):223–47.

43. Tomoaki Imamura, Hiroo Ide, and Hideo Yasunaga (2007) History of Public Health Crises in Japan, *Journal of Public Health Policy* 28(2):221–37.

44. The risks of ingesting mercury must be weighed against the benefits of including fish as part of a healthy diet. Moreover, some fish contain more mercury than others. For more background, see E. Oken et al. (2005) Maternal Fish Consumption, Hair Mercury, and Infant Cognition in a U.S. Cohort, *Environmental Health Perspectives* 113(10):1376–80. For dietary guidance see http://water.epa.gov/scitech/swguidance/fishshellfish/outreach/advice_index.cfm.

45. Quoted in Environmental News Service, Appeals Court Rejects EPA Mercury Cap-and-Trade Rule, February 8, 2008. http://www.ens-newswire.com/ens/feb2008/2008-02-08-01.asp.

46. Neela Banerjee, Obama Faces a Battle on Air Rules, *Los Angeles Times*, December 22, 2011.

47. For an overview of climate policies see Anita Engels (2013) Assessing Carbon Policy Experiments, *Global Environmental Politics* 13(3):138–43; and Barry G. Rabe (2008) States on Steroids: The Intergovernmental Odyssey of American Climate Policy, *Review of Policy Research* 25(2):105–28.

48. Because the global REDD+ program is still in the early stages, and requires the design of new rulemaking systems in dozens of countries, analysts are just beginning to understand the associated challenges. See, for example, Jacob Phelps, Edward L. Webb, and Arun Agrawal (2010) Does REDD+ Threaten to Recentralize Forest Governance?,

Science 328(5976):312–13; and William D. Sunderlin, et al. (2014) How Are REDD+ Proponents Addressing Tenure Problems? Evidence from Brazil, Cameroon, Tanzania, Indonesia, and Vietnam, *World Development* 55:37–52.

49. Critiques of market-based regulatory tools are plentiful, and range from philosophical objections to the "commodification" of nature to worries about the social and environmental effects. See, for example, Murat Arsel and Bram Büscher (2012) Nature™ Inc.: Changes and Continuities in Neoliberal Conservation and Market-based Environmental Policy, *Development and Change* 43(1):53–78.

50. Ronald Coase op. cit., pp. 18–19.

51. Kevin P. Gallagher and Lyuba Zarsky, *The Enclave Economy: Foreign Investment and Sustainable Development in Mexico's Silicon Valley*, MIT Press, Cambridge, MA, 2007.

52. The Argentina case is discussed in Sebastian Gailiani, Paul Gertler, and Ernesto Schargrodsky (2005) Water for Life: The Impact of the Privatization of Water Services on Child Mortality, *Journal of Political Economy* 113(1):83–120. The contrasting Colombian case is analyzed in Claudia Granados and Fabio Sánchez (2014) Water Reforms, Decentralization and Child Mortality in Colombia, 1990–2005, *World Development* 53:68–79. For critical overviews of privatization in the water sector, see Karen J. Bakker, *Privatizing Water: Governance Failure and the World's Urban Water Crisis*, Cornell University Press, Ithaca, NY, 2010; and Jessica Budds and Gordon McGranahan (2003) Are the Debates on Water Privatization Missing the Point? Experiences from Africa, Asia and Latin America, *Environment and Urbanization* 15(2):87–114.

CHAPTER 6

1. Sheila Jasanoff, Heaven and Earth: The Politics of Environmental Images, pp. 31–52 in Sheila Jasanoff and Marybeth Long Martello (eds.), *Earthly Politics: Local and Global in Environmental Governance*, MIT Press, Cambridge, MA, 2004.

2. Keith Bakx (1987) Planning Agrarian Reform: Amazonian Settlement Projects, 1970–86, *Development and Change*, 18:533–55.

3. Stephen G. Perz et al. (2013) Trans-boundary Infrastructure and Land Cover Change: Highway Paving and Community-level Deforestation in a Tri-national Frontier in the Amazon, *Land Use Policy* 34:27–41.

4. The importance of national governments in efforts to promote sustainability is described in John Barry and Robin Eckersley (eds.), *The State and the Global Ecological Crisis*, MIT Press, Cambridge, MA, 2005; and Paul F. Steinberg (2005) From Public Concern to Policy Effectiveness: Civic Conservation in Developing Countries, *Journal of International Wildlife Law & Policy* 8(4):341–65.

5. Hans P. Binswanger (1991) Brazilian Policies That Encourage Deforestation in the Amazon, *World Development* 19(7):821–29; Cynthia Simmons et al. (2010) Doing It for Themselves: Direct Action Land Reform in the Brazilian Amazon, *World Development* 38(3):429–44; and Philip M. Fearnside (2001) Soybean Cultivation as a Threat to the Environment in Brazil, *Environmental Conservation* 28(1):23–38.

6. Lester M. Salamon (2010) Putting the Civil Society Sector on the Economic Map of the World, *Annals of Public and Cooperative Economics* 81(2):167–210; and Lester M. Salamon, Helmut K. Anheier, Regina List, Stefan Toepler, and S. Wojciech Sokolowski and Associates, *Global Civil Society: Dimensions of the Nonprofit Sector*, Kumarian Press, West Hartford, CT, 1999.

7. Russell J. Dalton, Steve Recchia, and Robert Rohrschneider (2003) The Environmental Movement and the Modes of Political Action, *Comparative Political Studies* 36(7):743–71.

8. John Clark, *Democratizing Development: The Role of Voluntary Organizations*, Kumarian Press, West Hartford, CT, 1991.

9. Those vying for a piece of Antarctica include Australia, Norway, France, Chile, Argentina, New Zealand, and the United Kingdom. A map delineating the dispute is available on the website of the Australian Antarctic Division at http://www.antarctica.gov.au/about-antarctica/people-in-antarctica/who-owns-antarctica.

10. See the *Journal of Democracy* for high quality and up-to-date analyses of trends in democratization worldwide.

11. The Economist Intelligence Unit, *Democracy Index 2012: Democracy at a Standstill*, The Economist Intelligence Unit Ltd., London, 2013.

12. The data on growth in the number of independent countries are from Monty G. Marshall and Benjamin R. Cole, *Global Report 2009: Conflict, Governance, and State Fragility*, Center for Systemic Peace, George Mason University, Fairfax, VA, 2009; and The World Bank, *World Development Report 1997: The State in a Changing World*, Oxford University Press, New York, 1997, p. 21.

13. Michael L. Ross, *The Oil Curse: How Petroleum Wealth Shapes the Development of Nations*, Princeton University Press, Princeton, NJ, 2012; and Michael L. Ross (2001) Does Oil Hinder Democracy? *World Politics* 53(3):325–61. For a rejoinder, see Stephen Haber and Victor Menaldo (2011) Do Natural Resources Fuel Authoritarianism? A Reappraisal of the Resource Curse, *American Political Science Review* 105(1):1–26. On the importance of social rules in mediating the relationship among oil, politics, and economic development, see Pauline Jones Luong and Erika Weinthal, *Oil Is Not a Curse: Ownership Structure and Institutions in Soviet Successor States*, Cambridge University Press, New York, 2010.

14. Shannon McClelland (2002) Indonesia's Integrated Pest Management in Rice: Successful Integration of Policy and Education, *Environmental Practice* 4(4):191–95.

15. The history of political support for environmental policy in Brazil is described in Kathryn Hochstetler and Margaret E. Keck, *Greening Brazil: Environmental Activism in State and Society*, Duke University Press, Durham, NC, 2007.

16. Kathryn Hochstetler, Democracy and the Environment in Latin America and Eastern Europe, pp. 199–230 in Paul F. Steinberg and Stacy VanDeveer (eds.), *Comparative Environmental Politics: Theory, Practice, and Prospects*, MIT Press, Cambridge, MA, 2012.

17. Paul F. Steinberg, *Environmental Leadership in Developing Countries: Transnational Relations and Biodiversity Policy in Costa Rica and Bolivia*, MIT Press, Cambridge, MA, 2001.

18. These figures report gross national income per capita in 2010, adjusted to reflect the purchasing power of a dollar in each country. The resulting Purchasing Power Parity data are from Table 1.1 of The World Bank, *World Development Indicators 2012*, Washington, DC, 2012.

19. In 2005, the federal budget for environmental programs in Brazil was $1.2 billion. From Hochstetler and Keck, op. cit., Table 1.2, p. 41. For an excellent discussion of the challenges facing very small countries, see Godfrey Baldacchino (1993) Bursting the Bubble: The Pseudo-Development Strategies of Microstates, *Development and Change* 24(1):29–51.

20. For details on "Pride Campaigns" inspired by the Saint Lucia model, see www.rareconservation.org.

21. The challenge of promoting sustainability in systems characterized by chronic instability is further explored in Paul F. Steinberg, Welcome to the Jungle: Policy Theory and Political Instability, pp. 255–84 in Steinberg and VanDeveer, op. cit.

22. These figures are calculated from data in *Polity IV: Regime Authority Characteristics and Transitions Datasets 1800–2010*, Center for International Development and Conflict Management, University of Maryland, College Park. The analysis includes only countries with populations over 500,000.

23. Ibid.

24. This calculation uses data from A. S. Banks, *Cross-National Time-Series Data Archive*, Databanks International, Jerusalem, Israel, 2011; and US Central Intelligence Agency, *CIA World Factbook*, 2011.

25. The figures on armed conflicts are from Lotta Themnér and Peter Wallensteen (2013) Armed Conflicts, 1946–2012, *Journal of Peace Research* 50(4):509–21. Figures on the lifespan of democracies are from Adam Przeworski (2005) Democracy as an Equilibrium, *Public Choice* 123(3–4):253–73. Przeworski's national income figures refer to Purchasing Power Parity dollars which, as we saw earlier, takes account of variation among countries in the cost of living.

26. Susan Rose-Ackerman, *Corruption and Government: Causes, Consequences, and Reform*, Cambridge University Press, New York, 1999.

27. Edward Dommen (1997) Paradigms of Governance and Exclusion, *Journal of Modern African Studies* 35(3):485–94.

28. On the measurement of corruption, see Daniel Kaufmann, Aart Kraay, and Massimo Mastruzzi, *The Worldwide Governance Indicators: Methodology and Analytical Issues*, World Bank Policy Research Working Paper No. 5430, World Bank Institute, Washington, DC, 2010. Maps and charts comparing corruption levels are available at http://info.worldbank.org/governance/wgi/index.aspx#home.

29. Paul Robbins (2000) The Rotten Institution: Corruption in Natural Resource Management, *Political Geography* 19:423–43.

30. Brian Z. Tamanaha (2008) Understanding Legal Pluralism: Past to Present, Local to Global (Julius Stone Address), *Sydney Law Review* 30:375–411.

31. Clifford Geertz, *After the Fact: Two Countries, Four Decades, One Anthropologist*, Harvard University Press, Cambridge, MA, 1995. Quote p. 22.

32. Robin R. Sears and Miguel Pinedo-Vasquez (2011) Forest Policy Reform and the Organization of Logging in Peruvian Amazonia, *Development and Change* 42:609–31.

33. Peter Dauvergne, *Shadows in the Forest: Japan and the Politics of Timber in Southeast Asia*, MIT Press, Cambridge, MA, 1997.

34. Charles Benjamin illustrates the diversity of relationships that are possible when formal and informal rules collide. See Charles E. Benjamin (2008) Legal Pluralism and Decentralization: Natural Resource Management in Mali, *World Development* 36(11):2255–76.

35. For an introduction to the concept of social capital, see Michael Woolcock (1998) Social Capital and Economic Development: Toward a Theoretical Synthesis and Policy Framework, *Theory and Society* 27(2):151–208. Relations of reciprocity assume many forms, from mutual self-help associations to more hierarchical relationships involving patrons and their social networks. See James Scott (1972) Patron-client Politics and Political Change in Southeast Asia, *American Political Science Review* 66(1):91–113; and Peter Dauvergne, op. cit.

36. For more information on community forest management in the Sierra Gorda, see www.sierragorda.net and the PBS documentary *Mexico: The Business of Saving Trees,* 2008.

37. Mexico's leadership in community forestry is described in chapter 4 and in Camille Antinori and David Barton Bray (2005) Community Forest Enterprises as Entrepreneurial Firms: Economic and Institutional Perspectives from Mexico, *World Development* 33(9):1529–43.

38. PROPER denotes the Program for Pollution Control, Evaluation, and Rating.

39. The history of Indonesia's PROPER program is described in Shakeb Afsah, Allen Blackman, Jorge H. Garcia, and Thomas Sterner, *Environmental Regulation and Public Disclosure: The Case of PROPER in Indonesia*, Resources for the Future Press/Routledge, New York, 2013. See also Hemamala Hettige, Mainul Huq, Sheoli Pargal, and David Wheeler (1996) Determinants of Pollution Abatement in Developing Countries: Evidence from South and Southeast Asia, *World Development* 24(12):1891–1904.

40. The importance of political expertise for bringing about improvements in environmental quality is discussed in Paul F. Steinberg (2003) Understanding Policy Change in Developing Countries: The Spheres of Influence Framework, *Global Environmental Politics* 3(1):11–32.

41. John Bongaarts and Steven Sinding (2011) Population Policy in Transition in the Developing World, *Science* 333:574–76.

42. Judith Lipp (2007) Lessons for Effective Renewable Electricity Policy from Denmark, Germany and the United Kingdom, *Energy Policy* 35:5481–95.

43. Data are from the US Department of Energy, Carbon Dioxide Information Analysis Center, Oak Ridge National Laboratory. Available at http://cdiac.ornl.gov/trends/emis/den.html. Accessed October 14, 2013.

44. The experience with coastal management in the Philippines is described in Angel C. Alcala and Garry R. Russ (2006) No-take Marine Reserves and Reef Fisheries Management in the Philippines: A New People Power Revolution, *Ambio* 35(5):245–54; Alan T. White, Catherine A. Courtney, and Albert Salamanca (2002) Experience with Marine Protected Area Planning and Management in the Philippines, *Coastal Management*, 30:1–26; and Miriam C. Balgos (2005) Integrated Coastal Management and Marine Protected Areas in the Philippines: Concurrent Developments, *Ocean & Coastal Management* 48(11–12):972–95.

45. For a lively (and sympathetic) discussion of the pathologies of the Indian bureaucracy, see Gurcharan Das, *India Grows at Night: A Liberal Case for a Strong State*, Penguin Books, New York, 2012.

46. Alasdair Roberts (2010) A Great and Revolutionary Law? The First Four Years of India's Right to Information Act, *Public Administration Review* 70(6):925–33. The impact of the law is evaluated in Leonid Peisakhin and Paul Pinto (2010) Is Transparency an Effective Anti-corruption Strategy? Evidence from a Field Experiment in India, *Regulation and Governance* 4:261–80. Figures on the number of information requests filed by citizens are from Indian government statistics posted at http://rti.gov.in/.

47. Robert H. Bates, *Markets and States in Tropical Africa: The Political Basis of Agricultural Policies*, University of California Press, Berkeley, 1981. The political drivers of agricultural policy decisions also feature in C. Peter Timmer (ed.), *Agriculture and the State: Growth, Employment, and Poverty in Developing Countries*, Cornell University Press, Ithaca, NY, 1991.

48. Michael L. Ross, *Timber Booms and Institutional Breakdown in Southeast Asia*, Cambridge University Press, New York, 2001. Deforestation rates are documented in

Navjot S. Sodhi, Lian Pin Koh, Barry W. Brook, and Peter K. L. Ng (2004) Southeast Asian Biodiversity: An Impending Disaster, *TRENDS in Ecology and Evolution* 19(12):654–60.

49. Clark C. Gibson, *Politicians and Poachers: The Political Economy of Wildlife Policy in Africa*, Cambridge University Press, New York, 1999. Quote p. 156.

50. Gibson, op. cit., p. 158.

51. James C. Scott, *Seeing Like a State: How Certain Schemes to Improve the Human Condition Have Failed*, Yale University Press, New Haven, CT, 1998.

52. State-society conflicts in natural resource management have been documented extensively by researchers in fields including rural sociology, anthropology, geography, and agrarian studies. See, for example, Nancy Lee Peluso, *Rich Forests, Poor People: Resource Control and Resistance in Java*, University of California Press, Berkeley, 1992; and Madhav Gadgil and Ramachandra Guha, *Ecology and Equity: The Use and Abuse of Nature in Contemporary India*, Routledge, New York, 1995.

53. Ann Hironaka (2002) The Globalization of Environmental Protection: The Case of Environmental Impact Assessment, *International Journal of Comparative Sociology* 43(1):65–78.

54. David Vogel (2003) The Hare and the Tortoise Revisited: The New Politics of Consumer and Environmental Regulation in Europe, *British Journal of Political Science* 33:557–80. Quote p. 557.

55. Edward A. Parson, *Protecting the Ozone Layer: Science and Strategy*, Oxford University Press, New York, 2003.

56. R. Daniel Kelemen and David Vogel (2010) Trading Places: The Role of the United States and the European Union in International Environmental Politics, *Comparative Political Studies* 3(4):427–56. Quote p. 5. The shift in environmental leadership from the United States to Europe is documented in several insightful chapters within Andreas Duit (ed.), *State and Environment: The Comparative Study of Environmental Governance*, MIT Press, Cambridge, MA, 2014.

57. Elizabeth R. DeSombre (1995) Baptists and Bootleggers for the Environment: The Origins of United States Unilateral Sanctions, *Journal of Environment & Development* 4(1):53–75; and Kenneth A. Oye and James H. Maxwell (1994) Self-Interest and Environmental Management, *Journal of Theoretical Politics* 6(4):593–624.

58. Theda Skocpol, Naming the Problem: What It Will Take to Counter Extremism and Engage Americans in the Fight against Global Warming, paper prepared for the Symposium on the Politics of America's Fight against Global Warming, Harvard University, February 14, 2013.

59. Riley E. Dunlap, Chenyang Xiao, and Aaron M. McCright (2001) Politics and Environment in America: Partisan and Ideological Cleavages in Public Support for Environmentalism, *Environmental Politics* 10(4):23–48.

CHAPTER 7

1. José Delfín Duarte stepped down from his position with the water association subsequent to my visit with him in 2011. I maintained his day-in-the-life profile here for purposes of illustration.

2. Multilevel governance is the focus of a large research literature. Examples include Liliana B. Andonova and Ronald B. Mitchell (2010) The Rescaling of Global Environmental Politics, *Annual Review of Environment and Resources* 35:255–82; Harriet

Bulkeley and Michele Betsill (2005) Rethinking Sustainable Cities: Multilevel Governance and the "Urban" Politics of Climate Change, *Environmental Politics* 14(1):42–63; and Henrik Selin, *Global Governance of Hazardous Chemicals: Challenges of Multilevel Management*, MIT Press, Cambridge, MA, 2010.

3. According to Bach and Newman, "To this day, the FDA has in practice no formal authority to control market access and has comparatively little expertise in cosmetics, relying instead on industry-led ingredient review." See p. 685 in David Bach and Abraham L. Newman (2010) Governing Lipitor and Lipstick: Capacity, Sequencing, and Power in International Pharmaceutical and Cosmetics Regulation, *Review of International Political Economy* 17(4):665–95. The Environmental Working Group maintains a database with research and guidelines on cosmetics products at http://www.ewg.org/skindeep/.

4. For an excellent overview of the processes surrounding environmental treaties, see Daniel Bodansky, *The Art and Craft of International Environmental Law*, Harvard University Press, Cambridge, MA, 2011.

5. James N. Rosenau and Ernst-Otto Czempiel (eds.), *Governance without Government: Order and Change in World Politics*, Cambridge University Press, New York, 1992.

6. The relation between warfare and the rise of modern European states is analyzed in Charles Tilly, *Coercion, Capital, and European States, AD 990–1990*, Blackwell, Cambridge, MA, 1990.

7. For an analysis of the causes of European wars, see William R. Thompson (2003) A Streetcar Named Sarajevo: Catalysts, Multiple Causation Chains, and Rivalry Structures, *International Studies Quarterly* 47(3):453–74.

8. Louis L. Snyder, *The War: A Concise History*, Simon and Schuster, New York, 1960. Quote p. 502.

9. Data on the destruction wrought by World War II come from Snyder, op. cit.; and Robert Goralski, *World War II Almanac, 1931–1945: A Political and Military Record*, Random House, New York, 1987.

10. This history of Jean Monnet and the rise of the European Union draws principally on François Duchêne, *Jean Monnet: The First Statesman of Interdependence*, Norton, New York, 1994; Desmond Dinan, *Europe Recast: A History of the European Union*, Lynn Rienner, Boulder, CO, 2004; and Paul Magnette, *What Is the European Union? Nature and Prospects*, Palgrave Macmillan, New York, 2005.

11. Quoted in Duchêne, op. cit., p. 27.

12. Duchêne, op. cit.

13. League of Nations, International Labour Office, *The Third International Labour Conference, October–November 1921*, with a Foreword by Viscount Burnham, International Labour Office, Geneva, 1922.

14. Chiang Kai-shek's observation is reported in Duchêne, op. cit., p. 62.

15. For a discussion of Monnet's extraordinary influence in prompting the Americans to ramp up war production, see pp. 67–72 of Frederic J. Fransen, *The Supranational Politics of Jean Monnet: Ideas and Origins of the European Community*, Greenwood Press, Westport, CT, 2001.

16. Donald Kladstrup and Petie Kladstrup, *Wine and War: The French, the Nazis, and the Battle for France's Greatest Treasure*, Broadway Publishers, New York, 2002.

17. Reprinted in Pascal Fontaine (ed.), *Jean Monnet, A Grand Design for Europe*, Office for Official Publications of the European Communities, Luxembourg, 1988, p. 41.

Accessible through the University of Pittsburgh Archive of European Integration at http://aei.pitt.edu. I have modified the English translation of the quote to remove the repeated word in the phrase "with all that that implies."

18. The American influence in the post-war period is described in Desmond Dinan, *Europe Recast: A History of the European Union*, Lynne Rienner Publishers, Boulder, CO, 2004.

19. Monnet's lack of interest in environmental issues cannot be attributed to the historical period in which he lived. There was, in fact, an active transnational conservation movement after the war. In September 1948, President Eisenhower inaugurated the Inter-American Conference on the Conservation of Renewable Resources, held in Denver, Colorado, and attended by representatives from throughout the Americas. The following month, several dozen conservation leaders gathered in Fontainebleau, on the outskirts of Paris, to launch the International Union for the Conservation of Nature, the world's first global environmental organization. It seems likely that Monnet would have known about the Fontainebleau conference, given its proximity and high political profile, but there is no evidence to suggest that he took an interest in the environmental dimensions of international cooperation.

20. Barry Eichengreen, Institutions and Economic Growth: Europe After World War II, chapter 2 in Nicholas Crafts and Gianni Toniolo (eds.), *Economic Growth in Europe Since 1945*, Cambridge University Press, New York, 1996.

21. For a history of green parties in Western Europe, see Michael O'Neill, Political Parties and the "Meaning of Greening" in European Politics, pp. 171–95 in Paul F. Steinberg and Stacy D. VanDeveer (eds.), *Comparative Environmental Politics: Theory, Practice, and Prospects*, MIT Press, Cambridge, MA, 2012.

22. Herbert Kitschelt, *The Logics of Party Formation: Ecological Politics in Belgium and West Germany*, Cornell University Press, Ithaca, NY, 1989.

23. At the prodding of the United States, in the mid-1980s NATO announced plans to deploy tactical nuclear weapons on the European continent as a deterrent against Soviet aggression. The proposal prompted public outrage and large protests in West Germany and throughout Western Europe, fanning the flames of anti-nuclear sentiment and creating a fertile political environment for emergent green parties. See Anon., Hundreds of Thousands Protest Missiles in Europe: Urge U.S. to Match Soviet Halt, *Los Angeles Times*, April 8, 1985.

24. John Dryzek, David Downs, Hans-Kristian Hernes, and David Schlosberg, *Green States and Social Movements: Environmentalism in the United States, United Kingdom, Germany, and Norway*, Oxford University Press, New York, 2003.

25. The impact of the new voting rule on the diffusion of policy innovations in Europe is described in Simon Bulmer and Stephen Padgett (2005) Policy Transfer in the European Union: An Institutionalist Perspective, *British Journal of Political Science* 35(1):103–26. Qualified majority voting in Europe is described in greater detail at http://www.eurofound.europa.eu/areas/industrialrelations/dictionary/definitions/qualifiedmajorityvoting.htm

26. Magnette, op. cit., p. 2.

27. David Benson and Andrew Jordan, Environmental Policy, pp. 358–74, in Michelle Cini and Nieves Pérez-Solórzano Borragán (eds.), *European Union Politics*, 3rd edition, Oxford University Press, New York, 2010. Quote p. 359. Further background on European Union environmental policy is provided in Andrew Jordan and Camilla Adelle (eds.), *Environmental Policy in the EU: Actors, Institutions and Processes*, Routledge,

London, 2012; JoAnn Carmin and Stacy D. VanDeveer (eds.), *EU Enlargement and the Environment: Institutional Change and Environmental Policy in Central and Eastern Europe*, Routledge, London, 2005; and Christopher Knill and Duncan Liefferink, *Environmental Politics in the European Union: Policy Making, Implementation and Patterns of Multi-Level Governance*, Manchester University Press, Manchester, 2007.

28. The environmental rules that must be adopted by EU members are described at http://www.europarl.europa.eu/enlargement/briefings/17a2_en.htm.

29. REACH is the Regulation on Registration, Evaluation, Authorisation and Restriction of Chemicals. For insight into how reformers overcame the political opposition and made REACH the new law of the land in Europe, see Henrik Selin (2007) Coalition Politics and Chemicals Management in a Regulatory Ambitious Europe, *Global Environmental Politics* 7(3):63–93.

30. Katja Biedenkopf (2012) Hazardous Substances in Electronics: The Effects of European Union Risk Regulation on China, *European Journal of Risk Regulation* 3(4):477–87.

31. Tanja A. Börzel (2002) Member State Responses to Europeanization, *JCMS: Journal of Common Market Studies* 40(2):193–214.

32. Roberto J. Serrallés (2006) Electric Energy Restructuring in the European Union: Integration, Subsidiarity and the Challenge of Harmonization, *Energy Policy* 34:2542–51.

33. Anon., ETS, RIP? The Failure to Reform Europe's Carbon Market Will Reverberate Round the World, *The Economist*, April 20, 2013.

34. G. Pe'er et al. (2014) EU Agricultural Reform Fails on Biodiversity, *Science* 344(6188):1090–92.

CHAPTER 8

1. I have used a pseudonym to protect the identity of the landowner who provided this interview.

2. Jeffrey J. Ryan (2004) Decentralization and Democratic Instability: The Case of Costa Rica, *Public Administration Review* 64(1):81–91.

3. Andrew Nickson, *Where Is Local Government Going in Latin America? A Comparative Perspective*, Working Paper No. 6, Swedish International Centre for Local Democracy, Visby, Sweden, 2011.

4. Anthony Bebbington and his colleagues document how authoritarian regimes use decentralization to better monitor and control local populations. See Anthony Bebbington, Leni Dharmawan, Erwin Fahmi, and Scott Guggenheim (2006) Local Capacity, Village Governance, and the Political Economy of Rural Development in Indonesia, *World Development* 34(11):1958–76.

5. Jonathan Rodden (2004) Comparative Federalism and Decentralization: On Meaning and Measurement, *Comparative Politics* 36(4):481–500. Quote p. 481.

6. The historical tendency for states to centralize power is described in James Manor, *The Political Economy of Democratic Decentralization*, World Bank, Washington, DC, 1999 (see especially pp. 13–25); and Merilee S. Grindle and John W. Thomas, *Public Choices and Policy Change: The Political Economy of Reform in Developing Countries*, Johns Hopkins University Press, Baltimore, 1991.

7. Ronald L. Watts (1998) Federalism, Federal Political Systems, and Federations, *Annual Review of Political Science* 1:117–37.

8. On the role of centralization in postcolonial India, see especially the chapter Foundations of India's Development Strategy: The Nehru-Mahalanobis Approach, pp. 7–18 in Sukhamoy Chakravarty, *Development Planning: The Indian Experience*, Oxford University Press, New York, 1987.

9. The impact of international trade and multinational corporations on the people and environment of developing countries is described in Richard P. Tucker, *Insatiable Appetite: The United States and the Ecological Degradation of the Tropical World*, University of California Press, Berkeley, 2000.

10. The history of populism in Latin America is traced in Michael L. Conniff (ed.), *Latin American Populism in Comparative Perspective*, University of New Mexico Press, Albuquerque, 1982.

11. Many environmental policies in these regions reflect national and European Union agreements. A growing body of local legislation, however, has resulted from independent initiatives taken by local governments.

12. Teresa Garcia-Milà and Therese J. McGuire, Fiscal Decentralization in Spain: An Asymmetric Transition to Democracy, pp. 208–26 in Richard M. Bird and Robert D. Ebel (eds.), *Fiscal Fragmentation in Decentralized Countries: Subsidiarity, Solidarity and Asymmetry*, Edward Elgar, Cheltenham, 2007.

13. J. David Tábara, A New Climate for Spain: Accommodating Environmental Foreign Policy in a Federal State, pp. 161–84 in Paul G. Harris (ed.), *Europe and Global Climate Change: Politics, Foreign Policy and Regional Cooperation*, Edward Elgar, Cheltenham, 2007.

14. Jean-Claude Thoenig (2005) Territorial Administration and Political Control: Decentralization in France, *Public Administration* 83(3):685–708.

15. Is it realistic to expect governments to give up some of their responsibilities? The research suggests a qualified "yes." In a review of the published literature on the behavior of US agencies, James Wilson finds that the image of government bureaucrats seeking to protect and expand their "turf" is greatly oversimplified; there are many examples of agencies willingly reducing the scope of their authority to advance other aims. See James Q. Wilson, *Bureaucracy: What Government Agencies Do and Why They Do It*, BasicBooks, New York, 1989.

16. The first of the oil shocks came about when a group of Arab countries cut off oil exports to the United States, Japan, and Western Europe to protest their support for Israel in the Yom Kippur War. The second resulted from disruptions to the oil supply following the Iranian Revolution. These events sent the global economy into a tailspin, devastating the economies of those developing countries that did not have the good fortune of sitting atop accessible oil reserves.

17. In Argentina, an estimated 20,000 citizens were "disappeared" by the government during the 1970s and early 1980s, meaning they were kidnapped, tortured, and often killed for opposing the regime.

18. Merilee S. Grindle, *Audacious Reforms: Institutional Invention and Democracy in Latin America,* Johns Hopkins University Press, Baltimore, 1991; and Kathleen O'Neill (2003) Decentralization as an Electoral Strategy, *Comparative Political Studies* 36(9):1068–91. The political forces driving similar reforms in Europe are explored in Jason Sorens (2009) The Partisan Logic of Decentralization in Europe, *Regional and Federal Studies* 19(2):255–72. Some authors have posited that decentralization in developing countries was driven by the demands of international lenders such as the International Monetary Fund, which required reduced spending by central governments

as a prerequisite for new loans. In Latin America, the research suggests that these demands were not a significant cause of decentralization. See Alfred P. Montero and David J. Samuels, The Political Determinants of Decentralization in Latin America: Causes and Consequences, pp. 3–32 in Montrero and Samuels (eds.), *Decentralization and Democracy in Latin America*, University of Notre Dame Press, Notre Dame, IN, 2004. Beginning in the 2000s, a growing number of developing countries (including Brazil, India, and South Africa) actually rejected the "Washington Consensus" model of economic growth, with its emphasis on scaling back the state, in favor of China's hybrid model in which a strong state steers a market-oriented economy.

19. The political snowball effect of local governments testing out their new powers, then demanding more, was especially pronounced in countries that handed over real decision-making power to locals at the outset, rather than beginning the decentralization process with more cautious fiscal and administrative reforms. See Tulia G. Falleti (2005) A Sequential Theory of Decentralization: Latin American Cases in Comparative Perspective, *American Political Science Review* 99(3):327–46.

20. Jesse Ribot, *Democratic Decentralization of Natural Resources: Institutionalizing Popular Participation*, World Resources Institute, Washington, DC, 2002.

21. Wolfram H. Dressler, Christian A. Kull, and Thomas C. Meredith (2006) The Politics of Decentralizing National Parks Management in the Philippines, *Political Geography* 25:789–816.

22. Maria Carmen Lemos and João Lúcio Farias de Oliveira (2004) Can Water Reform Survive Politics? Institutional Change and River Basin Management in Ceará, Northeast Brazil, *World Development* 32(12):2121–37; and Margaret Wilder and Patricia Romero Lankao (2006) Paradoxes of Decentralization: Water Reform and Social Implications in Mexico, *World Development* 34(11):1977–95. The new rules governing water management in Costa Rica are described in chapter 7.

23. Anne M. Larson (2003) Decentralisation and Forest Management in Latin America: Towards a Working Model, *Public Administration and Development* 23:211–26.

24. As part of its declared War on Poverty, the administration of President Lyndon Johnson went so far as to fund hundreds of local legal assistance clinics that helped citizens to sue their city governments.

25. Garrett Hardin (1968) The Tragedy of the Commons, *Science* 162(3859): 1243–48.

26. Hardin, op. cit. Quotes p. 1244.

27. Aristotle, *The Politics*, Book II, translated by T. A. Sinclair, Penguin Books, Baltimore, 1962. Quotes pp. 58 and 61.

28. H. Scott Gordon (1954) The Economic Theory of a Common-Property Resource: The Fishery, *Journal of Political Economy* 62(2):124–42. Quote p. 135.

29. The videogame *Law of the Jungle* can be downloaded at The Social Rules Project website at www.rulechangers.org. Part of a genre called "serious games," it combines entertainment value with in-depth educational content on the institutional dimensions of tropical deforestation.

30. In his original article, Hardin advocates "mutual coercion, mutually agreed upon by the majority of the people affected." That sounds a lot like social rules arrived at through legitimate means. But Hardin was unclear about what role, if any, local communities might play in such an arrangement. He suggested that rules need not emanate from "arbitrary decisions of distant and irresponsible bureaucrats." But in later work he stated that the choice was between private property rights and state socialism,

completely overlooking local common-property institutions. See Garret Hardin, Political Requirements for Preserving Our Common Heritage, pp. 310–17 in Howard P. Brokaw (ed.), *Wildlife and America*, Council on Environmental Quality, Washington, DC, 1978. Quotes are from Hardin 1968, op. cit., p. 1247.

31. S. V. Ciriacy-Wantrup and Richard C. Bishop (1975) Common Property as a Concept in Natural Resource Policy, *Natural Resources Journal* 15:713–27. Quotes pp. 713 and 719.

32. Ostrom relates this history in her best-known work, Elinor Ostrom, *Governing the Commons: The Evolution of Institutions for Collective Action*, Cambridge University Press, New York, 1990.

33. Elinor Ostrom, *Public Entrepreneurship: A Case Study in Ground Water Basin Management*, Ph.D. dissertation, University of California, Los Angeles, 1965.

34. Thrainn Eggertsson (1992) Analyzing Institutional Successes and Failures: A Millennium of Common Mountain Pastures in Iceland, *International Review of Law and Economics* 12(4):423–37.

35. Ostrom, op. cit., quote p. 82.

36. Arun Agrawal, Local Institutions and the Governance of Forest Commons, pp. 313–40 in Paul F. Steinberg and Stacy D. VanDeveer, eds., *Comparative Environmental Politics: Theory, Practice, and Prospects*, MIT Press, Cambridge, MA, 2012.

37. Fikret Berkes (2003) Alternatives to Conventional Management: Lessons from Small-Scale Fisheries, *Environments* 31(1):5–20; and Fisheries and Aquaculture Department, Food and Agriculture Organization of the United Nations, *The State of World Fisheries and Aquaculture 2010*, Rome, 2010.

38. Calculated from Table 1.1 of Jonathan B. Mabry, The Ethnography of Local Irrigation, in Jonathan B. Mabry (ed.), *Canals and Communities: Small-scale Irrigation Systems*, University of Arizona Press, Tucson, 1996.

39. Agrawal, op. cit.

40. United Nations Department of Economic and Social Affairs, Population Division, *Population World Urbanization Prospects: The 2011 Revision*, United Nations, New York, 2012.

41. The imbalance between urban and rural studies has produced an odd skew in political science research: Studies of national environmental policy typically feature industrialized countries, whereas research on local environmental governance (emphasizing the commons) more often focuses on developing countries.

42. Arrangements for governing common-pool resources can be found in urban settings. Ostrom's original research on shared water resources, for example, involved California cities and towns. But urban rulemaking differs in fundamental respects from the typical common-property arrangement. Cities are run by formal governments that possess regulatory powers and resources specified in state and national laws.

43. Erik Nelson, Michinori Uwasu, and Stephen Polasky (2006) Voting on Open Space: What Explains the Appearance and Support of Municipal-Level Open Space Conservation Referenda in the United States?, *Ecological Economics* 62:580–93.

44. Jeffrey M. Sellers and Anders Lidström (2007) Decentralization, Local Government, and the Welfare State, *Governance: An International Journal of Policy, Administration, and Institutions* 20(4):609–32.

45. Alan DiGaetano and Elizabeth Strom (2003) Comparative Urban Governance: An Integrated Approach, *Urban Affairs Review* 38(3):356–95.

46. Matthew Potoski (2001) Clean Air Federalism: Do States Race to the Bottom?, *Public Administration Review* 61(3):335–42. Quote p. 339.

47. Potoski, op. cit. Quote p. 339.

48. Kent E. Portney and Jeffrey M. Berry (2010) Participation and the Pursuit of Sustainability in U.S. Cities, *Urban Affairs Review* 46(1):119–39.

49. For research upending the myth that poor people and poor countries care less about the environment, see Riley Dunlap and Richard York, The Globalization of Environmental Concern, pp. 89–112 in Steinberg and VanDeveer, op. cit.; and Paul F. Steinberg, Environmental Privilege Revisited, chapter 2 in *Environmental Leadership in Developing Countries: Transnational Relations and Biodiversity Policy in Costa Rica and Bolivia*, MIT Press, Cambridge, MA, 2001.

50. Krister P. Andersson, Clark C. Gibson, and Fabrice Lehoucq (2006) Municipal Politics and Forest Governance: Comparative Analysis of Decentralization in Bolivia and Guatemala, *World Development* 34(3):576–95.

51. For a sample of the research investigating why some local governments promote sustainability more vigorously than others, see George A. González (2002) Local Growth Coalitions and Air Pollution Controls: The Ecological Modernisation of the US in Historical Perspective, *Environmental Politics* 11(3):121–44; Aidan While, Andrew E. G. Jonas and David Gibbs (2004) The Environment and the Entrepreneurial City: Searching for the Urban "Sustainability Fix" in Manchester and Leeds, *International Journal of Urban and Regional Research* 28(3):549–69; Daniel Press, *Saving Open Space: The Politics of Local Preservation in California*, University of California Press, Berkeley, 2002; Harriet Bulkeley and Michele Betsill (2005) Rethinking Sustainable Cities: Multilevel Governance and the "Urban" Politics of Climate Change, *Environmental Politics* 14(1):42–63; and Robert F. Young (2010) The Greening of Chicago: Environmental Leaders and Organisational Learning in the Transition Toward a Sustainable Metropolitan Region, *Journal of Environmental Planning and Management* 53(8):1051–68.

52. James Madison, Federalist No. 10, in Alexander Hamilton, James Madison, and John Jay, *The Federalist* (with Letters of Brutus), Terence Ball (ed.), Cambridge University Press, New York, 2003. Quote p. 46.

53. The diffusion of environmental policy innovations among local and state governments within the United States, has been studied by Barry Rabe among others. See Barry G. Rabe (2008) States on Steroids: The Intergovernmental Odyssey of American Climate Policy, *Review of Policy Research* 25(2):105–28.

54. Piers Blaikie, *The Political Economy of Soil Erosion in Developing Countries*, Longman Scientific & Technical (with John Wiley & Sons, Inc., New York), Essex, UK, 1985. Quote p. 88.

55. The design of this graphic is based on a figure originally published in Inter-American Development Bank, *Economic and Social Progress in Latin America, 1996 Report: Special Section*, IADB, Washington, DC, 1997. The data have been updated based on figures from Robert Daughters and Leslie Harper, Fiscal and Political Decentralization Reforms, pp. 213–61 in Eduardo Lora (ed.), *The State of State Reform in Latin America*, Inter-American Development Bank, Washington, DC, 2007; and Christopher Sabatini (2003) Decentralization and Political Parties, *Journal of Democracy* 14(2):138–150.

56. Jesse C. Ribot (1999) Decentralisation, Participation and Accountability in Sahelian Forestry: Legal Instruments of Political-Administrative Control, *Africa* 69(1):23–65.

57. Anne M. Larson and Jesse C. Ribot (2004) Democratic Decentralisation through a Natural Resource Lens: An Introduction, *European Journal of Development Research* 16(1):1–25.

58. Jesse Ribot, 2002, op. cit. Quote pp. 1–2.

59. Stefano Pagiola (2008) Payments for Environmental Services in Costa Rica, *Ecological Economics* 65:712–24.

60. Steinberg, 2001, op. cit.

61. The amount of carbon stored in forests is analyzed in Yude Pan et al. (2011) A Large and Persistent Carbon Sink in the World's Forests, *Science* 333:988–92. Data on carbon emissions resulting from deforestation are from Mary L. Tyrrell, Mark S. Ashton, Deborah Spalding, and Bradford Gentry (eds.), *Forests and Carbon: A Synthesis of Science, Management, and Policy for Carbon Sequestration in Forests*, Yale School of Forestry & Environmental Studies, New Haven, CT, 2009.

62. There are no peer-reviewed publications on GRILA's history, but the events described here are corroborated by news accounts and unpublished reports written by participants in the process. Latin America's role in pushing the world to include forest conservation under the climate change treaty is described by officials from Mexico and Colombia in Manuel Estrada Porrua and Andrea García-Guerrero, A Latin American Perspective on Land Use, Land-Use Change, and Forestry Negotiations under the United Nations Framework Convention on Climate Change, pp. 209–22 in Charlotte Streck, Robert O'Sullivan, Toby Janson-Smith, and Richard Tarasofsky (eds.), *Climate Change and Forests: Emerging Policy and Market Opportunities,* Brookings Institution Press, Washington, DC, and Chatham House, London, 2008.

63. For an overview of the international program to combat global warming through the protection of forests, see Arild Angelsen (ed.), *Realising REDD: National Strategy and Policy Options*, Center for International Forestry Research, Bogor, Indonesia, 2009.

CHAPTER 9

1. For insight into the historical roots of the US environmental movement, including its relation to earlier social movements, see Robert Gottlieb, *Forcing the Spring: The Transformation of the American Environmental Movement*, Island Press, Washington, DC, 2005.

2. *New York Times*, April 23, 1970. The graph updates and uses a design originally employed in Nazli Choucri (ed.), *Global Accord: Environmental Challenges and International Responses*, MIT Press, Cambridge, MA, 1993. Laws corresponding to the acronyms used in the figure are as follows: AEA (Atomic Energy Act), CAA (Clean Air Act), CAAA (Clean Air Act Amendments), CERCLA (Comprehensive Environmental Response, Compensation and Liability Act), CPS (Consumer Product Safety Act), CWA (Clean Water Act), CZMA (Coastal Zone Management Act), DWPA (Deepwater Port Act), EPAA (Environmental Programs Assistance Act), ESA (Endangered Species Act), ESECA (Energy Supply and Environmental Coordination Act), FCMA (Fisheries Conservation and Management Act), FDC (Federal Food, Drug, and Cosmetic Act), FIFRA (Federal Insecticide, Fungicide, and Rodenticide Act), FLPMA (Federal Land Policy Management Act), FMA (Forest Management Act), FQPA (Food Quality Protection Act), FWCA (Fish and Wildlife Coordination Act), MBTA (Migratory Bird Treaty Act), MMPA (Marine Mammal Protection Act), NCA (Noise Control Act), NEPA (National Environmental Policy Act), NFMA (National Forest Management Act), NHPA (National Historic Preservation Act), NPA (National Park System Organic Act), NWPA

(Nuclear Waste Policy Act), ODA (Ocean Dumping Act), ODBA (Ocean Dumping Ban Act), OPA (Oil Pollution Act), OSHA (Occupational Safety and Health Act), PTSA (Port and Tanker Safety Act), PWSA (Ports and Waterways Safety Act), RA (Reclamation Act), RCRA (Resource Conservation and Recovery Act), RHA (Rivers and Harbors Act), SARA (the Superfund Amendments and Reauthorization Act), SCA (Social Conservation Act), SMCRA (Surface Mining Control and Reclamation Act), SWDA (Safe Water Drinking Act), TGA (Taylor Grazing Act), TSCA (Toxic Substances Control Act), WA (Wilderness Act), WPCA (Water Pollution Control Act), WPCAA (Water Pollution Control Act Amendments), WRA (National Wildlife Refuge Administration Act), WRPA (Water Resources Planning Act), and WSRA (Wild and Scenic Rivers Act).

3. The number and coverage of protected areas is from the International Union for the Conservation of Nature at http://worldparkscongress.org/about/what_are_protected_areas.html. For more background, see Lisa Naughton-Treves, Margaret Buck Holland, and Katrina Brandon (2005) The Role of Protected Areas in Conserving Biodiversity and Sustaining Local Livelihoods, *Annual Review of Environment and Resources* 30:219–52.

4. James Meadowcroft, Greening the State?, pp. 63–87 in Paul F. Steinberg and Stacy D. VanDeveer, *Comparative Environmental Politics: Theory, Practice, and Prospects*, MIT Press, Cambridge, MA, 2012.

5. Langdon Winner, *Autonomous Technology: Technics-out-of-control as a Theme in Political Thought*, MIT Press, Cambridge, MA, 1977. See also Langdon Winner, *The Whale and the Reactor: A Search for Limits in an Age of High Technology*, University of Chicago Press, Chicago, 1986.

6. John Tierney, Use Energy, Get Rich and Save the Planet, *New York Times*, April 20, 2009.

7. See David I. Stern (2004) The Rise and Fall of the Environmental Kuznets Curve, *World Development* 32(8):1419–39. The name is a play on the original Kuznets curve, from a paper published by Simon Kuznets in 1955 in which he documented an apparent rise and then decline in inequality as economies grow.

8. Adam B. Jaffe, Richard G. Newell, and Robert N. Stavins (2005) A Tale of Two Market Failures: Technology and Environmental Policy, *Ecological Economics* 54(2–3):164–74.

9. Robert D. Putnam, *Bowling Alone: The Collapse and Revival of American Community*, Simon & Schuster, New York, 2000.

10. Thomas H. Sander and Robert D. Putnam (2010) Still Bowling Alone? The Post-9/11 Split, *Journal of Democracy* 21(1):9–16.

11. The impact of Internet use on political participation is analyzed in Kay Lehman Schlozman, Sidney Verba, and Henry E. Brady (2010) Weapon of the Strong? Participatory Inequality and the Internet, *PS: Politics and Society* 8(2):487–509.

12. Michael F. Maniates (2001) Individualization: Plant a Tree, Buy a Bike, Save the World?, *Global Environmental Politics* 1(3):31–52. Quote p. 41.

13. Andrew Szasz, *Shopping Our Way to Safety: How We Changed from Protecting the Environment to Protecting Ourselves*, University of Minnesota Press, Minneapolis, 2009. Quotes p. 195 and 201.

14. Douglass C. North, *Institutions, Institutional Change and Economic Performance*, Cambridge University Press, Cambridge, 1990. Quote p. vii.

15. Defense outlays are from the International Institute for Strategic Studies, *The Military Balance 2013*, London, 2013. The comparison of expenditures as a percentage

of the economy is from The World Bank, online data table, available at http://data.worldbank.org/indicator/MS.MIL.XPND.GD.ZS. Data on tax collection as a percentage of Gross Domestic Product are from OECD statistics available at http://stats.oecd.org/Index.aspx?QueryId=21699.

16. According to the US Department of Education, in 2012 there were 13 million full-time college students in the United States. If we assume that the average tuition and fees total $10,000 per year, that's a cost of $130 billion per year. Twenty percent of the current $700 billion defense budget is $140 billion—more than enough to cover tuition and fees for each of these students every year. Note that the $700 billion figure understates the true cost of US military commitments because it does not include veteran's benefits.

17. Rebecca U. Thorpe (2010) The Role of Economic Reliance in Defense Procurement Contracting, *American Politics Research* 38(4):636–75.

18. Roderick F. Nash, *Wilderness and the American Mind*, Yale University Press, New Haven, CT, 2001 (orig. 1967). Quote p. 67.

19. Lily Tsai (2007) Solidarity Groups, Informal Accountability, and Local Public Goods Provision in Rural China, *American Political Science Review* 101(2):355–72.

20. Ronald Inglehart (1995) Public Support for Environmental Protection: Objective Problems and Subjective Values in 43 Societies, *PS: Political Science and Politics* 28(1):57–72. See also Paul R. Abramson, Critiques and Counter-Critiques of the Postmaterialism Thesis: Thirty-four Years of Debate, paper prepared for the Global Cultural Changes Conferences, Leuphana University, Lüneburg, Germany and University of California, Irvine, March 11, 2011.

21. Carolyn Sachs, Dorothy Blair, and Carolyn Richter (1987) Consumer Pesticide Concerns: A 1965 and 1984 Comparison, *Journal of Consumer Affairs* 21(1):96–107. The shift in attitudes toward pesticides in the United States has endured—see Riley E. Dunlap and Curtis E. Beus (1992) Understanding Public Concerns About Pesticides: An Empirical Examination, *The Journal of Consumer Affairs* 26(2):418–38.

22. Paul F. Steinberg, *Environmental Leadership in Developing Countries: Transnational Relations and Biodiversity Policy in Costa Rica and Bolivia*, MIT Press, Cambridge, MA, 2001.

23. James C. Scott, *Weapons of the Weak: Everyday Forms of Peasant Resistance*, Yale University Press, New Haven, CT, 1985.

24. Arthur Grube, David Donaldson, Timothy Kiely, and La Wu, *Pesticides Industry Sales and Usage: 2006 and 2007 Market Estimates*, US Environmental Protection Agency, Office of Pesticide Programs, Office of Chemical Safety and Pollution Prevention, Washington, DC, 2011.

25. David Wheeler (2001) Racing to the Bottom? Foreign Investment and Air Pollution in Developing Countries, *Journal of Environment & Development* 10:225–45.

26. Peter Dauvergne, *Shadows in the Forest: Japan and the Politics of Timber in Southeast Asia*, MIT Press, Cambridge, MA, 1997; Michael J. Watts (2005) Righteous Oil? Human Rights, the Oil Complex, and Corporate Social Responsibility, *Annual Review of Environment and Resources* 30:373–407; and Emily McAteer and Simone Pulver (2009) The Corporate Boomerang: Shareholder Transnational Advocacy Networks Targeting Oil Companies in the Ecuadorian Amazon, *Global Environmental Politics* 9(1):1–30.

27. Glen Dowell, Stuart Hart, and Bernard Yeung (2000) Do Corporate Global Environmental Standards Create or Destroy Market Value?, *Management Science* 46(8):1059–74. Quote p. 1066.

28. The World Bank, *Greening Industry: New Roles for Communities, Markets, and Governments*, Oxford University Press, New York, 2000.

29. David Vogel, *The Market for Virtue: The Potential and Limits of Corporate Social Responsibility*, Brookings Institution Press, Washington, DC, 2005.

30. Paul J. DiMaggio and Walter W. Powell (1983) The Iron Cage Revisited: Institutional Isomorphism and Collective Rationality in Organizational Fields, *American Sociological Review* 48(2):147–60.

31. Ronie Garcia-Johnson, *Exporting Environmentalism: U.S. Multinational Chemical Corporations in Brazil and Mexico*, MIT Press, Cambridge, MA, 2000. For insights into the environmental behavior of domestic firms in developing countries, see Simone Pulver (2007) Introduction: Developing-Country Firms as Agents of Environmental Sustainability?, *Studies in Comparative International Development* 42(3–4):191–207.

32. Stephen D. Parkes et al. (2013) Understanding the Diffusion of Public Bikesharing Systems: Evidence from Europe and North America, *Journal of Transport Geography* 31:94–103.

33. James G. Lewis, *The Forest Service and the Greatest Good: A Centennial History*, Forest History Society, Durham, NC, 2005.

34. Samuel P. Hays, *Conservation and the Gospel of Efficiency: The Progressive Conservation Movement, 1890–1920*, Harvard University Press, Cambridge, MA, 1959.

35. Char Miller, quoted in *The Greatest Good: A Forest Service Centennial Film*.

36. Lewis, op. cit., p. xiv.

37. Herbert Kauffman, *The Forest Ranger: A Study in Administrative Behavior*, Johns Hopkins Press, Baltimore, 1960. Quote pp. 4–5.

38. The relationship between changing ideas and changing rules in American forests is explored in Miles Burnett and Charles Davis (2002) Getting out the Cut: Politics and National Forest Timber Harvests, 1960–1995, *Administration & Society* 34(2):202–28.

39. The quote is from the *Use Book* of 1907, a pocket-sized summary of regulations guiding forest rangers' activities. From Lewis, op. cit., p. 50.

40. Alexander Hamilton, addressing the New York constitutional ratification convention, June 28, 1788. In Harold C. Syrett and Jacob E. Cooke (eds.), *The Papers of Alexander Hamilton*, Volume V, Columbia University Press, New York, 1962. Quote p. 118.

41. Thomas Paine, *Rights of Man: Being an Answer to Mr. Burke's Attack on the French Revolution*, London, 1791. Quote p. 9. Italics in original.

42. Thomas Jefferson, letter to James Madison, September 6, 1789. In Philip B. Kurland and Ralph Lerner, *The Founders' Constitution*, University of Chicago Press, Chicago, Document 23, Papers 15:392–97. Online edition available at http://press-pubs.uchicago.edu/founders/.

43. For a discussion of the resilience of the rules we create for biodiversity conservation, see Paul F. Steinberg (2009) Institutional Resilience amid Political Change: The Case of Biodiversity Conservation, *Global Environmental Politics* 9(3):61–81.

CHAPTER 10

1. This scenario of children designing mutually binding rules for a game requires the inclusion of older children in the mix, and this in itself reveals something about the power of human instincts to create rules. Toddlers cannot improvise and articulate their own rules, and so their free-form play is a more chaotic affair. They understand that they are supposed to follow rules (take turns, no hitting), but the only decision they

face is whether to comply or rebel. Developmental psychologists Donna Weston and Elliot Turiel of the University of California at Berkeley report that beginning around age six, children become aware that some social rules are up for debate—"changeable, as relative to the social context, and as determined by consensus." In other words, as soon as humans are able, we make rules, debate rules, and support or defy them. See Donna R. Weston and Elliot Turiel (1980) Act-Rule Relations: Children's Concepts of Social Rules, *Developmental Psychology* 16(5):417–24.

2. Elinor Ostrom and Sue Crawford distinguish among three levels of rules: operational rules that govern day-to-day management decisions, like how much water to draw from an irrigation channel; collective choice rules that specify who gets to participate in decision making; and constitutional choice rules that spell out what it takes to change collective choice rules (Elinor Ostrom and Sue Crawford, Classifying Rules, pp. 186–216 in Elinor Ostrom, *Understanding Institutional Diversity*, Princeton University Press, Princeton, NJ, 2005).

I have collapsed these into two levels—rules and super rules—both for simplicity and because I find their distinction between collective choice and constitutional choice to be somewhat arbitrary. In practice, rules are embedded within other rules like the layers of an onion. Consider an agency ruling that establishes acceptable levels of toxic formaldehyde in consumer products. In the United States, this is affected by a higher-level rule, the Toxic Substances Control Act, which empowers the EPA to make that decision. The implementation of the law is, in turn, affected by the Administrative Procedure Act, which requires the EPA to make its decision available for public review. The Administrative Procedure Act was itself shaped by the US Constitution, which granted lawmakers the right to create it. A rule's "super" quality is therefore a matter of degree, depending on how many rulemaking processes it affects.

3. The rise of the Greens in Western Europe is discussed in Michael O'Neill, Political Parties and the "Meaning of Greening" in European Politics, pp. 171–95 in Paul F. Steinberg and Stacy D. VanDeveer (eds.), *Comparative Environmental Politics: Theory, Practice, and Prospects*, MIT Press, Cambridge, MA, 2012.

4. The tendency of winner-takes-all voting systems to produce two-party systems is known as Duverger's Law, after the French sociologist who first proposed the explanation in 1951. Exceptions to Duverger's Law can be found where small parties enjoy concentrated pockets of support in particular regions of a country (as is seen in Canada, the United Kingdom, and India), and are therefore able to defeat the major parties in those regions and achieve some presence in the legislature. For an overview see Kenneth Benoit (2006) Duverger's Law and the Study of Electoral Systems, *French Politics* 4(1):69–83.

5. Germany's voting rules are described in Thomas Gschwend, Ron Johnston, and Charles Pattie (2003) Split-Ticket Patterns in Mixed-Member Proportional Election Systems: Estimates and Analyses of their Spatial Variation at the German Federal Election, 1998, *British Journal of Political Science* 33:109–27.

6. For an accessible overview of the strengths and weaknesses of various voting systems, see Ian Stewart (2010) Why Voting Is Always Unfair, *New Scientist* 206(2758):28–31.

7. On the electoral success of far-right parties in contemporary Europe, see William Wheeler, Europe's New Fascists, *New York Times*, November 17, 2012, p. SR4; and Anon., European Fascism: A Movement Grows in Hungary, *Boston Globe*, May 18, 2013. For an evaluation of their impact on policy, see Michael Minkenberg (2013) From

Pariah to Policy-Maker? The Radical Right in Europe, West and East: Between Margin and Mainstream, *Journal of Contemporary European Studies* 21(1):5–24.

8. Timothy Doyle and Adam Simpson (2006) Traversing More than Speed Bumps: Green Politics under Authoritarian Regimes in Burma and Iran, *Environmental Politics* 15(5):750–67. Quotes p. 755.

9. In August 2011, protesters in the city of Tabriz in northwestern Iran were beaten and arrested after demanding that the government take measures to save Lake Urmia, which is disappearing as a result of policies encouraging unsustainable surface water use in drought conditions. See Robert Mackey, Protests in Iran Over Disappearing Lake, *New York Times*, August 30, 2011. See also Amnesty International's coverage of the May 2011 arrest of Farzad Haghshenas, a member of the environmental group Sabzchia (The Green Mountain Society) at http://ua.amnesty.ch/urgent-actions/2011/06/195-11?ua_language=en.

10. Peter Bachrach and Morton S. Baratz (1962) Two Faces of Power, *American Political Science Review* 56(4):947–52.

11. Bachrach and Baratz, op. cit. Quote p. 952.

12. Kent E. Portney and Jeffrey M. Berry, Neighborhoods, Neighborhood Associations, and Social Capital, pp. 2–43 in Susan A. Ostrander and Kent E. Portney (eds.), *Acting Civically: From Urban Neighborhoods to Higher Education*, Tufts University Press/University Press of New England, Lebanon, NH, 2007. Quote p. 25.

13. For an overview of experiences with participatory budgeting in Latin America, see Yves Cabannes (2004) Participatory Budgeting: A Significant Contribution to Participatory Democracy, *Environment and Urbanization* 16(1):27–46.

14. Frank R. Baumgartner and Bryan D. Jones (1991) Agenda Dynamics and Policy Subsystems, *Journal of Politics* 53(4):1044–74. Quote p. 1047.

15. Wayne B. Gray and Jay P. Shimshack (2011) The Effectiveness of Environmental Monitoring and Enforcement: A Review of the Empirical Evidence, *Review of Environmental Economics and Policy* 5(1):3–24. Quote p. 5.

16. Senator Edmund Muskie, September 21, 1970, quoted in Jeffrey G. Miller, *Citizen Suits: Private Enforcement of Federal Pollution Control Laws*, Wiley Law Publications, Hoboken, NJ, 1987, pp. 4–5.

17. In the American legal system, the Attorney General is the country's highest law enforcement official and carries the ultimate responsibility for pursuing violations of federal laws.

18. James R. May (2003), Now More Than Ever: Trends in Environmental Citizen Suits at 30, *Widener Law Review* 10(1):1–48. Quote pp. 3–4.

19. Lesley K. McAllister, *Making Law Matter: Environmental Protection and Legal Institutions in Brazil*, Stanford Law Books, Stanford, CA, 2008. See also Bernardo Mueller (2010) The Fiscal Imperative and the Role of Public Prosecutors in Brazilian Environmental Policy, *Law & Policy* 32(1):104–26.

20. Alex Aylett (2010) Conflict, Collaboration and Climate Change: Participatory Democracy and Urban Environmental Struggles in Durban, South Africa, *International Journal of Urban and Regional Research* 34(3):478–95. Quote p. 488.

21. Lynton K. Caldwell (1963) Environment: A New Focus for Public Policy?, *Public Administration Review* 23(3):132–39. Quotes pp. 136 and 139.

22. Per-Olof Busch, Helge Jörgens, and Kerstin Tews (2005) The Global Diffusion of Regulatory Instruments: The Making of a New International Environmental Regime, the *ANNALS of the American Academy of Political and Social Science* 598:146–67.

23. See the remarks of Chief Justice Hughes in *Panama Refining Co. v. Ryan*, 293 U.S. 388 (1935). For a history of efforts to increase transparency in the US government, see Walter Gellhorn (1986) The Administrative Procedure Act: The Beginnings, *Virginia Law Review* 72(2):219–33.

24. Administrative Procedure Act, 5 USC Chapter 5 § 552b.

25. Data on transparency are from Table 1 of Colin Bennett (1997) Understanding the Ripple Effects: The Cross-National Adoption of Policy Instruments for Bureaucratic Accountability, *Governance: An International Journal of Public Policy and Administration* 10(3):213–33; and Tero Erkkilä, Transparency and Nordic Openness: State Tradition and New Governance Ideas in Finland, pp. 348–72 in Stephan A. Jansen, Eckhard Schröter, and Nico Stehr (eds.), *Transparenz: Multidisziplinäre Durchsichten durch Phänomene und Theorien des Undurchsichtigen*, VS Verlag/Springer, Heidelberg, Germany, 2010. See also Jeannine E. Relly and Meghna Sabharwal (2009) Perceptions of Transparency of Government Policymaking: A Cross-national Study, *Government Information Quarterly* 26(1):148–57.

26. The history and impact of transparency initiatives are discussed in Aarti Gupta and Michael Mason (eds.), *Transparency in Global Environmental Governance: Critical Perspectives*, MIT Press, Cambridge, MA, 2014.

27. For an analysis of the challenge of implementing greater openness in post-communist countries, written by two practitioners close to the process, see Tatiana R. Zaharchenko and Gretta Goldman (2004) Accountability in Governance: The Challenge of Implementing the Aarhus Convention in Eastern Europe and Central Asia, *International Environmental Agreements: Politics, Law, and Economics* 4:229–51.

28. The two US Supreme Court decisions that emasculated campaign finance reform were *Citizens United v. the Federal Elections Commission* (2010) and *McCutcheon v. the Federal Elections Commission* (2014).

29. For an astute analysis of the tension between democracy and capitalism and how this has shifted over time in the United States, see particularly chapter 4 of Robert B. Reich, *Supercapitalism: The Transformation of Business, Democracy, and Everyday Life*, Alfred A. Knopf, New York, 2007.

30. Thomas R. Rochon and Daniel A. Mazmanian (1993) Social Movements and the Policy Process, *ANNALS of the American Academy of Political and Social Science* 528:75–87. Quote p. 83.

CHAPTER 11

1. Diane Wedner, The Lawn Arm of the Law, *Los Angeles Times*, September 6, 2008.

2. Interview with Matthew Lyons, Director of Planning and Conservation, Long Beach Water Department, June 6, 2014. The law prohibiting homeowners associations from interfering with water-wise landscaping is California Assembly Bill 1061.

3. Athena Mekis, Long Beach Water Department Announces Completion of Its 500th Water-conserving Lawn, *Signal Tribune Newspaper*, November 11, 2011.

4. Daron Acemoglu and James A. Robinson, *Why Nations Fail: The Origins of Power, Prosperity and Poverty*, Crown Publishers, New York, 2012.

5. If your Google Scholar search produces a surplus of publications, leaving you unsure where to start, try the following trick. Do a general Google search for the issue that concerns you, combined with the word "syllabus." This will help to identify

readings that have been vetted by experienced teachers sharing the cream of the crop. This approach works best for issues that attract reasonably broad public interest (broad enough to be featured in a college course), and have been around for at least a few years, allowing time for research and publication.

6. The challenges involved in bridging the worlds of research and practice are explored in Paul F. Steinberg (ed.), Is Anyone Listening? The Impact of Academic Research on Global Environmental Practice, special issue of *International Environmental Agreements: Politics, Law, and Economics*, Kluwer Academic Publishers/Springer, vol. 5(4) December, 2005.

7. Leonie Huddy, Lilliana Mason, and Lene Aaroe, Measuring Partisanship as a Social Identity, Predicting Political Activism, paper presented at the annual meeting of the International Society for Political Psychology, San Francisco, CA, July 7–10, 2010.

8. For a discussion of strategies to ensure the durability of environmental reforms, see Paul F. Steinberg (2009) Institutional Resilience amid Political Change: The Case of Biodiversity Conservation, *Global Environmental Politics* 9(3):61–81.

9. On the relation between social rules for the environment and economic growth, see Leena Lankoski, *Linkages between Environmental Policy and Competitiveness*, OECD Environment Working Papers, No. 13, Organisation for Economic Cooperation and Development (OECD) Publishing, Paris, 2010; and Stefan Ambec, Mark A. Cohen, Stewart Elgie, and Paul Lanoie, *The Porter Hypothesis at 20: Can Environmental Regulation Enhance Innovation and Competitiveness?*, Resources for the Future, Washington, DC, 2011.

10. Paul F. Steinberg, *Environmental Leadership in Developing Countries: Transnational Relations and Biodiversity Policy in Costa Rica and Bolivia*, MIT Press, Cambridge, MA, 2001.

11. Daniel Brockington and James Igoe (2006) Eviction for Conservation: A Global Overview, *Conservation and Society* 4(3):424–70.

12. Richard Rose (1991) What Is Lesson-Drawing?, *Journal of Public Policy* 11(1):3–30.

13. Everett M. Rogers, *Diffusion of Innovations*, 5th ed., Free Press, New York, 2003.

14. Process expertise and related categories of site-specific political know-how are described in Steinberg, 2001, op. cit.

15. Margaret E. Keck and Kathryn Sikkink, *Activists Beyond Borders: Advocacy Networks in International Politics*, Cornell University Press, Ithaca, NY, 1988; and Alison Brysk, *From Tribal Village to Global Village: Indian Rights and International Relations in Latin America*, Stanford University Press, Stanford, CA, 2000.

16. James Q. Wilson, *Bureaucracy: What Government Agencies Do and Why They Do It*, BasicBooks, New York, 1989.

Index

"f" refers to figure; "n" to notes

Aarhus Convention on Access to Information, 259
Acemoglu, Daron, 105, 293n19, 314n4
acidification of the ocean, 43. *See also* coral reefs
acid rain, 108, 115–17, 119
acquis communautaire (EU), 180
activism, 16, 134, 203. *See also* advocacy; social movements, impact of
activists
    Bolivian, 14
    at Chicago wade-in event, 21, 22f1.1
    civil rights, 118
    community, 18
    environmental justice, 118
    German, 176
    hazardous waste, 261
    human rights, 164
    Iranian, 251
    Irwin, June, 6–7, 11, 55, 263 (*See also* pesticide(s))
    Karen ethnic group (Burma), 251
    for Louisiana's "Cancer Alley," 118
    South African, 256
    transnational, 231
    Yadana gas pipeline and Burmese, 251
advocacy. *See also* lobbyists; rulemakers
    Brower, David, 223
    Carson, Rachel, 17, 20, 223, 225
    Colburn, Theo, 17
    groups, 259, 268
    Holdren, John, 17
    Muir, John, 223, 235, 237
    Myers, Norman, 17
    Nogales, Juan, 138–39
    pesticide movement led by Cesar Chavez and the United Farm Workers, 118, 225
    public, 17, 21 (*See also* activism)
    Sagan, Carl, 17, 283n18
    Schneider, Stephen, 17
    Thoreau, Henry David, 223
    Transparency International, 259
    US Forest Service, 83, 234–35, 238f9.4, 239, 268
African National Congress, 271
agriculture. *See also* green revolution; pesticide(s); pesticide regulation
    investment in, 187
    national economic policies and, 175

318   Index

agriculture (*continued*)
   pesticides and, 8
   property rights reform and, 77
   sustainable, 175
   tax exemptions for, 133
   US Department of Agriculture, 51, 212, 257–58
air pollution. *See also* Air Pollution Control Act; Clean Air Act
   acid rain, 108
   in American cities, 163
   annual deaths worldwide from, 43
   emissions, regulation of, 190
   in Los Angeles, 211, 241
   Los Angeles Bureau of Smoke Control, 212
   from marine transport, 23
   scrubbing technology to reduce, 216
   Senate Subcommittee on Air and Water Pollution, 254
   urban, 5, 211
Air Pollution Control Act (California), 212
air quality. *See also* air pollution; Clean Air Act
   cap-and-trade program, 116
   local, 114
   monitoring, 148
   regulations, 108
   rules, 23, 216
   standards, 113, 202
   urban, in Italy, 100
American Academy of Pediatrics, 121
American Chemical Society, 23
American Society for Testing and Materials (ASTM), 38–40, 253
anti-corruption campaigns, 142
Antinori, Camille, 85, 290n54, 299n37
apartheid-era structures, 271
Appalachian Regional Reforestation Initiative, 93
Ascher, William, 150
ASTM. *See* American Society for Testing and Materials
Audubon Society, 91
authoritarian
   governments, 250
   military juntas (Madagascar), 136
   monarchies (Saudi Arabia), 136
   regimes, 136–37, 137f6.2, 137nf6.2, 138, 140, 185, 191, 224, 251, 303n4
   rule, 136, 205
   single-party systems (China), 136
Aylett, Alex, 256

Bachrach, Peter, 252
Balcones Canyonlands Preserve, 43
bald eagle, 65
Bangladesh, 148–49, 188
bankruptcy laws, 25f2.2, 103
Baratz, Morton, 252
Bardach, Eugene, 58–59, 286n29
Bates, Robert, 150, 299n47
Battle of Waterloo (1815), 29
BBS. *See* North American Breeding Bird Survey
A Beautiful Mind (film), 196
Benavides, Felipe, 139
Berry, Jeffrey, 203, 253, 294n26, 307n48, 313n12
Berry, Wendell, 203, 253, 278, 307n48, 313n11
bicycle ridership in Portland, 42, 205, 264
bicycle lanes, 13, 205, 268
Bicycle Transportation Alliance (Portland), 264
Biedenkopf, Katja, 181, 303n30
Binswanger, Hans, 133, 296n5
biodiversity, 152, 182, 217, 237, 246, 257–58, 273. *See also* deforestation; Endangered Species Act
   conservation and Costa Rica, 217
   conservation and value creation, 273
   Convention on International Trade in Endangered Species, 153
   Forest Service (US), 237
   "green" agricultural practices (EU), 181–82
   loss and deforestation, 217
   oceanic, 22
   in Peru, 69–70, 82
Biodiversity Unit (EU), 161–62
birth control, 148–49
Blackstone, William (Sir), 66
Blaikie, Piers, 205, 307n54

Index   319

Bolivia
   Brazil vs. Bolivia, conservation rules, 131, 132f6.1, 133–34
   decentralization, 192
   deforestation, 131–32, 132f6.1, 133–34
   dictatorship, 14
   endangered species, 138–39
   environmental news events, 228f9.3
   forest in, 14
   Kaa-Iya del Gran Chaco National Park, 273
   local activists, 14
   local decisions governing forests, 192, 203–4
   Noel Kempff National Park, 14
   rainforests, 14
   Santa Cruz de la Sierra, 14, 199–200, 220
   social change in, 14
Bonaparte, Napoleon, 28–30, 76, 186, 263
Bongaarts, John, 148, 299n41
boomerang effect, 277
Bray, David Barton, 85, 290n54, 299n37
Brazil
   Bolivia vs. Brazil, impact of conservation rules, 131, 132f6.1, 133–34
   Constitution, 256
   decentralization, 192
   deforestation, 131–32, 132f6.1, 133–34
   Ministério Público (Brazil), 256
   policy granting land to those who remove trees, 31
Brophy, Jim, 264
Brower, David, 223
By-law 270 (pesticide ban), 7, 11, 240, 268

Caldwell, Lynton, 256–57, 313n21
Canada
   Aboriginal peoples' property rights recognized, 81
   Appalachian mountains and wildlife species, 90
   cerulean warbler's migratory route, 90, 275
   democracy, established, 259
   federalist system, 186
   forest conservation, 210
   Hudson Bay Company of Canada, 169
   lawn care industry lawsuit, 8
   manufacturing plants for warplanes, 171
   migratory birds, ban on hunting or capturing of, 64
   NATO and global security, 222
   northern cod fishery, collapse of, 23
   pesticides, ban on nonessential, 7–9, 11, 25f2.2, 230, 263
   political system, lack of corruption, 143
Canadian Supreme Court, 8–9
cancer clusters, 5
cannon shot rule, 87
capacity
   to act, 139–41
   to act by letting go of responsibilities and power, 190
   to deliver social services, 186
   to perform rulemaking functions, 140
   to process information, 45, 129
   to respond in a coordinated fashion to environmental problems, 163
cap-and-trade program
   air quality, 116
   for carbon dioxide, 121
   environmental protection, 120–24, 156, 181
   environmental regulations, market-based, 120–24, 156, 181
   for forest based carbon, 123
   for mercury, 120–21, 295n45
carbon
   capture, 122
   footprint, 36
   markets, forest-based, 123
   storage, forest-based, 208
   tax, 58, 121, 123
carbon dioxide. *See also* cap-and-trade program; climate change; deforestation; forest; greenhouse gases
   acidification of the ocean, 43
   from air conditioners, 67
   atmospheric carbon dioxide traps heat, 267
   emissions, 27, 43, 107, 217
   emissions decline in Denmark, 149
   emissions from China, 122
   Emissions Trading System, 121–22

carbon dioxide (*continued*)
  payments to local landowners to reduce, 208
  photosynthesis and, 208
  power plants, new rules limiting emissions from, 122
  sink in forests, 84, 272
  sources of, 121
  tropical forest destruction and, 208
Carson, Rachel, 17, 20, 223, 225
cattle ranching, 132–133, 184, 193
cerulean warbler. *See also* endangered
  Atlantic Coast, Costa Rica, 70f4.1, 82–84
  Barro Colorado Island (Panama), 70f4.1, 83–84
  Boone County (West Virginia), 70f4.1, 89–93
  Callanga Valley (Peru), 69–74, 70f4.1
  Cerulean Warbler Reserve (Santander, Colombia), 70f4.1, 78–79
  Cerulean Warbler Technical Group, 83
  Doppler radar showing migratory birds arriving on the Texas coast, 90f4.2
  Gulf of Mexico, 70f4.1
  Lancandon Rainforest (Mexico), 70f4.1, 84
  Migratory Bird Treaty, 94
  migratory route, 70
  property rights, 70f4.1, 71, 74, 77–81, 83, 85, 87–89, 91–92, 94
  risk of extinction, 69
  Santa Catarina Ixtepeji (Mexico), 70f4.1, 84–86
  Sierra Nevada de Santa Marta (Colombia), 70f4.1, 79–82
  survey by BBS volunteers, 69
CFCs. *See* chlorofluorocarbons (CFCs)
Chavez, Cesar, 118, 225
A Chemical Reaction (film), 8
ChemLawn, 8
Chernobyl nuclear power plant, 177
Chesapeake Bay Foundation, 41
child labor, 148
China, 23, 25f2.2, 26, 122, 136–37, 152
Chinese boxes, 205–7
chlorofluorocarbons (CFCs), 153

Citizens for Alternatives to Pesticides, 8
citizenship, indigenous, 277
citizens' movements, 15
citizen suits, 255
city by-laws, 11
city codes, 59, 134
city planning guidelines, 268
Civil Code, 29, 76. *See also* Napoleonic code
civil law, 76
Civil Rights era, 21, 192
Clean Air Act (1970). *See also* air pollution; air quality of 1970, 96, 113, 155, 255, 263
  amendments of 1977, 114
  amendments of 1990, 115, 124, 152, 154, 156, 272
Clean Water Act (1970), 30, 91, 152, 155, 255, 263. *See also* water
climate change. *See also* carbon dioxide; fossil fuels; global warming; greenhouse gases
  3 R's (roles, rights, responsibilities), 27
  about, 5, 15, 101, 115, 166
  auto industry lawsuit and, 156
  Earth Summit in Rio de Janeiro, 179
  EPA and, 156
  European Union and, 169, 179
  Figueres, Christiana, 209
  global effort to address, 210
  Group of 77 and, 209
  market-based solution to, 122
  multilevel world view of, 163
  science of, 122
  super rules and UN procedures, 247
  transnational nature of, 133
  treaty, 154, 208–9
coal-fired power plants and mercury, 120–21
Coalition for Sensible Pesticide Policy, 9
coastal access
  in Ireland, 20
  in Scotland, 20
  in the US, 20–21
coastal reserves (Philippines), 149
cod fishery, Canada's northern, 23
coffee, 78
Colburn, Theo, 17

collective action (coordinated social
    action)
  about, 37, 47–48
  free-rider behavior, 48, 50, 55
Colombian Constitution (1991), 81
colonial law, 20
commercial rules, 144
common-pool resources, 306n42.
    See also the commons; tragedy
    of the commons
the commons
  about, 193, 195–96, 200
  Digital Library of the Commons, 197
  freedom of, 194
  local, 200
  tragedy of, 193–96
Communities for a Better Environment, 118
community
  activist, 18, 55 (*See also* Irwin, June)
  enterprises, 86
  forestry (Mexico), 85–86, 299n37
  harvesting rights (Peru), 75
  institutions, 144
  leaders, 16, 124
  organizations, 134, 149
  peer pressure, 145
  property, local, 75
Community Forest Enterprises
    (Mexico), 86
community gardens, 134, 246, 268, 273, 278
complementarity, 59
compliance (legal)
  citizen suits, 255
  PROPER program (Indonesia), 153,
    299n38–39
  South Africa empowers citizens, 256
Congress of Europe, 174
conservation groups, 48, 57, 164, 276
Conservation International, 82, 265
constitutions
  about, 229, 239–40, 268
  Brazilian Constitution, 256
  Colombian Constitution (1991), 81
  crisis, 136
  guarantees, 19
  national, 205, 240
  reform, 192
  rules, 23, 134

  structure, 140
  US Constitution, 173, 246
  US Senate, role of, 261
contracts
  about, 31, 59, 103, 229
  car insurance, 24
  for community water management, 162
  corporate, 28
  federal defense, 222
  harvesting, in Ixtepeji, 86
  legal, 274
  mutually binding, 196
  patents and labor, 215
  private, 11
  purchasing, 268
  renewal periods, 240
  rental, 27
Convention on International Trade
    in Endangered Species, 153
copyright law, 25f2.2
coral reefs, 13, 43. *See also* acidification
    of the ocean; oceans
corruption, 51, 104, 142–43. *See also*
    lobbyists
Corzo, Pati Ruiz, 146
Costa Rica
  biodiversity conservation, 217
  cerulean warbler, 70f4.1, 82–84
  decentralization in, 192
  environmental news events, 228f9.3
  environmental services of farmers, 13,
    84, 207
  forest in, 13, 84, 183–84, 207–10
  national parks in, 208
  Payment for Ecosystem Services
    program, 207–8, 272
  tropical forests of, 183
  Water and Sewer Institute, 162
  water associations, rules for, 163
  water management, 161–62
CTG Energetics, 41
Cuba, 23
cultural norms, 8, 28, 223, 225, 229, 246
Culture Moves (Rochon), 226–27

dams, 17, 188
da Silva, Lula, 138
Dauvergne, Peter, 145

decentralization. *See also* tragedy of the commons
  in Bolivia, Honduras, Guatemala, and Nicaragua, 192
  in Costa Rica, 192
  defined, 185–86
  democracy and, 205, 207
  dictatorships and, 192
  in Europe, 182, 189–91
  European Union, 182, 190–91, 277
  global trend toward, 157, 164, 182, 185–86, 190–92, 277
  in Latin America, 191–92, 205, 206f8.2
  laws, 182, 192
  laws in Japan, South Korea, Australia and New Zealand, 192
  local control of forests, 198
  local control over environmental decisions, 192
  local environmental outcomes, 204
  in Mexico, Brazil, and Costa Rica, 192
  in the Philippines, 192
  power-sharing paradox, 277
  tragedy of the commons and local control, 193
  urban governments and new environmental responsibilities, 199–200, 202
  in West Africa, 207
Deepwater Horizon rig (BP's), 89
DeFalco, Paul, 113–14
Defenders of Wildlife, 91
deforestation. *See also* biodiversity; carbon dioxide; forest
  of the Amazon, 145
  in Brazil and Bolivia, 131–32, 132f6.1, 133–34
  citizen protests, 227
  of eastern United States and Canada, 275
  encomienda system, 73
  environmental news events, 228f9.3
  in Malaysia, 231
  property rules and, 74
  Reducing Emissions from Deforestation and Forest Degradation (REDD+), 210, 295–96n48
  relation to economic growth, 217
  soybean production and, 134
  in United States, 234

democracy
  about, 135–38, 137f6.2, 152
  in Australia, Canada, and the Netherlands, 259
  authoritarian rule and fight for, 205, 206f8.2
  autocratic rule vs., 136
  average lifespan of, 140
  in Burma, 251
  decentralization and, 205, 207
  dictatorships and, 192
  environmental concerns of, 248
  fossil fuel dependency, 136–37
  in Greece, 176–77
  Green parties, 176–78, 180, 247–48
  growth in, 136, 137f6.2
  in India, 149–50
  in Indonesia, 147–48
  innovations in urban democracy, 253
  Latin America, election of local mayors, 206f8.2
  local, 192
  Madison, James, 204
  newly established countries vs., 188
  in the Philippines, 149
  pro-democracy movement, 137
  reversals from, 136
  in Southeast Asia, 151
  in Spain, 190
  transparency and, 259
  in the US, 260
  world's wealthy, 222, 247
democratic
  accountability, 142, 224
  countries, 105
  dialogue, 18, 54
  environmental protection, 138
  experiment in the US, 186
  forms of government, 136
  government, 140
  government, authoritarian vs., 140
  institutions, 105, 207
  natural resources, 138
  participation, 113
  reforms, 85, 186, 205, 262
  rights, 52
  South Africa, 271
  transitions in countries, 136

Democratic Party (US), 212, 248
design standards, 11, 19, 59, 181, 229, 232, 268
de Gaulle, Charles (General), 172–73, 176
De Soto, Hernando, 103
dictatorship(s). *See also* authoritarian, regimes
  Bolivia, 14
  decentralization and, 192
  Earth is ruled by nations, 152
  in Indonesia, 147–48
  Marcos (Philippines), 192
  in Peru, 139
  rulemaking under, 14
  Samuel Doe's (Liberia), 128
Diffusion of Innovations (Rogers), 274, 315n13
Digital Library of the Commons, 197
Directive on End-of-Life Vehicles (EU), 181
Disabilities Act (1990), 24
Doe, Samuel, 127–28
Dowell, Glen, 230–31, 310n27
Downs, Anthony, 30
Doyle, Timothy, 250–51, 313n8
Drinking Water Protection Zone (Austin Texas), 43
droughts, 43, 123, 264–65. *See also* water, scarcity
Dryzek, John, 177, 302n24
Duarte, José Delfín, 161–63
Dunlap, Riley, 154
Dupont Deepwater plant, 96
Dutch flower industry, 109–10

Earth Day, 113, 118, 185, 212, 214f9.2
Earth Summit (Rio de Janeiro), 179, 208
economic growth and institutions. *See* institutional economics
ecosystem services. *See* Payment for Ecosystem Services program
educational opportunities for girls, 148
effectiveness (physical change). *See also* European Union
  bicycle ridership in Portland, 42, 205, 264
  cap-and-trade, 120–24, 156, 181
  Directive on End-of-Life Vehicles (EU), 181
  Emissions Trading System (EU), 121–22
  government effectiveness and political and economic instability, 140
  lead, 12, 120, 170, 255
  local social capital, mobilization of, 147
  pesticide levels lower in Canadian waterways, 8
  pollution emissions program PROPER (Indonesia), 147
  REACH policy (EU), 181
  smog-forming ozone, progress in combating, 213f9.1
  tetraethyl lead, 96
ejidos (Mexican property institutions), 85–86, 290n55
electoral systems, 179, 268
Elliot, Michael, 7
Emergency Planning and Community Right-to-Know Act (1986), 153
Emissions Trading System (EU), 121–22
endangered. *See also* cerulean warbler; extinction
  animals and acts of sabotage, 229
  bird, 12, 67
  cerulean warbler, 32, 70f4.1, 91, 275–76
  Convention on International Trade in Endangered Species, 153
  elephants in Liberia and Sierra Leone, 129
  "Red Data" books on endangered species (Russian), 130
  Saint Lucia Parrot, 140
  species, government initiatives to protect, 227
  species, illegal trade of, 166
  species and rulemaking structures, 162
  species in Bolivia, 138–39
Endangered Species Act (1970), 30, 91, 152, 155, 246, 258. *See also* biodiversity
energy. *See also* fossil fuel(s)
  activation, 49, 49f3.1
  building construction and operation requirements, 39
  carbon footprint, 36
  consumption, 26
  efficiency, 36, 39
  efficiency in Japan, 138
  entrepreneurs, 107
  Hampton Roads Energy Company (Virginia), 114
  incentives, 133

energy (*continued*)
  LEED-certified buildings, 41
  markets, competitive, 107
  nuclear, 177
  peregrine falcons and warm air currents, 86
  pricing policies (Dutch), 149
  renewable, 13, 149
  solar, 12, 52, 107
  sources, traditional, 149
  technology, alternative, 17, 164, 175, 215
  windows, energy-efficient, 40, 46
environment (environmental). *See also* market-based environmental regulations; policy (policies), environmental
  concerns of democracies, 248
  decisions, local control over, 192
  decisions, race or class bias in, 118
  imperative of regulating use of oceans, 88
  justice, 118
  law and policy, 107
  laws, 91, 107, 155, 208, 214f9.2, 231, 255–56
  movement, American, 118
  National Environmental Management Act (South Africa), 256
  National Environmental Policy Act (1970), 153, 156, 257
  National People of Color Environmental Leadership Summit, 118
  news events (Costa Rica and Bolivia), 228f9.3
  outcomes, local, 204
  policies, 91, 95, 98, 100–101, 111, 145, 150, 154, 156, 175, 180–81, 256–57, 268
  problems and property rights, 94
  protection and cap-and-trade program, 120–24, 156, 181
  responsibilities, urban governments and new, 199–200, 202
  science, 12, 227
  services of farmers (Costa Rica), 84, 207
  standards in Mexico, 15
  stewardship, 12, 40, 135, 179, 251
  summits, global, 135
  treaties and amendments, 165, 165f7.1

Environmental Defense Fund, 120
environmental groups
  about, 41, 50, 138, 203
  German, 248
  Iranian, 251
  US, 10, 114–15, 118, 134, 238, 248, 255
environmentalism, 42, 156, 220, 251, 278
Environmental Kuznets Curve, 217
environmental protection. *See also* sustainability; US Environmental Protection Agency
  bottom-line approaches to, 123–25
  cap-and-trade, 120–24, 156, 181
  carbon markets, forest-based, 123
  environmentalists and efficient regulation, 112
  equity and, 118
  in the US, 89, 96, 100, 118
environmental regulations, 18
  costs imposed on industry, 108
  for diesel industry, 108
  efficient regulation and lower costs, 112
  market-based, 100–101, 111, 115, 117, 119, 121–24, 292n5, 296n49 (*See also* cap-and-trade)
  opponents of, 100
  smart regulation, 110
environmental stewardship, 12, 40, 135, 179, 251
EPA. *See* US Environmental Protection Agency
Erlich, Paul, 218
EU directive on urban wastewater, 181
European Court of Justice, 176, 266
European Parliament, 176, 178, 250
European Union (EU)
  Biodiversity Unit, 161–62
  decentralization, 182, 189–91, 277
  Directive on End-of-Life Vehicles, 181
  domestic laws of new members are adjusted to EU policies, 181
  Ecolabel program, 179
  economic policy, 176
  Emissions Trading System, 121–22
  environmental policy, 180–82
  environmental problems, 163
  European Council, 166, 178
  international cooperation, 164

invasive species, rules for control of, 162
local governments rulemaking
    power, 277
Maastricht Treaty, 179
Monnet, Jean, 167–72, 172f7.2, 173–79, 182, 187, 264
national economic policies and agriculture, 175
polluter pays principle, 179
priorities of European heads of state, 162, 176
proportional representation, 248–49, 249f10.1, 250
REACH policy, 181
rulemaking power and local EU governments, 277
Single European Act, 178
urban wastewater treatment, 181
water quality regulations, 182
Exporting Environmentalism (Garcia-Johnson), 15
externalities, 105–7, 111. *See also* spillover effects
externality (externalities), 105–7, 111
extinction. *See also* endangered
    of bird species, 69
    crisis, global, 237, 283n18
    efforts to save the cerulean from, 276
    of freshwater fish species, 43
    irreversibility of, 240
    of mammals, 6
    species, 5–6
    species at risk of, 246
    of wild vicuñas, 138
Exxon Valdez spill in Alaska (1989), 89

FeatherFest in Galveston, Texas, 89
Federal-Aid Highway Act (1956), 24
federalism
    in Brazil and Germany, 186
    in Canada, 9, 186
    in Great Britain, 186
    in the US, 164, 173, 204
Federalist Papers, 204
Federal Register, 258–59
Figueres, Christiana, 209
First Indigenous Congress (Colombia), 81
flaring from oil rigs, 88

Food and Drug Administration rules, 25f2.2
food production, 152, 218, 268.
    See also overfishing; sustainable, agriculture
forest. *See also* carbon dioxide; deforestation
    Amazonian rainforests, 16
    Appalachian mountains, 90–91, 93
    Appalachian Regional Reforestation Initiative, 93
    ASTM standards for sustainability harvesting, 39
    in Bolivia, 14
    in Borneo, 151
    in Burma, 251
    carbon, global market for, 122–23
    carbon capture, 122
    carbon dioxide, sources of, 121
    in Colombian Andes, 77–79
    conservation and GRILA, 209–10
    conservation and industrialized countries, 208
    conservation blocked by Group of 77, 209
    conservation easements, 93
    conservation in Dominican Republic, 140
    conservation in Spain, 190
    in Costa Rica, 13, 84, 183–84, 207–10
    decentralized forest management in West Africa, 207
    deforestation, prohibitions against, 27
    Environmental Kuznets Curve and, 217
    homesteading laws (US), 212, 215
    in Honduras, 150
    Kempff, Francisco, 14
    law in Peru, 145
    local citizens and forest management in Guatemala and Bolivia, 203–4
    local communities control 800 million acres of forests worldwide, 198
    local community management of shared forest resources in Mexico and Colombia, 195
    local decisions governing forests of Bolivia, Honduras, Guatemala, and Nicaragua, 192

forest (*continued*)
   national forests (US), 155, 234
   ownership of, 66, 198
   Payment for Ecosystem Services program (Costa Rica), 207–8, 272
   "peace park" in Gola rainforest, 141
   Peruvian rainforests, 12, 91–92
   policy and sustainability, 141–42
   property-rights and land management agencies, 151
   property rules in West Virginia, 92
   rainforest in Panama, 83–84
   rainforest of Santa Catarina Ixtepeji (Mexico), 84–86
   Reducing Emissions from Deforestation and Forest Degradation (REDD+), 210, 295–96n48
   in Russia, 130
   Shadows in the Forest, 145
   Sierra Gorda Biosphere Reserve (Mexico), 145–46, 148
   tropical forests of Southeast Asia, 151
   tropical forests of southern Peru, 69–77
   US Forest Service, 83, 234–35, 238f9.4, 239, 268
The Forest Ranger (Kaufmann), 236, 311n37
Forest Stewardship Council, 102
fossil fuel(s). *See also* climate change; energy; global warming; mountaintop removal
   carbon dioxide from, 121
   cerulean's flight path and, 77
   dependency on, 54, 136–38
   emissions absorbed by forests, 208
   environmental costs of, 12
   global warming and, 138
   industry, 91, 107, 241
   property owner of, 92
   reducing, 46
   rules promoting extraction of, 76
   solar energy vs., 12, 107
Franklin, Benjamin, 96
Franz Ferdinand (Archduke of Austria), 168
freedom and rules, 24–26
free-rider behavior, 48–50, 55, 145, 272
free speech, right to, 68, 260

garbage can model of organizational choice, 49–51
Garcia-Johnson, Ronie, 15
gay marriage, 26
Geertz, Clifford, 144, 298n31
German common law, 144
glaciers, melting, 5
global warming, 5, 12. *See also* climate change; fossil fuel(s)
   bill sponsored by John McCain, 156
   cap-and-trade, 121
   carbon taxes, 121
   combating, 84
   environmental services, 84, 207
   oil and fossil fuel dependency, 138
   Payment for Ecosystem Services program (Costa Rica), 207–8, 272
   Reducing Emissions from Deforestation and Forest Degradation (REDD+), 210
   Republican Party and, 156
   sea levels rise and, 5, 107
   US scientists and, 154
Google Scholar, 47, 269, 314–15n5
Gordon, H. Scott, 194–95, 305n28
Great Louisiana BirdFest, 89
Green Building Council, 39–41, 46, 233–34, 253, 284n4
green consumerism, 5, 220
greenhouse gases, 107, 121, 208. *See also* carbon dioxide; climate change; sulfur dioxide
Greenland ice sheets, 100
green parties, 176–78, 180, 247–48
Greenpeace, 118
green revolution, 218–19. *See also* agriculture, sustainable
Green States and Social Movements (Dryzek et al.), 177, 302n24
GRILA. *See* Latin American Initiative Group
Grindle, Merilee, 191, 304–5n18
Group of 77, 209

Hamilton, Alexander, 239, 311n40
Hamilton, Alice, 96, 291n1
Hammond, Merryl, 8

Hampton Roads Energy Company
    (Virginia), 114
Hardin, Garrett, 193–96, 200, 305n25–26,
    305n30
hat-making trade, 120
hazardous waste, 181, 261, 295n42
heat waves, 123
Herrold-Menzies, Melinda, 15
Hill Country Conservancy, 43
Hochstetler, Kathryn, 138, 297n15–16,
    297n19
Holdren, John, 17
"How One Woman Has Created a
    Biosphere" (film), 146
Hudson Gazette, 7
human rights
    about, 26, 36, 130, 231, 273
    abuses, 129, 191, 231
    activists, 164

Illinois Central Railroad Company, 20
impact of policies and rules. *See also*
    effectiveness
    of American rules governing oil
        extraction, 88–89
    Brazil vs. Bolivia, impact of new rules
        in, 131, 132f6.1, 133–34
    of Danish energy pricing policies on
        developing new technologies, 149
    of environmental impact assessments,
        153, 257–58
    of environmental services on
        landowners in Costa Rica, 84, 207
    of the European Union, 157
    of India's Right to Information Law, 149
    of LEED on the environment, 41,
        233–34
    of allowing then banning leaded
        gasoline, 98–100
    Paul DeFalco and new rules on toxic
        lead, 113–14
    of political decentralization, 157
    of property rights shift (Colombia), 81
    water quality, 118, 182
implementation, 8–9, 93, 123, 254. *See also*
    compliance; effectiveness; PROPER
    program
indigenous peoples, 79–81, 85, 277

informal institutions, 19, 27–28, 75–77,
    143–45, 224–25
institutional economics, 45–46, 103.
    See also North, Douglas
institutional isomorphism, 311n30
institutional resilience, 239–40, 270–71,
    311n43, 315n8. *See also* stability and
    change
institutions. *See also* social rules;
    constitutions; markets
    3 R's (roles, rights, responsibilities), 27
    American federalist, 173
    common property, based on, 195
    the commons (*See* the commons)
    community-level, 195, 144, 188, 224
    colonial, 27, 88, 187
    contracts (*See* contracts)
    defined, 11, 28, 37
    democratic, 105, 136–7
    design standards, 11, 19, 59, 181, 229, 232
    European Union, 180
    ideas, relation to, 217, 224–29
    informal (*See* informal institutions)
    international, 80, 164–66
    laws (*See* laws)
    markets for buying and selling of
        goods and services, 103
    national and international, 80, 146
    non-market, relation to market, 104
    norms (*See* norms)
    older local, 144–45
    policies (*See* policy (policies))
    power and rules, 247
    property (*See* property rights)
    private (*See* rulemaking, private sector)
    religious, 29, 144
    resilience, 239–40, 270–71, 311n43,
        315n8
    stability and change, 31, 221–23
*Institutions, Institutional Change, and
    Economic Performance* (North), 104,
    293n16
insurance regulations, 103
Integrated Pest Management techniques, 138
International Court of Justice (UN), 166
International Criminal Court, 166
international institutions, 80, 164–66
International Labor Organization, 277

international organizations, 219, 277–78
international treaty on transparency, 259
INTERPOL, 166
Interstate Highway System, 28
Irwin, June, 6–7, 11, 55, 263

James, Allen, 9
Jefferson, Thomas, 239–40
Johnson, Charles, 87
Jospin, Lionel, 106
Justinian, Roman Emperor, 20

Kaa-Iya del Gran Chaco National Park (Bolivia), 273
Kaufmann, Daniel, 143, 298n28
Kaufmann, Herbert, 236, 311n37
Kaunda, Kenneth, 151
Keck, Margaret, 277, 315n15
Kempff, Francisco, 14
Kempff, Noel, 14
Kempff, Rolando, 14
King, Martin Luther, Jr., 272–73
Kogi Indians (Colombia), 79–82

Land Reform Act (2003, Scotland), 20
land rights, traditional, 80, 91, 273, 277
landscape ordinances, water-efficient, 265
land trusts, 93
Latin American Initiative Group (GRILA), 208–10
Lawn-to-Garden Incentive Program, 265
Law of Se (Colombia), 80
Law of the Jungle (videogame), 195
Law of the Sea (UN), 87–88, 290–91n58, 290n58
laws. *See also* Clean Air Act; policy; rule(s); social rules; super rule(s)
  Administrative Procedure Act, 258
  Americans with Disabilities Act, 24
  antipollution laws, 254
  bankruptcy laws, 25f2.2, 103
  for battered women, 134
  Bill of Rights, 31
  by-laws, 11, 240, 268
  California law forbidding homeowner associations from forcing their members to keep their lawns, 264
  California law for cities to adopt water-efficient landscape ordinances, 265
  city by-laws, 11
  city ordinance, 240
  Civil Code, 76
  civil law systems, 76
  Clean Water Act, 30, 91, 152, 155, 255, 263
  colonial law, 20, 87–88
  commercial rules, 144
  for conservation easements, 93
  constitutions, 268
  copyright law, 25f2.2
  customs agreed upon by merchants, 144
  decentralization laws, 182, 192
  Directive on End-of-Life Vehicles, 181
  election laws, 248–50
  Emergency Planning and Community Right-to-Know Act, 153
  Endangered Species Act, 30, 91, 152, 155, 246, 258
  environmental laws, 91, 107, 155, 208, 214f9.2, 231, 255–56
  Federal-Aid Highway Act, 24
  federal law on voting at motor vehicle offices, 58
  Federal Register, 258–59
  feudal law, 29
  Forest and Wildlife Law, 75–76, 145
  Freedom of Information Act, 153, 259
  German common law, 144
  homesteading laws, 212–13
  Land Reform Act (Scotland), 20
  Law of Se (Colombia), 80
  Law of the Mother (Colombia), 80
  Law of the Sea, 87–88, 290–91n58, 290n58
  laws vs. the actual rules in play, 144
  legal pluralism, (*See* legal pluralism)
  Millennial Laws (Shaker), 102
  Montreal Protocol, 23, 154–56
  Napoleonic Code, 29
  National Environmental Management Act (South Africa), 256
  National Environmental Policy Act (1970), 153, 156, 257
  patent, 23, 103, 108, 109f5.3, 215
  Persistent Organic Pollutants treaty, 154

pesticide laws (Canada), 7–9, 11, 25f2.2, 230, 263
pesticide laws (US), 9–10, 65, 212
pollution laws (Mexico), 147
preemption laws, 10, 281–82n8
prohibition laws, 154–55
property laws (19th-century US), 64. *See also* property rights
Right to Information law (India), 149–50, 153
Roman Catholic Church dictates, 144
Roman law, 29, 144
rule of law, 104, 143
rules of courts created by manors, guilds, and royalty, 144
Single European Act, 178
sunshine laws, 259–60
Surface and Mining Control and Reclamation Act (1977), 93
trademark, 103
unwritten rules codified in the law, 28
US Coastal Zone Management Act (1972), 21
weights and measures, standardized, 103
lead
  Ethyl Corporation advertisement promoting the use of leaded gasoline, 97f5.1
  human health, impact on, 96
  League of Nations resolution banning lead from interior paints, 13
  Natural Resources Defense Council and EPA lawsuit over lead, 255
  in older homes (US), 12
  Paul DeFalco and new rules on toxic lead, 113–14
  rule changes and environmental, 99f5.2
  tetraethyl, 95–96
  toxicity at low doses and long-term environmental persistence, 120
Leadership for Energy and Environmental Design (LEED), 41–42, 44, 46, 233–34, 263, 284n4–5
League of Nations, 13, 169–70
LEED. *See* Leadership in Energy and Environmental Design
Leeson, Peter, 87, 290n57
legal pluralism, 298n30, 298n34

legitimacy
  of ASTM, 40
  crisis of, 191
  of effective rules, 145
  of the Mexican state and its ruling party, 85
  of national governments, 190
  of national policy, 145
  of the old regime and its rules, 141
Liberia, 127–30, 140–41
libertarianism, 24–26, 102, 128, 176, 217–19
lobbyists. *See also* advocacy; corruption; rulemakers
  auto industry, 156
  Coalition for Sensible Pesticide Policy, 9
  National Agricultural Chemical Association, 9
  pesticide industry, 9–10, 230
  Responsible Industry for a Sound Environment (RISE), 9
  US Chamber of Commerce, 9
lobster fishing community (Maine), 22
local government, 8, 134, 149, 163, 184–86, 188, 190, 192, 201–03. *See also* decentralization
The Logic of Collective Action (Olson), 47, 285n15
Long Beach Water Department, 265
Los Angeles Bureau of Smoke Control, 212

Maastricht Treaty, 179
Madison, James, 204, 239–40
Making Law Matter (McAllister), 256
Maniates, Michael, 5, 220
Maoist China, 137
Mao's War on Nature (Shapiro), 137
market-based environmental regulations
  about, 100–101, 111, 115, 117, 119, 121–24, 292n5, 296n49
  cap-and-trade, 120–24, 156, 181
  carbon taxes, 58, 121, 123
  pollution permits, 112, 117, 121, 294n33
market failures, 105, 107
markets
  American and European, 189
  for buying and selling of goods and services, 103, 122
  cap-and-trade and forest based carbon, 123

markets (*continued*)
    competitive energy, 107
    for forest-based carbon storage, 208
    government rules and, 103, 105, 123–24, 133
    natural resources valuation and, 133
    non-market institutions support of, 104
    poverty alleviation and, 18
    price fluctuation on international, 142
    regulation vs., 95, 100–101, 104, 187
    within rules, 110–11
    wealth-creating capacity of, 106
Markets and States in Tropical Africa (Bates), 150, 299n47
Mazmanian, Daniel, 261–62, 314n30
McAllister, Lesley, 256
McGinty, Dalton, 8
medicines, traditional, 123
mercury, 120–21, 278–79, 295n44–295n45
methyl chloroform, 110
metropolitan area, pedestrian-friendly, 270
Mexican government, 191
Mexico City earthquake (1982), 191
migratory birds, 64. *See also* cerulean warbler
Migratory Bird Treaty, 94, 308n2
Millennial Laws (Shaker), 102
Minamata Disease (Japan), 120
mining code of conduct, 273
mining rules, 266. *See also* mountaintop removal
Ministério Público (Brazil), 256
Mississippi Flyway Birding Festival, 89
Monnet, Jean, 167–72, 172f7.2, 173–79, 182, 187, 264
Montreal Protocol, 23, 154–56
mountaintop removal, 91–92, 266, 291n64
Muir, John, 223, 235, 237
multi-level governance
    about, 303n27
    Chinese boxes, 205, 206f8.2, 207
    think vertically, 163, 181, 185, 276–78
multinational
    agreements. *See* climate change treaty, Law of the Sea, Migratory Bird Treaty, Montreal Protocol, Persistent Organic Pollutants treaty
    corporations, 15, 77, 92, 124, 133, 145, 166, 189, 230–32

municipal trading initiative (Japan), 121
Myers, Norman, 17
The Mystery of Capital: Why Capitalism Triumphs in the West and Fails Everywhere Else (De Soto), 103

Napoleonic code, 29. *See also* civil law
Nash, Roderick F., 223, 310n18
National Agricultural Chemical Association, 9
National Environmental Management Act (South Africa), 256
National Environmental Policy Act (1970), 153, 156, 257
national parks. *See also* protected areas
    in Costa Rica, 208
    Kaa-Iya del Gran Chaco National Park (Bolivia), 273
    Manu National Park (Peru), 69, 71, 74–75, 82
    Noel Kempff National Park (Bolivia), 14
    in Panama, 83–84
    Peruvian national park system, 70
    in the Philippines, 192
    rights of local communities, 273
    in Liberia and Sierra Leone, 129
    Ulla-Ulla National Fauna Reserve, 139
    in the US, 155, 212–13, 222–23
    in the world, 212
National People of Color Environmental Leadership Summit, 118
Natural Resources Defense Council, 41, 114, 255, 294–95n36
The Nature Conservancy, 93
New Public Management, 51
Niebuhr, Reinhold, 52–54, 286n21–22
Noel Kempff National Park, 14
Nogales, Juan, 138–39
nonprofit organizations, 50, 81, 100, 134, 208, 240, 246
normative research, 16–18
norms
    authoritarian rule, 205
    cross-cultural differences in, 225
    cultural, 8, 25f2.2, 28, 223, 225, 229, 246
    and institutions, 217, 224–29

of the Kogi Indians, 276
political instability, 140
social, 11, 224, 227, 229
North, Douglass, 11, 104, 221
North Alabama Birding Festival, 89
North American Breeding Bird Survey (BBS), 68
North American Free Trade Agreement, 24
northern cod fishery (Canada), 23
North Korea, 136
nuclear energy, 177
nuclear freeze movement, 261
nuclear weapons proliferation, 16

oceans. *See also* coral reefs
　acidification of, 43
　currents, 19
　environmental imperative of regulating use of, 88
　Law of the Sea, 87–88, 290–91n58, 290n58
　oil drilling and spills, 5, 16, 88–89
　overfishing, 23, 194
　Panama Canal and, 83
　public access to, 20–21
　social rules and, 22
　Truman Proclamation, 87
　vested interests in, 133
oil drilling, offshore, 16
oil refinery near wildlife refuge, 114
oil spills, 5, 88–89
Olazábal, Claudia, 161–63
Olson, Mancur, 47, 285n15
Ontario Government, 8
open-access publications, 269
organizational behavior, 49–52
Ostrom, Elinor, 196–98, 282n13, 306n32–33, 306n42, 312n2
overfishing, 43, 149, 194, 268. *See also* food production
ozone layer, 23
ozone treaty, 156. *See also* Montreal Protocol

Panama Canal, 83
Parker, Ted, 82–83
patents, 23, 103, 108, 109f5.3, 215
Payment for Ecosystem Services program (Costa Rica), 207–8, 272

Peace of Westphalia (1648), 167
"peace park" in Gola rainforest, transboundary, 141
peer pressure, 145, 224
peer-to-peer learning, 274
Persistent Organic Pollutants treaty, 154
Peruvian Amazon, 75, 77
Perz, Stephen G., 132, 296n3
pesticide(s). *See also* anti-pesticide activists; pesticide regulation
　By-law 270 (banning nonessential pesticides), 7
　Canadian ban on nonessential, 7–9, 11, 25f2.2, 230, 263
　Citizens for Alternatives to Pesticides, 8
　Coalition for Sensible Pesticide Policy, 9
　golf course aesthetic, 8
　industry, 7–11, 263
　levels lower in Canadian waterways, 8
　movement led by Cesar Chavez and the United Farm Workers, 118, 225
　policy (Dutch), 109
　public perception of danger, 226
　reform community, 10
　school cafeterias, pesticide-free produce in, 54
Pesticide Action Network International, 9–10
pesticide regulation. *See also* Irwin, June; pesticide(s); toxic substances
　DDT ban, 23–24
　Hudson, Quebec ban all nonessential, 6–9
　Indonesia's ban of organophosphate pesticides, 138
　Ontario ban of common pesticides, 8–9
　preemption rules of US states forbidding local control of pesticides, 10f1.1
　in the US, 9–10, 65, 212
Pinchot, Gifford, 235–7
Pinedo-Vasquez, Miguel, 76
policy (policies). *See also* laws; rule(s); super rule(s)
　about, 11, 31, 247
　agricultural, 133, 219, 257
　climate, 277
　conservation, 130, 140

policy (policies) (*continued*)
   Danish energy pricing, 149
   of decentralization in EU, 182
   of deforestation (Brazil), 132, 132f6.1, 133
   economic (EU), 176
   environmental, 91, 95, 98, 100–101, 111, 145, 150, 154, 156, 175, 180–81, 256–57, 268
   EU guidelines encompass dozens of national policies, 180
   EU policies, domestic laws of new members are adjusted to, 181
   foreign, 52, 59, 155, 179, 261
   granting land to those who "improve" it by removing the trees (Brazil), 31
   grassroots activism and government policy, 134
   immigration, 250
   on invasive, European, 162
   of low prices for food staples, 150
   to manage forests, empowering local Mexican communities, 146, 148
   management practices, general policies translated into, 236
   national, 25f2.2, 133, 144–45, 185
   national economic and agriculture (EU), 175
   National Environmental Policy Act, 153, 156, 257
   on pesticide use (Dutch), 109
   pollution control in Indonesia, 148
   priorities of European heads of state, 162, 176
   REACH (EU), 181
   Reagan-era, 261
   that hurt the countryside, 150
   timber operators are encouraged to cut and sell wood quickly, 151
   urban planning, 134, 184
   urban wastewater treatment (EU), 181
   wildlife (sub-Saharan Africa), 151
   workplace, 259
policymakers, 13, 18, 52, 59, 115, 118, 121–22, 148, 152, 155, 270, 277
political
   instability, 31, 140
   patronage, 51, 133, 142
   scientists, 13, 15, 131, 136, 166, 250, 253

The Political Economy of Soil Erosion in Developing Countries (Blaikie), 205, 307n54
political engagement, 21, 266
polluter pays principle, 179
pollution laws (South Africa), 256
pollution permits, 112, 117, 121, 294n33
Pollution Prevention Pays initiative, 272
The Population Bomb (Erlich), 218
population growth, 5
   in Bangladesh, 148–49
   birth control and, 148
   in China, 137
   in cities in developing countries, 200
   The Population Bomb, 218
Portney, Kent, 203, 253, 294n26, 307n48, 313n12
Potoski, Matthew, 202, 304n46
power-sharing paradox, 277
Pralle, Sarah, 9
preemption rules, 9–10, 10f1.1, 263, 281–82n8
principles for action
   rule the earth, 267–68
   bridge research and action, 269–70
   coalitions, building unconventional, 270–71
   public value, creating, 271–73
   beg, borrow, or steal, 273–75
   process expertise, cultivating, 275–76
   think vertically, 276–78
   recycling, keep, 278–79
private property. *See* property rights
PROPER program (Indonesia), 153, 299n38–39
Property and Persuasion (Rose), 67, 287n6
property rights. *See also* rule(s); social rules; super rule(s)
   of indigenous peoples in Canada, 81
   along the cerulean warbler's migratory route, 70f4.1
   Callanga Valley (Manu), 71–72
   in Caribbean waters, 87
   cerulean warbler migratory route, 71, 74, 77–81, 83, 85, 87–89, 91–92, 94
   in Colombia, 77–79, 81
   conservation easements and, 92
   access to resources vs. ownership, 92

corporate behavior and, 96
defined, 65–68
incentives to care for, 67
indigenous peoples', 80–81, 85
in Manu National Park, 71
in Mexico, 85–86
modern farming and private, 194
of Panamanian forest, 83–84
in Peru, 74, 103
political creations, 68
pollution permits and, 112
private property vs. government regulation, 67
right to possession *(direito de posse)* (Brazil), 134
to the sea and underlying seabed, 88
possession vs. ownership, 66
social rules, 19, 59, 65, 83
subsurface, 91–92
sustainability and, 65, 100
water, 71
proportional representation, 248
protected areas, 14, 149, 212.
See also national parks
public transit, 200, 268
public trust doctrine, 20
public opinion, 30, 203, 223, 226, 239
Putnam, Robert, 219–20n9

race-to-the-bottom hypothesis, 202
radiation, 23, 228
Radio Zamaneh (Iran), 251
REACH policy (EU), 181
recycling, 6, 35–36, 57, 181, 219–20, 265–66, 268, 278
Reducing Emissions from Deforestation and Forest Degradation (REDD+), 210, 295–96n48
Regional Strategy for Climate Change (Andalusia), 190
renewable energy, 13, 149, 164, 175, 215
Republican Party (US), 155–56, 212, 248, 249f10.1
Responsible Industry for a Sound Environment (RISE), 90
Ribot, Jesse, 207, 305n20, 307n56–57, 308n58
Right to Information Law (India), 149–50

RISE. *See* Responsible Industry for a Sound Environment
Roberts, Bartholomew, 87
Robinson, James, 105, 293n19, 314n4
Rochon, Thomas, 226–27, 261–62, 314n30
Rodrik, Dani, 104, 293n18
Rogers, Everett, 274, 315n13
Roman Catholic Church, 144
Roman law, 144
Rose, Carol, 67, 287n6
Rose, Richard, 274, 315n12
Rose-Ackerman, Susan, 142, 298n26
Ross, Michael, 138, 151, 299–300n48
rule(s). *See also* laws; policy; property rights; social rules; super rule(s)
on air quality, 23, 216
beach access, restricting, 31
benefits for society, generate, 30
Brazil vs. Bolivia, impact of new rules in, 131, 132f6.1, 133–34
cannon shot, 87
Catholic Church's Catechism 2415, 29
China's market growth and entry into WTO, 23
commercial, 144
constitutional, 23, 134
of courts created by manors, guilds, and royalty, 144
crop subsidies, US, 31
decentralization. (*See* decentralization)
for deforestation, 26
for domestic air quality, 23
for economic growth and political change, 28
environmental (EU), 181–82
on environmental activists, impact of, 250
fishing practices, governing offshore, 23
fossil fuel, promoting extraction of, 76
freedom and, 24–26
governing oil extraction, impact of, 88–89
Kogi craft rules (Colombia), 80
landowners and timber companies, guiding decisions of, 27
of law, 104, 143
markets vs., 95, 100–101, 104
for nature protection in Siberia, 15
of the old regime, 141

rule(s) (*continued*)
   in play vs. laws in force, 144
   and power, 33, 152, 156–57, 189, 202, 205, 252
   power plants, rules limiting carbon dioxide emissions from, 122
   preemption, 9–10, 10f1.1, 263
   private sector, 31, 266
   regulations, combine formal written and unwritten, 28
   rulemaking, governing, 247
   shape our rivers, our skies, and energy we use, 26
   social rules interlock and intertwine, 28
   on toxic lead, 113–14
   transparency, 259–60
   unwritten rules codified in the law, 28
   voting, 19, 248, 249f10.1, 253
   for water associations, 163
rule of law, 104, 143
ruling coalitions, 271
rummaging, 58

safety codes, 246
Sagan, Carl, 17, 283n18
sanitation services, 43. *See also* wastewater
Santa Cruz de la Sierra (Bolivia), 14, 199–200, 220
satisficing, 45
Save Barton Creek Association, 43
Say, Jean-Baptiste, 272
Schneider, Stephen, 17
Schuman, Robert, 175
Scientists and Engineers for Responsible Technology, 16
Scott, James, 152, 229, 298n35, 300n51, 310n23
sea levels, rising, 5, 67, 107, 123
Senate Subcommittee on Air and Water Pollution, 254
Shadows in the Forest (Dauvergne), 145
Shapiro, Judith, 137
Sierra Gorda (Mexico), 146
Sikkink, Kathryn, 277, 315n15
Silent Spring (Carson), 17, 225
Simpson, Adam, 250–51, 313n8
Sinai Peninsula, 59
Sinding, Steven, 148, 299n41

Singer, Joseph, 67
Single European Act, 178
Six-Day War (1967), 59
Smith, Adam, 66
social capital, 145, 147, 298n35
reputations, 224
taboos, 27, 148
social change
   about, 13–15, 26, 32, 36–37, 210, 212
   in Bolivia, 14
   inner workings of, 215
   major shifts within a single human lifetime, 26
   rules, new vs. old, 223
   social stability and, 221
social movements. *See also* activism
   about, 136, 216, 260–61
   Green States and Social Movements, 177
   impact of, 15, 113, 155, 204, 261
social rules. *See also* institutions; laws; property rights; rule(s); super rule(s)
   defined, 11–12
   durability, 29
   escape notice, 12
   human rights, protect, 26
   long-term prosperity, promote, 26
   property rights, 19, 59, 65, 83
   shape our everyday existence, 18
   for tankers dumping oil at sea, 22
Social Rules Project, 16, 305n29, x
Society for Testing and Materials, 25f2.2
soil erosion
   The Political Economy of Soil Erosion in Developing Countries (Blaikie), 205, 307n54
solar energy, 12, 52, 107
solar water heating systems (Spain), 190
South Africa, 256, 270–71
sovereignty, national, 27, 131, 167, 173
Soviet Union, 26, 130, 136, 140, 187
species conservation, 15, 228.
   See also cerulean warbler; biodiversity
species extinction, 5
spillover effects, 105–6. *See* externality
Spratly Islands, 135
stability and change, 31, 221–23.
   See also institutional resilience

Index 335

standard operating procedures, 12, 230–32
standards of evidence, 53, 268
Stavins, Robert, 116, 295n39
Steel and Coal Community (EU), 175
Suharto, 138, 188
sulfur dioxide. *See also* greenhouse gases
   emissions, 108, 115–17, 119, 216, 272
   permits, 119
   reduction, 119
   scrubber technology, 108
Sunshine Act (US), 259
sunshine laws, 259–60
super rule(s). *See also* policy; property rights; rule(s); social rules
   of ASTM, 253–54
   changes are commonplace, 253
   citizen suit, 255
   defined, 178, 246–47, 312n2
   determine participation, 247, 251–52, 254, 268
   election laws, 248–49, 249f10.1, 250
   environmental impact assessment, 257
   for environmental law enforcement, 256
   transparency, 258–59
Surface Mining Control and Reclamation Act (1977), 93
sustainability. *See also* environmental protection
   choice, individual and social, 266
   coalitions for, 271
   corporate rules and, 102
   democratic reform and, 262
   economic development vs. the environment, 273
   LEED buildings, 41–42
   nation-state and, 133
   political systems, unsustainable, 131
   political will for, 139
   property rights and, 65, 100
   social and political choice, 266–67
   strategic environmental planning and, 258
sustainable
   agriculture, 175
   buildings, 39, 253
   communities in Mexico and Colombia, 195
   development, 133, 146, 271
   harvesting agreements in oceans, 22
   forestry, 75, 235–6
Szasz, Andrew, 220, 225n40

tax codes, 134
Taylor, Charles, 127–30
technical standards, 268
terrorist network, relies on rules, 246
think vertically, 163, 181, 185, 276–78
Thoreau, Henry David, 223
Thorpe, Rebecca, 222, 310n17
3M Corporation, 102, 272
3-mile limit, traditional, 87
Tierney, John, 216, 309n6
Timber Booms and Institutional Breakdown in Southeast Asia (Ross), 151, 299–300n48
toxic substances. *See also* air pollution; lead; pesticide(s); pesticide regulation; sulfur dioxide
   acid rain pollution, 108
   dumping in Ecuador by multinational firms, 231
   hazardous waste, 181, 261, 295n42
   industrial chemicals, 41, 270
   landfills, 35
   mercury, 120–21, 278–79, 295n44–295n45
   solvents, 233
   tetraethyl lead, 96
   US Toxics Release Inventory, 153
   waste, 13, 39, 106, 123, 270
tradable permit program (US), 100. *See also* cap-and-trade
trademark, 103
tragedy of the commons, 193–96. *See also* the commons; decentralization
Transparency International, 259
transparency rules, 259–60
transportation, 221
   bicycle ridership in Portland, 42, 205, 264
   Bicycle Transportation Alliance (Portland), 264
   infrastructure, 133
   network, 187
   sector and carbon dioxide, 208

treaties
- 3 R's (roles, rights, responsibilities), 27
- banning lead in exterior house paint, 170
- citizens' group and, 23
- climate change, 154, 208–10
- Convention on International Trade in Endangered Species, 153
- global environmental summits and treaty-signing, 135
- growth in number over time, 165, 165f7.1
- Law of the Sea, 87–88, 290–91n58, 290n58
- Maastricht Treaty, 179
- Migratory Bird Treaty, 64, 94, 308n2
- Montreal Protocol (1987), 23, 154–56
- Peace of Westphalia (1648), 167
- Persistent Organic Pollutants treaty, 154
- Reducing Emissions from Deforestation and Forest Degradation (REDD+), 210, 295–96n48
- tankers banned from dumping excess oil at sea, 22
- trade in vicuñas, treaty banning international, 139
- on transparency, international, 259
- US fails to translate ratified treaties into domestic law, 154
- US shunned several major treaties, 154
- watershed, treaty protecting Panamanian, 83–84
- wildlife treaty challenged by Missouri, 64

Treaty of Paris, 175
Treaty of Rome, 175–76
tropical forests
- about, 5, 123
- of Costa Rica, 183
- local stewardship of, 203
- of South America, 69, 77
- of Southeast Asia, 151

Truman Proclamation, 87
Tsai, Lily, 224, 231
Tukey, Paul (filmmaker), 8
2,4-D (pesticide), 7

Ulla-Ulla National Fauna Reserve (Bolivia), 139
ultraviolet radiation, 23

United Church of Christ's Commission for Racial Justice, 118
United Farm Workers, 118, 225
United Nations (UN)
- Conference on the Human Environment, 153
- Directorate General for the Environment., 161–62
- Framework Convention on Climate Change (*see* climate change, treaty)
- humanitarian agencies, 165
- Intergovernmental Panel on Climate Change, 154

United States (US). *See also* US Environmental Protection Agency
- Administrative Procedure Act (1946), 258
- Centers for Disease Control, 7, 96, 292n6
- Chamber of Commerce, 9
- Coastal Zone Management Act (1972), 21
- Congress, 113–15, 122, 153, 155, 218, 234, 246, 248, 254
- Constitution, 173, 246
- defense spending, 221
- Democratic Party, 212, 248
- Department of Agriculture, 51, 212, 257–58
- Federal Register, 258–59
- Food and Drug Administration, 23, 25f2.2
- Forest Service, 83, 234–35, 238f9.4, 239, 268
- Freedom of Information Act (1966), 153, 259
- Minerals Management Service, 89
- national parks, 222–23
- pesticide laws, 9–10, 65, 212
- political discourse, divisive, 270
- ratified treaties, US failed to translate treaties into domestic law, 154
- Republican Party, 155–56, 212, 248, 249f10.1
- scientists and global warming, 154
- Supreme Court, 9, 20–21, 64, 227, 268
- Toxics Release Inventory, 153
- winner-takes-all voting method, 248, 249f10.1, 312n4

"Up and Down with Ecology: The Issue- Attention Cycle" (Downs), 30

urban
    air quality in Italy, 100
    democracy, 253
    drinking water systems, 125
    governance, 36
    megacities, 199
    planning, 134, 199, 274
    waste management, 115, 125
urbanization, 148
US Environmental Protection Agency (EPA)
    Clean Air Act lawsuit, 114
    climate change, 156
    lawsuit over lead, 255
    Office of Environmental Justice, 118
    ozone-depleting chemicals banned, 110
    tetraethyl lead, 96

value creation, 56–59, 272–73
Védrine, Hubert, 179
venue shaping, 252–53
venue shopping, 252–53
Vogel, David, 153–54, 293n15, 300n54, 300n56
voter registration at motor vehicle offices, 58
voting
    alternative, 250
    assemblies, 198
    by consensus, 178
    in European Council, 178
    European Economic Community, 178
    instant-runoff, 250
    proportional representation, 248–49, 249f10.1, 250
    qualified majority (EU), 178
    requirements, 268
    rights, 188
    rules, 19, 248, 249f10.1, 253
    winner-takes-all voting method, 248, 249f10.1, 312n4

wade-in protests, 21
Walmart, 11
war (armed conflict), 140, 168–69
wastewater, urban, 181, 217
water. *See also* Clean Water Act
    building code requirements for, 39
    Chile's plan to flood the Patagonian wilderness, 135
    citizens' common right to, 21
    conservation, 43, 264–65
    conservation and transfer program in Southern California, 115
    Costa Rican water associations, rules for, 163
    Deepwater Horizon oil spill, 89
    drinking water systems and waste treatment, urban, 125
    Dutch flower industry and, 109–10
    Exxon Valdez oil spill in Alaska (1989), 89
    fish species extinction, 43
    Forest Service and tree planting on watersheds, 234–39
    Great Lakes, multinational agreements to conserve, 273
    groundwater depletion and water management, 196–98
    industrial pollution in local waterways, 134, 227
    Law of the Sea, 87–88, 290–91n58, 290n58
    LEED-certified buildings, 41
    management in Costa Rica, 161–62
    management of water resources, local government, 192
    paper pulp and EPA water effluent standards, 25f2.2
    Payment for Ecosystem Services program, Costa Rica's, 207–8, 272
    pesticides levels in Ontario's waterways, 8
    pollution, 163, 254
    property rights, 71
    public access to safe water, 125, 131
    public trust doctrine, 20
    quality, 118, 182
    rights in West Virginia, 92–93
    shortage, 109–10, 196–98, 228–29, ix (*See also* droughts)
    territorial waters, 88
    treatment, 215
    user fees, 84
    water management in Egypt, 205
    water quality regulations (EU), 182

water (*continued*)
    waterways for fishing and recreation, 105
    Working for Water program (South Africa), 135
watershed, Panamanian, 83–84
The Wealth of Nations (Smith), 66
weights and measures, standardized, 103
West Basin Water Association (Indiana), 196–97
Western Climate Initiative (California), 122
West Virginia Land Trust, 93
White House Council on Environmental Quality (US), 155
Why Governments Waste Natural Resources (Ascher), 150
Why Nations Fail (Acemoglu and Robinson), 105, 293n19, 314n4
Wilderness and the American Mind (Nash), 223, *310n18*
win-win solutions, 37, 59, 110, 179, 272
Women's Society Against Environmental Pollution (Iran), 251
World Trade Organization, 23
World Values Survey, 225

Yadana gas pipeline (Burma), 251
Yinchu, Ma, 137

Zapata, Emiliano, 85
zero-sum game, 58–59